Tony Allan

This book follows Tony Allan's journey through the maze of water management and global awareness of the risks of mismanagement, tracing the trajectory of the hydraulic mission from a time of infinite possibilities—*fiat panis* and piped water—to risk aversion, loss of confidence and political will, and post-water societies.

Part I, spanning fifteen chapters, explores Tony Allan's formative years at SOAS, his groundbreaking work in remote sensing, and his Middle Eastern studies. It examines his influential contributions to understanding the political economy of water, the concept of virtual water, and the intricacies of transboundary water disputes. His keen insights into power dynamics shaped his perspective on regional agriculture, environmental issues, and cross-border water management.

Part II investigates two decades of profound societal and economic transformation. It highlights the expanding influence of the private sector and showcases Tony Allan's transdisciplinary collaboration with diverse stakeholders, from institutions to communities. The narrative follows his strategic pivot from conventional water issues to food systems—the primary water consumer—culminating in the comprehensive *Oxford Handbook of Food, Water and Society.*

Through sixteen detailed chapters, this book examines the mixed legacy of privatization and the complex politics of the water-energy-food nexus. It tackles contentious issues of sustainability and governance, ultimately advocating the rebalancing of power dynamics within food systems and broader society.

This book offers critical insights for scholars, policymakers, and practitioners in the fields of water management, agriculture, and the environment. Most of the chapters in this volume were originally published in *Water International.*

David Dent graduated from University of Durham, UK. A soil surveyor, he has dug in every continent. Turning to land evaluation, land use planning, and the science-policy interface, he has advised national governments, international organisations, and multinationals. He is former Director of ISRIC-World Soil Information.

Martin Keulertz is Adjunct Assistant Professor of Food Security at the American University of Beirut, Lebanon.

Michael Gilmont is a Research Affiliate at the Institute for Science, Innovation and Society at the University of Oxford, focusing on interdisciplinary hydrological, political and economic approaches to water management. Subsequent to work on this book, he has been an adviser to the UN Environment Programme.

Routledge Special Issues on Water Policy and Governance

Edited by: Cecilia Tortajada *(IJWRD) – Institute of Water Policy, Lee Kuan Yew School of Public Policy, NUS, Singapore, and* **James Nickum** *(WI) – International Water Resources Association, France*

Most of the world's water problems, and their solutions, are directly related to policies and governance, both specific to water and in general. Two of the world's leading journals in this area, the *International Journal of Water Resources Development* and *Water International* (the official journal of the International Water Resources Association), contribute to this special issues series, aimed at disseminating new knowledge on the policy and governance of water resources to a very broad and diverse readership all over the world. The series should be of direct interest to all policy makers, professionals and lay readers concerned with obtaining the latest perspectives on addressing the world's many water issues.

Source-to-Sea Management
Edited by Josh Weinberg, Qinhua Fang, Sarantuyaa Zandaryaa, Greg Leslie and James Nickum

Strengthening Cooperation over Transboundary Groundwater Resources
Edited by Gabriel Eckstein and Alice Aureli

Wicked Problems of Water Quality Governance
Edited by James E. Nickum, Raya Marina Stephan and Henning Bjornlund

Groundwater
Recent Advances in Interdisciplinary Knowledge
Edited by Raya Marina Stephan, James E. Nickum and Philippus Wester

Public Banks, Public Water
Exploring the Links in Europe
Edited by Thomas Marois and David A McDonald

Smart Water Management
Truly Intelligent or Just Another Pretty Name?
Edited by Henning Bjornlund, Stephanie Kuisma, James E. Nickum and Raya Marina Stephan

Tony Allan
An Intellectual Journey
Edited by David Dent, Martin Keulertz and Michael Gilmont

For more information about this series, please visit: www.routledge.com/Routledge-Special-Issues-on-Water-Policy-and-Governance/book-series/WATER

Tony Allan

An Intellectual Journey

Edited by
David Dent, Martin Keulertz and Michael Gilmont

IWRA

Routledge
Taylor & Francis Group
LONDON AND NEW YORK

First published 2026
by Routledge
4 Park Square, Milton Park, Abingdon, Oxon, OX14 4RN

and by Routledge
605 Third Avenue, New York, NY 10158

Routledge is an imprint of the Taylor & Francis Group, an informa business

British Library Cataloguing in Publication Data
A catalogue record for this book is available from the British Library

ISBN13: 978-1-032-97144-5 (hbk)
ISBN13: 978-1-032-97145-2 (pbk)
ISBN13: 978-1-003-59240-2 (ebk)

DOI: 10.4324/9781003592402

Typeset in Minio Pro
by codeMantra

Publisher's Note
The publisher accepts responsibility for any inconsistencies that may have arisen during the conversion of this book from journal articles to book chapters, namely the inclusion of journal terminology.

Contents

Citation Information viii
Notes on Contributors xii

Introduction: A life exploring blind corners: Tony Allan's legacy 1
David Dent, Martin Keulertz and Michael Gilmont

Part I
Foundations, innovations and water politics: Tony Allan's early legacy **9**

1 'Tony here!' Reflections on Professor Tony Allan 11
Stephen F. Lintner

2 Scalable Water Balances from Earth Observations (SWEO): results from 50 years of remote sensing in hydrology 13
Tim Hessels, Jeffrey C. Davids and Wim Bastiaanssen

3 Professor Tony Allan and Libya 34
Salem Maiar

4 Of intellectual friendship in *fin-de-siècle* London 37
Chibli Mallat

5 Recollections of a peacemaker 41
Munther J. Haddadin

6 Sanctioned discourse and the power of hegemonic imaginings 43
Charles Tripp

7 Water wars, conflict and cooperation – how the virtual water concept helped change the discourse 48
Anders Jägerskog and Jan Lundqvist

8 How virtual water saved the Middle East from water wars 52
Greg Shapland

9 From zero-sum to variable-sum on the Nile 56
John Waterbury

10 Egypt's water balancing act 59
Stephen Brichieri-Colombi

11 Contested baselines and transboundary water resources management, with illustrations from the Nile 81
Dale Whittington

12 Water and complex problemsheds in Karamoja, Uganda 99
Alan Nicol, Liza Debevec and Samuel Okene Ayaru

13 *Ozymandias* in the desert: irrigation in Saudi Arabia 116
Elie Elhadj

14 Locating the channel and other tales from the river bank: constants and change in river boundary delimitation 126
Richard Schofield

15 Power plus: Tony Allan's contributions to understanding transboundary water arrangements 148
Mark Zeitoun, Ana Elisa Cascão, Marwa Daoudy, Francesca Greco, Naho Mirumachi and Jeroen Warner

Part II
Beyond water: private enterprise, food security, and systemic change **163**

16 I remember Tony 165
Carl Hausmann

17 The role of the private sector in sustainable development 167
Rabi H. Mohtar

18 The private sector and water services: a reflection 176
David Lloyd Owen

19 Water governance and system coordination across diverse risk-management cultures 181
Brendan Bromwich, Damian Crilly and Jyoti Banerjee

20 Chronic crisis: 30 years on from the Dublin Principles and still no market to value water 192
Martin Keulertz and Phil Riddell

21 When the virtual water runs out: local and global responses to addressing unsustainable groundwater consumption 204
Iman Haqiqi, Chris J. Perry and Thomas W. Hertel

22 The problem with water footprints outside of irrigated drylands 229
Mark Mulligan

23 Virtual water, international relations and the new geopolitics of food 252
Eckart Woertz

24 The role of virtual-water decoupling in achieving food–water security: lessons from Egypt, 1962–2013 262
Ahmed Tayia, Alexandra M. Collins and Michael Gilmont

25 Unexpected bright spots: how the pandemic, climate change and biodiversity loss are shaping the evolution of the nexus 284
Nathanial Matthews, Bart Schoonbaert and Elizabeth Burlon

26 Tony Allan: a magic toolbox of theoretical frameworks, a never-ending story 291
Francesca Greco

27 Accountants will save the world! 295
Andrew Ross

28 Irrigated agriculture: more than 'big water' and 'accountants will [not] save the world' 299
Bruce A. Lankford

29 Farmers will save the world! 309
Brian Chatterton

30 Crossed wires: public regulation and private action for water stewardship and sustainable farming 313
Peter Newborne

31 How decisions are made by politicians through the advocacy of peer reviewed research: the lens of advocacy coalition theory 331
Anthony J. Colman

32 The cost of food. Consequences of not valuing soil and water and the people who manage them 336
Tony Allan and David Dent

Index 351

Citation Information

Most chapters in this book were originally published in *Water International*, volume 47, issues 6 and 7 (2022). When citing this material, please use the original page numbering for each article, as follows:

Chapter 1
'Tony here!' Reflections on Professor Tony Allan
Stephen F. Lintner
Water International, volume 47, issue 6 (2022), pp. 864–865

Chapter 2
Scalable Water Balances from Earth Observations (SWEO): results from 50 years of remote sensing in hydrology
Tim Hessels, Jeffrey C. Davids and Wim Bastiaanssen
Water International, volume 47, issue 6 (2022), pp. 866–886

Chapter 3
Professor Tony Allan and Libya
Salem Maiar
Water International, volume 47, issue 6 (2022), pp. 887–889

Chapter 4
Of intellectual friendship in fin-de-siècle *London*
Chibli Mallat
Water International, volume 47, issue 6 (2022), pp. 890–893

Chapter 5
Recollections of a peacemaker
Munther J. Haddadin
Water International, volume 47, issue 6 (2022), pp. 894–895

Chapter 6
Sanctioned discourse and the power of hegemonic imaginings
Charles Tripp
Water International, volume 47, issue 6 (2022), pp. 896–900

Chapter 7

Water wars, conflict and cooperation – how the virtual water concept helped change the discourse
Anders Jägerskog and Jan Lundqvist
Water International, volume 47, issue 6 (2022), pp. 901–904

Chapter 8

How virtual water saved the Middle East from water wars
Greg Shapland
Water International, volume 47, issue 6 (2022), pp. 905–908

Chapter 9

From zero-sum to variable-sum on the Nile
John Waterbury
Water International, volume 47, issue 6 (2022), pp. 909–911

Chapter 10

Egypt's water balancing act
Stephen Brichieri-Colombi
Water International, volume 47, issue 6 (2022), pp. 912–933

Chapter 11

Contested baselines and transboundary water resources management, with illustrations from the Nile
Dale Whittington
Water International, volume 47, issue 6 (2022), pp. 934–951

Chapter 12

Water and complex problemsheds in Karamoja, Uganda
Alan Nicol, Liza Debevec and Samuel Okene Ayaru
Water International, volume 47, issue 6 (2022), pp. 952–968

Chapter 13

Ozymandias *in the desert: irrigation in Saudi Arabia*
Elie Elhadj
Water International, volume 47, issue 6 (2022), pp. 969–978

Chapter 14

Locating the channel and other tales from the river bank: constants and change in river boundary delimitation
Richard Schofield
Water International, volume 47, issue 6 (2022), pp. 979–1000

Chapter 15

Power plus: Tony Allan's contributions to understanding transboundary water arrangements
Mark Zeitoun, Ana Elisa Cascão, Marwa Daoudy, Francesca Greco, Naho Mirumachi and Jeroen Warner
Water International, volume 47, issue 6 (2022), pp. 1001–1015

Chapter 16
I remember Tony
Carl Hausmann
Water International, volume 47, issue 7 (2022), pp. 1021–1022

Chapter 17
The role of the private sector in sustainable development
Rabi H. Mohtar
Water International, volume 47, issue 7 (2022), pp. 1023–1031

Chapter 18
The private sector and water services: a reflection
David Lloyd Owen
Water International, volume 47, issue 7 (2022), pp. 1032–1036

Chapter 19
Water governance and system coordination across diverse risk-management cultures
Brendan Bromwich, Damian Crilly and Jyoti Banerjee
Water International, volume 47, issue 7 (2022), pp. 1037–1047

Chapter 20
Chronic crisis: 30 years on from the Dublin Principles and still no market to value water
Martin Keulertz and Phil Riddell
Water International, volume 47, issue 7 (2022), pp. 1048–1059

Chapter 21
When the virtual water runs out: local and global responses to addressing unsustainable groundwater consumption
Iman Haqiqi, Chris J. Perry and Thomas W. Hertel
Water International, volume 47, issue 7 (2022), pp. 1060–1084

Chapter 22
The problem with water footprints outside of irrigated drylands
Mark Mulligan
Water International, volume 47, issue 7 (2022), pp. 1085–1107

Chapter 23
Virtual water, international relations and the new geopolitics of food
Eckart Woertz
Water International, volume 47, issue 7 (2022), pp. 1108–1117

Chapter 24
The role of virtual-water decoupling in achieving food–water security: lessons from Egypt, 1962–2013
Ahmed Tayia, Alexandra M. Collins and Michael Gilmont
Water International, volume 47, issue 7 (2022), pp. 1118–1139

Chapter 25

Unexpected bright spots: how the pandemic, climate change and biodiversity loss are shaping the evolution of the nexus
Nathanial Matthews, Bart Schoonbaert and Elizabeth Burlon
Water International, volume 47, issue 7 (2022), pp. 1140–1146

Chapter 26

Tony Allan: a magic toolbox of theoretical frameworks, a never-ending story
Francesca Greco
Water International, volume 47, issue 7 (2022), pp. 1147–1150

Chapter 27

Accountants will save the world!
Andrew Ross
Water International, volume 47, issue 7 (2022), pp. 1151–1154

Chapter 28

Irrigated agriculture: more than 'big water' and 'accountants will [not] save the world'
Bruce A. Lankford
Water International, volume 47, issue 7 (2022), pp. 1155–1164

Chapter 29

Farmers will save the world
Brian Chatterton
Water International, volume 47, issue 7 (2022), pp. 1165–1168

Chapter 30

Crossed wires: public regulation and private action for water stewardship and sustainable farming
Peter Newborne
Water International, volume 47, issue 7 (2022), pp. 1169–1186

Chapter 31

How decisions are made by politicians through the advocacy of peer reviewed research: the lens of advocacy coalition theory
Anthony J. Colman
Water International, volume 47, issue 7 (2022), pp. 1187–1191

Chapter 32

The cost of food. Consequences of not valuing soil and water and the people who manage them
Tony Allan and David Dent
Regenerative Agriculture. Springer, Cham.
https://doi.org/10.1007/978-3-030-72224-1_1

Notes on Contributors

Tony Allan, Formerly King's College and SOAS, London, UK.

Samuel Okene Ayaru, Consultant, Kaabong, Uganda.

Jyoti Banerjee, North Star Transition, Richmond, UK.

Wim Bastiaanssen, Department of Water Management, Faculty of Civil Engineering and Geosciences, Delft University of Technology, the Netherlands; IrriWatch, Wageningen, the Netherlands.

Stephen Brichieri-Colombi, formerly Geography Department, King's College London, UK.

Brendan Bromwich, Water Consultancy Division, Mott MacDonald, Cambridge, UK.

Elizabeth Burlon, Accenture, London, UK.

Ana Elisa Cascão, Independent researcher, Lisbon, Portugal.

Brian Chatterton, Independent Researcher.

Alexandra M. Collins, Centre for Environmental Policy, Imperial College London, UK.

Anthony J. Colman, Water Security Centre, University of East Anglia, Norwich, UK; Columbia University, USA; University of Cape Town, South Africa.

Damian Crilly, Water Strategy and Partnerships, Environment Agency, UK.

Marwa Daoudy, School of Foreign Service (SFS) and Center for Contemporary Arab Studies (CCAS), Georgetown University, Washington, DC, USA.

Jeffrey C. Davids, Davids Engineering, Chico, USA; California State University, Chico, USA.

Liza Debevec, IWMI, East Africa Office, Addis Ababa, Ethiopia.

David Dent, Chestnut Tree Farm, Norfolk, UK.

Elie Elhadj, Former Chief Executive Officer, Arab National Bank, Saudi Arabia.

Michael Gilmont, Institute for Science, Innovation and Society (Insis), University of Oxford, UK.

Francesca Greco, Independent Researcher, Nicosia, Cyprus.

Munther J. Haddadin, Former Minister of Water and Irrigation, Jordan; Former Chairman of the Board of Trustees, University of Science and Technology, Jordan.

Iman Haqiqi, Department of Agricultural Economics, Purdue University, West Lafayette, USA.

Carl Hausmann, Cereol and Bunge.

Thomas W. Hertel, Department of Agricultural Economics, Purdue University, West Lafayette, USA.

Tim Hessels, Department of Water Management, Faculty of Civil Engineering and Geosciences, Delft University of Technology, the Netherlands; Institute for Water Education, UN-IHE Delft, the Netherlands; IrriWatch, Wageningen, the Netherlands.

Anders Jägerskog, World Bank, Washington, DC, USA; School of Global Studies, University of Gothenburg, Sweden.

Martin Keulertz, American University of Beirut, Lebanon.

Bruce A. Lankford, School of International Development, University of East Anglia, Norwich, UK.

Stephen F. Lintner, Department of Geography, King's College London, UK.

Jan Lundqvist, Stockholm International Water Institute, Sweden.

Salem Maiar, UK Representative, Man-made River Authority.

Chibli Mallat, Mallat Law Offices, Beirut, Lebanon; Law School, University of Utah, Salt Lake City, USA.

Nathanial Matthews, Department of Geography, King's College London, UK; The Global Resilience Partnership, London, UK.

Naho Mirumachi, Department of Geography, King's College London, UK.

Rabi H. Mohtar, Department of Biological and Agricultural Engineering, Zachry Department of Civil and Environmental Engineering, Texas A&M University, USA; Faculty of Agricultural and Food Sciences, American University of Beirut, Lebanon.

Mark Mulligan, Department of Geography, King's College London, UK.

Peter Newborne, Independent Researcher and Consultant, UK.

Alan Nicol, IWMI, East Africa Office, Addis Ababa, Ethiopia.

David Lloyd Owen, Envisager, Wales, UK.

Chris J. Perry, Independent Researcher, London, UK.

Phil Riddell, Riddell Associates, London, UK.

Andrew Ross, Global Garden Ltd, London, UK.

Richard Schofield, Department of Geography, King's College London, UK.

Bart Schoonbaert, Arup, London, UK.

Greg Shapland, Visiting Senior Fellow, Middle East Centre, London School of Economics, UK.

Ahmed Tayia, Centre for Environmental Policy, Imperial College London, UK.

Charles Tripp, SOAS, University of London, UK.

Jeroen Warner, Social Sciences Group, Wageningen University, the Netherlands.

John Waterbury, Professor Emeritus of Politics and International Affairs, Princeton University, USA.

Eckart Woertz, German Institute for Global and Area Studies (GIGA), University of Hamburg, Germany.

Dale Whittington, Departments of Environmental Sciences & Engineering and City & Regional Planning, University of North Carolina, USA.

Mark Zeitoun, Geneva Water Hub, Graduate Institute of Geneva, Switzerland.

Introduction: A life exploring blind corners: Tony Allan's legacy

David Dent, Martin Keulertz and Michael Gilmont

John Anthony (Tony) Allan graduated with a first class honours in geography from Durham University in 1958, into a world very different from today. Much was expected of governments that held the commanding heights of their economies and of the United Nations and its agencies. In those days, we all believed in planning. And it was appreciated at the highest level that, to meet the challenges of the day, we needed intelligence at the point of decision about the location of resources, their condition, and how this was changing. So, there was plenty of work for geographers.

For his national service in 1959–61, Tony Allan was posted to Aden and Oman to lead a survey team which, at the time, entailed demanding fieldwork with optical instruments. Advances in aerial photography and photogrammetry soon carried our viewpoint from six feet above ground to six or even ten thousand feet; the stereoscope provided the three-dimensional perspective of an eagle; and false-colour film sensitive to the infrared waveband revealed details that even eagles cannot see. Then, in 1972, the first Earth resources satellite offered repeated global mapping at a resolution of 80m. Tony, who had joined the School of Oriental and African Studies (SOAS) in 1966, was in the thick of it. But all the satellite could see was light reflected from the Earth's surface: when it came to surveys of soil and water resources, pretty pictures of clouds, treetops, or desert sands have their limitations. Few of us imagined the advances to come and the brilliant tweaks of interpretation reported here by Wim Bastiannssen and his colleagues.

Through Durham, his national service, and his new home in SOAS, Tony Allan was drawn to the Middle East. With his lifelong friend Keith McLachlan and Edith Penrose, he edited *Libya: Agricultural Economy and Development* (1973), followed by further perceptive contributions to regional geography – but the chronic shortage of water claimed more and more of his attention. When it came to water, Tony considered himself an amateur but he could see the big picture: water obeys physical laws that operate whether we will or not, but what happens to that water is driven by societal laws that operate by consent – and can change. So, his *weltanschauung* grew to encompass not only geography and environmental sciences but also the subtleties of political economy.

Gifted with curiosity and imagination, the life scientific was as exciting as a ride through London in his Citroën 2CV. Tony's humanity shone through in his teaching and appreciation of the cardinal role of people, as individuals and societies, and the real-life outcomes of the issues he was investigating: the tangled web of transboundary water and water hegemony, and his espousal of the embedded water in crops, that is, the water consumed to produce the ultimate harvest – which he translated into the concept of *virtual water* (Allan 1993). He rehearsed the political economy of virtual water in the Middle East (Allan 2001, 2003) but all the World's a stage, and the centrality of the concept to global food trade and geopolitics was recognised by the award of the Stockholm Water Prize in 2008. The Nobel Peace Prize would have been no less appropriate: it takes 500–4 000 tons of water to produce a ton of grain, so countries short of water do well to import food rather than going to war over disputed water resources to grow their own. Moreover, it takes 4 500–15 500 tons of water to produce a ton of meat: Tony became a vegetarian.

His thinking was framed by five successive paradigms (Figure 1): (1) relatively small-scale *pre-modern* water development; (2) *the hydraulic mission* characterised by big engineering projects in the mid-20th century; then (3–5) from the 1970s, *environmental reflexivity, economic reflexivity*, and, in the new millennium, *political and institutional reflexivity*, recognising the need for fundamental changes in the management and governance of natural resources to achieve just and sustainable reallocation.

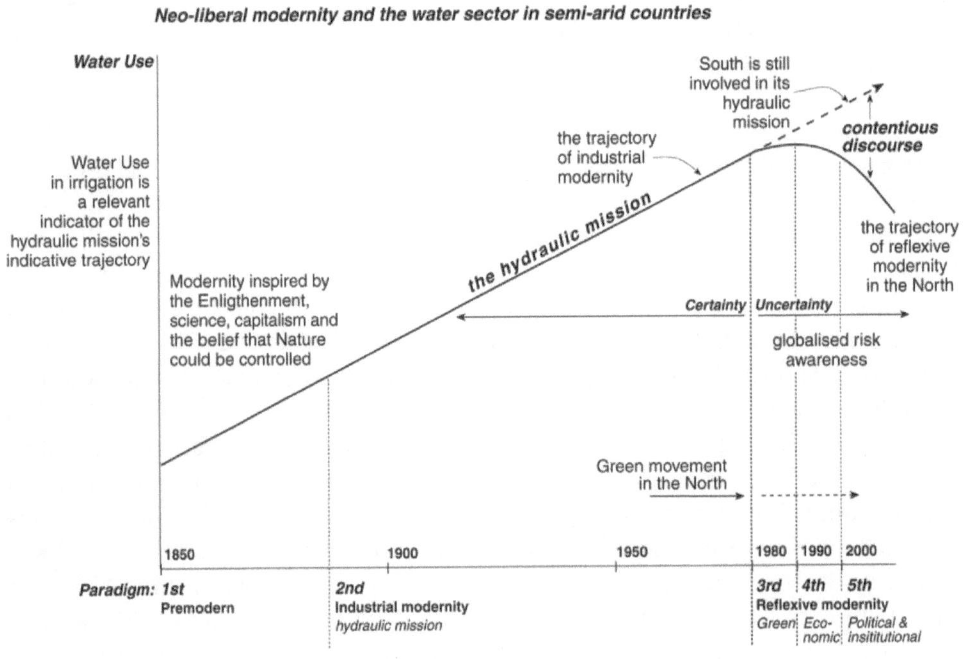

Figure 1. Five successive water paradigms. Source: Allan (2003).

Big engineering projects are still in vogue in the Global South while arguments rage amongst the rich and comfortable. Tony Allan adapted Mary Douglas's *ways of life analysis* (Douglas 1970) to map out the different camps, their rationales, and the tensions between them: *fatalistic* (essentially conforming and adopted by most of society most of the time), *hierachists* (conforming and controlling – .gov in Figure 2); *entrepreneurs* (not controlling, non-conforming – .com), and *ethicists* (controlling and non-conforming – .org). This frame helped to explain why resource managers continue to 'do the wrong thing extremely well' (Allan 2011); it also underscored the need for different voices to be heard and for negotiations between competing claims. The balance of forces, combined

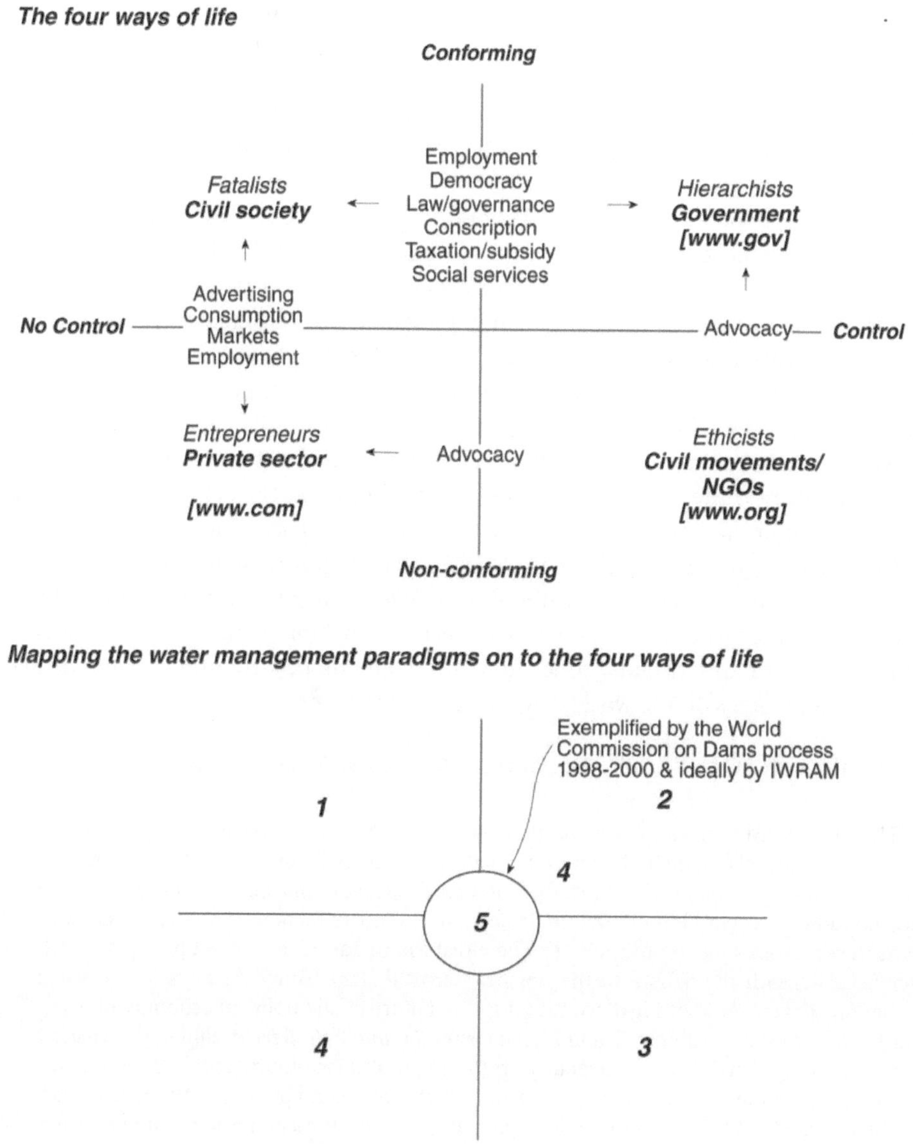

Figure 2. Ways of life analysis. Source: Allan (2003).

with social, political, and environmental shocks, also helped to explain why progress might regress after attempts to do the right thing and why it might take several cycles to achieve a sustained trajectory.

Observation of power in action informed Tony's work from local and national agriculture, water, and the environment, to transboundary waters, to the global food system. He tried to engage with all sides, deploying his capacity to communicate complex ideas – characteristically with one-liners such as 'water flows uphill to money and power' (Reisner 1986). There were no bounds to this engagement. One of the most telling tributes that follow is from Carl Hausmann, CEO of Bunge: 'I was first impressed by his gentle courtesy and rather quaint appearance. However, very quickly I said to myself: "He knows something that I must learn more about."'

Tony Allan's early years during the fatalistic era of the Cold War witnessed the power of government and the private sector to mobilise natural resources, and his work in remote sensing and geographical information systems was arguably at the technical frontier of the day. He lived through the rise of environmentalism, and unfettered neoliberalism, and saw the failure of each successive paradigm to achieve its desired ends. This led him to reflexivity. That is to say, the need for an alternative to the domination of one ideology or another, excluding all others – exemplified in one of his favourite arguments that solutions to water security lie beyond the water sector (Allan 2009, 2011), and by his late-career swerve from water to the food system in which society manages most of its water (Allan *et al.* 2019). He was a tireless convenor bringing together, in the London Water Group, extraordinary talents and diverse viewpoints. Tony's quest also threw down challenges for his beloved Labour Party and bred frustration with radicals whose political philosophy he shared but who lack the reflexivity to bridge parallel tracks and build a platform for power. And frustration with business leaders: These people know all about investment and operational risks; they know the reasons for market failure; they curse the darkness – but where is the candle to illuminate the reason why we have the current accounting rules and legal regime?' (Allan & Dent 2021). Surely, he underestimated the power of his own brief candle.

The contributions that follow appeared first in *Water International* **47** (7, 8) in 2022. Following Stephen Lintner's affectionate recollections in *Tony here!* we have arranged contributions from many friends, colleagues, and students in a rough order of the appearance of these topics on the world stage. In Part I we cover:

1966–90: the early SOAS YEARS – Regional geography and remote sensing

Tim Hessels, Geoff Davids and Wim Bastiannssen bring the water aspects of remote sensing up to date with *Scalable water balances from Earth observations: results from 50 years of remote sensing in hydrology.* Salem Maiar recalls *Professor Tony Allan and Libya* going back to the Libya-London University Joint Research Project, the country's metamorphosis to a republic, and its uncompromising commitment to The Great Manmade River Project piping groundwater from beneath the Sahara to the northern coastal strip. Tony was always concerned whether any government of Libya would grasp the nature of the political economy of water security and invest accordingly. Chibli Mallat writes: *Of intellectual friendship in fin-de-siècle London*, recalling Tony's subtle collegiality promoting multidisciplinary scholarship that initially focussed on the interwoven issues of land, water, people and politics of the Middle East but which turned out to have worldwide implications for law, food and water, war and peace.

The 1990s: The political economy of water and virtual water

Munther Haddadin's *Recollections of a peacemaker* record that Tony Allan became involved in the Middle East Peace Process in 1991 and the concept of virtual water, itself, has been a peacemaker. In *Sanctioned discourse and the power of hegemonic imaginings*, Charles Tripp observes that hegemony is exercised not only through the material power of those who make the rules and authorise what can and cannot be said but, also, though censorship invisible to those immersed in the discourse – that they take for granted and, so, are unaware of the implicit inequality and domination. Appreciation of this invisible power opens up an examination of the rules – spoken and unspoken, written and unwritten – that govern the content, boundaries and forms of sanctioned discourse, and the forces that may contest it.

Thirty-five years ago, water wars were not improbable. Greg Shapland comments on *How virtual water saved the Middle East from water wars.* In the event, Middle Eastern regimes concluded that it was less risky – and cheaper – to import the food they needed rather than go to war to secure a bigger share of local water resources so as to grow it themselves. Anders Jägerskog and Jan Lundqvist sum up this perspective in *Water wars, conflict and cooperation – how the virtual water concept helped change the discourse.* At the time of writing, we were patting ourselves on the back but rational behaviour cannot be taken for granted: Russia's grab for the Donbas and Dnieper, the renewed *intifada* against the sharp elbows of Israeli settlers on the West Bank of the Jordan and the subsequent wider conflicts have confounded us.

2001–10 Transboundary water wrangles

Egypt's water requirements now far outstrip its historical allocation of Nile waters. In *From zero-sum to variable-sum on the Nile,* John Waterbury comments that Egypt and Sudan might seem doomed to a zero-sum contest. The geography is immutable; the politics inscrutable; but the concept of virtual water embedded in food imports and exports has realised variable sum alternatives to head-to-head conflict. In knowledgeable detail, Stephan Brichieri-Colombi assesses *Egypt's water balancing act:* its implications, and measures that would enable Egypt to cohabit with its neighbours as they implement their own plans to exploit Nile waters. Egypt is grappling with its water scarcity more through actions in the wider economy than through integrated water resource management, and must continue to do so to reduce its vulnerability at the end of the supply line. In *Contested baselines and transboundary water resources management, with illustrations from the Nile,* Dale Whittington presents three kinds of problems that confront policy analysts in specifying dynamic baselines against which to assess alternative futures: *unexamined baselines,* a failure to grasp the need for a dynamic baseline; *uncertain baselines,* failure to agree on how to account for uncertainties in the forecast of the dynamic baseline; and *contested baselines,* a failure to agree on the baseline because of disagreements over whether the *status quo* or some alternative is a fair and reasonable basis for the baseline – such disagreement may arise from feelings of past injustices and the resulting perception of inequitable resource allocation. From the headwaters of the Nile basin, Alan Nicol, Liza Debevec and the late Samuel Okene Ayaru consider *Water and complex problemsheds in Karamoja, Uganda,* small-water issues that are, nonetheless, big issues in the region. Responding to growing pressure on resources, donors and the Ugandan government are investing in new surface-water resources but the power and political-economy issues rooted in resource-use relationships are rarely factored into water infrastructure development so its governance is not fit-for-purpose.

None of these issues concerns an absolute monarch. Elie Elhadj's commentary *Ozymandeas. Folly in the desert* analyses Saudi Arabia's experiment in grain self-sufficiency. Between 1980 and 2018, irrigation consumed around 700bcm of groundwater – most of it non-renewable. If the aquifers had supplied only drinking water and household water, some 4.4bcm could have been saved in 2018 alone; instead, in that year, almost 5.2bcm of virtual water was exported to neighbouring states, capping a 20-year trend. Elie scrutinises the cost in terms of water, money and environmental damage, how the decision to irrigate the desert was made, and whom might have benefitted.

In *Locating the channel and other tales from the river bank: constants and change in river boundary delimitation,* Richard Schofield considers wrangles along several international boundaries drawn along rivers. The very names – Shatt al-Arab, Jordan, Chobe, and Bahr al-Arab – spell trouble. Finally in Part 1, Mark Zeitoun leads a cohort of Tony Allan's former colleagues considering *Power plus: Tony Allan's contributions to understanding transboundary water arrangements.* A suite of theory and analytical frames within the field of political economy originated from Tony's mainstreaming of *power* as a determining factor in the control of transboundary water flows. These include hydro-hegemony, coexisting conflict and cooperation and, most recently, virtual water rivers – exemplified briefly through cases from around the globe and in greater depth through application to the Nile, and in a research agenda to carry them forward.

Part II spans two decades of unprecedented change in society and the global economy that the water sector and water science accommodated through a greatly increased role of the private sector and the evolution of the concept, practice, and politics of the water-energy-food nexus. Tony Allan's response was to engage with a broader and broader spectrum of people and institutions that, looking back, we can only describe as transdisciplinary. There was no let-up in his activism but he shifted his stance from mainstream consideration of water resources to investigation of the food system, far and away the principal consumer of water. This shift was marked by the publication of the *Oxford Handbook of Food, Water and Society* (Allan *et al.* 2019), which Tony would have preferred to call *Food-water and society.*

The introductory commentary by Carl Hausmann, CEO of the commodity trader Cereol when he first encountered Tony Allan, underscores two essential issues that Tony's evangelism carried to global decision-makers: first, trading a commodity means trading all the inputs, not only virtual water but virtual environment; secondly, an appreciation of the links between land, water, and people requires both sound science and humility.

The curate's egg

Milton Friedman was awarded the Nobel Prize for economics in 1976 and neoliberalism swept all before it. In *The role of the private sector in sustainable development*, Rabi Mohtar reflects that the private sector supplies most of our basic needs in the water, energy and food chains. These chains are our lifelines but they are knowingly responsible for the pollution of the land, water and air and for most greenhouse gas emissions. Achieving a business model for sustainable development requires concerned scientists and politicians to engage and work with the private sector – which requires better understanding of the challenges and constraints on commodity flow along supply chains. Writing before the public exposure of pollution of Britain's rivers and coastline by its clapped-out infrastructure, David Lloyd Owen discusses the pros and cons of *The private sector and water services.* Brendan Bromwich, Damien Crilly and Jynte Banerjee, respectively engineer, public administrator and academic, draw on cultural theory to advocate layered, collaborative *Water governance and coordination across diverse risk-management cultures*, so as to draw on the strengths of the public, private and voluntary sectors.

The water-energy-food nexus

Here, we consider the economic mobility of water, responses to unsustainable groundwater exploitation, the new geopolitics of food, and virtual-water decoupling in Egypt. In *Chronic crisis: 30-years on from the Dublin principles and still no market to value water,* Martin Keulertz and Phil Ridell step outside the box to propose ways to establish regulated water markets that will enable water to flow to its greatest economic, societal and environmental value. In *Local and global responses to unsustainable groundwater consumption,* Imam Haqiqi, Chris Perry and Thomas Hertel, grappling with the unbridled demands of irrigation, assess the impact of future restrictions on local and global food production, prices, local and distant land use and management. Optimistically, they conclude that local-to-global linkages and market responses will surely mitigate the impact on food security. Mark Mulligan's *The problem with water footprints outside irrigated drylands* is a shot across the bows of quantitative certainty that has been taken for granted, at least by politicians: in most places, agriculture consumes a good deal less water than the wild vegetation it has replaced.

In *Virtual water and the new geopolitics of food*, Eckart Woertz traces the evolution of global food politics from dumping of surplus agricultural commodities to today's asymmetric interdependence. In the aftermath of the Second World War, nobody thought of food imports as mitigating water scarcity – which they did – but the virtual-water paradigm switched attention from conflicting claims over water resources and questioned national aspirations of self-sufficiency that flew in the face of actual resource endowments. The global food trade has transmogrified production, processing, distribution and the provision of inputs to the food system that are now dominated by a handful of international corporations; an increasing share of exports now comes from tropical countries; and Russia has recaptured its historical position as a grain exporter.

Attempts to secure food supplies by investment in farmland abroad, commonly in countries with insecure land rights and problematic governance, created a political backlash; decades after the investments were announced, imports continued to come from established suppliers and, when states actually put money on the table, it was further up the value chain in food processing and distribution rather than in politically contentious farmland acquisition. Food security and virtual water considerations continue to play out in the new geopolitics of food after the global financial-food crisis of 2007/8. Imports suddenly seemed tenuous; and globalisation created asymmetric trade interdependence that can be weaponized. Almost 30 years after Tony Allan first introduced the virtual-water paradigm, it continues to surprise us. His unfinished paper on *Virtual water decoupling in Egypt, 1962–2013* is completed here by Ahmed Tayia, Alexandra Collins and Michael Gilmont. It traces Egypt's struggles to feed its burgeoning population by making the best use it can of Nile water while importing increasing amounts of staple commodities – which requires raising foreign exchange.

This section concludes with a commentary by Nate Matthews, Bart Schoonbaert and Elizabeth Burlon: *Unexpected bright spots: how the pandemic, climate change and biodiversity loss are shaping the evolution of the water-energy-food-climate nexus*. The nexus concept emerged early in the millennium and has been extended to embrace the unlikely bedfellows of ecosystems and finance. Fossil fuels might appear to be the least-cost option to meet burgeoning energy needs but they are water-intensive as well as generating greenhouse gas emissions. These aspects are accounted for by switching from pure-cost assessments to system-value frameworks based on the water-energy-food nexus. Virtual water unassumingly contributes to food and water security but, also, allows potentially detrimental impacts on the environment to be offshored by importers and consumers onto exporters and producers. Covid-19 perturbed these arrangements, but virtual-water thinking does not seem to have informed the immediate responses to the pandemic. However, water is finding a voice and there is growing awareness that water is the primary medium through which we will feel the effects of climate change.

Tony's final years of service: Under-priced food for under-paid people; water accounting, governance and the food system

Tony Allan's final years focused on 'under-priced food for under-paid people,' embracing water accounting and governance in the food system; it remains unfinished business. This section is introduced by Francesca Greco's *Never-ending story*. Tony was always seeking a more all-encompassing theoretical framework within which to explore the fundamental role of water in the environment and in society. But his most troubled thoughts were always about communication of these issues, in particular to journalists who carry messages to society.

Four contrasting commentaries take some of Tony Allan's contentions as their starting points. Andrew Ross takes *'Accountants will save the world'*: all things have a cost and if that cost is not accounted for in the ultimate price of goods, all manner of decisions will be flawed. This is certainly the case in the food system which consumes more than 90 per cent of fresh water and, in the absence of such accounting, we have a grotesquely distorted market that squanders land and water resources. Sixty per cent of the world's public companies have committed to net zero but how are they going to do it? Money is not the problem: the problem is a lack of propositions at institutional investment grade and scale of more than $US 1 billion. Bruce Lankford counters that accountants will not save the world: in *Irrigated agriculture is more than big water,* he argues that irrigation is poorly served when its hydrology is seen solely as *big*, or a subject of accounts. Brian Chatterton would certainly agree with Tony Allan that *'Farmers will save the world'* – if they are given the chance! He paints striking differences between conventional top-down research and farmers' hands-on pragmatism, and the rifts between ministries and mindsets. Peter Newborne cross-examines the words and actions of major retailers in *Crossed wires: public regulation and private action for water stewardship and sustainable farming*. Farmers need support if they are to farm sustainably. Whether major retailers intend to use their power in the food chain to give that support is not clear from the public reports and websites of eleven major retailers; the goal of alignment of price signals with environment seems remote. Meanwhile, the local planning regime operates independently from the national regulator. If we have to fall back on politicians to save the world, then Tony (AJ) Colman's *How politicians make decisions* is germane. Politicians want certainty. Not models, not probabilities. How, then, can we get them to act on what might be good research outcomes yet are not conclusive? Advocacy Coalition Theory sees changes in the beliefs of decision-makers at three levels: *deep core beliefs*, the normal attitude and behaviour of society; *policy core beliefs*, which can be changed by evidence and can lead to coalition formation; and *secondary beliefs*, subject to change over time, leading to fine tuning of reforms. Out of these beliefs come systems to support change to new coalitions in support of policy.

It seems fitting to give Tony Allan the last word. So, we conclude with his last paper, presented at the end of 2019 and published posthumously (Allan & Dent 2021). *The cost of food: Consequences of not valuing soil and water and the people who manage them* expresses his mantra of 'under-priced food for under-paid people.' Affordable food is a political imperative and, yet, there is nothing more expensive: the food system delivers cheap food but takes no account of the damage done along the way. The system is unsustainable – ecologically, environmentally and economically – and there are limits to growth. It will only be sustainable if the power relations across the food system can be redressed. If consumers paid the costs of maintaining environmental services, farmers in a viable commercial system might be able to remedy some of the ecological ills. *Vale*.

But we trust that this will not actually be the last word. Our authors still convene as the The Tony Allan Society. The two most striking contributions to our discussions in 2024 were from a team of Palestinians and Israelis and Jordanians installing water and sewerage in the disputed territory, striving to reconstitute the River Jordan from an open sewer and, probably, the only people from these countries still working together; and from a team in southern Ukraine working on the daily water supply to replace the half of Ukraine's fresh water obliterated by Russian aggression and working out how to reconstitute it all as soon as the war ends. There will always be need for environmental peacemaking and running repairs.

References

Allan JA (1993) Fortunately there are substitutes for water, otherwise our hydro-political futures would be impossible. 13–26 in *Priorities for water resources allocation and management. Proc. Natural Resources Advisors' Conference, Southampton, 1992.* (London: Overseas Development Administration)

Allan JA (2001) *The Middle-East water question. Hydropolitics and the global economy.* (London: IB Tauris)

Allan JA (2003) IWRM/IWRAM: A new sanctioned discourse? Occasional Paper 50. (London: SOAS Water Issues Study Group)

Allan JA (2009) Prioritising the processes beyond the water sector that will secure water for society – Farmers in political, economic and social contexts. *4th Marceliano Botin water workshop – Rethinking water and food security, Sept. 21-3.* https://rac.es/ficheros/doc/00725.pdf

Allan JA (2011) *Virtual water. Tackling the threat to our planet's most precious resource.* (London: IB Tauris)

Allan JA, Bromwich B, Keulertz M, Colman AJ (eds) (2019) *The Oxford handbook of food, water and society.* (New York: Oxford University Press)

Allan JA, Dent DL (2021) The cost of food: Consequences of not valuing soil and water and the people who manage them. 3–20 in *Regenerative agriculture. What's missing? What do we still need to know?* (Cham: Springer Nature Switzerland)

Allan JA, McLachlan KS, Penrose ET (eds) (1973) *Libya: Agriculture and economic development.* (London: Routledge)

Douglas M (1970) *Natural symbols: Explorations in cosmology.* (New York: Pantheon Books)

Reisner M (1986) *Cadillac desert: The American West and its disappearing water.* (New York: Viking Press)

Part I

Foundations, innovations and water politics: Tony Allan's early legacy

'Tony here!' Reflections on Professor Tony Allan

Stephen F. Lintner

'Tony Here!' – his signature greeting, always heralded an enlightening conversation, whether we spoke over the phone or met in person. Those conversations crossed the globe: in the mountains of West Virginia, Wadi Rum, the Dead Sea, along the Gulf or, very often, walking the streets of London, exploring farmers' markets or visiting Patisserie Valerie in Marylebone. Tony had a love of chocolate, a serious interest in other sweets and, on occasion, the Glenlivet Enigma. A vegetarian for many years, Tony was nevertheless forgiving of his meat-eating companions. Kind and generous to a fault, he often had bags of treats along with bottles of water to share with his friends and students.

A tall man, until his later years he walked swiftly in his long, thin, laced shoes, but he never lorded his height, always stooping to speak with those less lofty. And if you were not walking with him, you might be treated to a ride in his vintage Citroën 2CV (*Deux Chevaux*) – an opportunity to reflect on your mortality.

That generosity of spirit, having others' interests at heart, was reflected in his teaching: sharing the wisdom of many years of professional and intellectual achievements, engaging with students through his wit and reflections, encouraging them to think for themselves, pondering together with them on the big questions. He was a man of deep convictions and the gift of conveying these clearly to his audience. While he was at times frustrated by the challenges of bridging the gap between 'the academy' and other actors on the world stage, his determination and vision kept him on track and allowed him to significantly influence those he interacted with at many levels. His often-unorthodox views were matched with pragmatism, which was put to good use in his work as an adviser on policy and technical issues with non-academic parties, such as the World Bank.

Nearly a decade ago, he lured me to King's College London, where I greatly enjoyed our team teaching on water resources management from a combination of political, policy and technical perspectives. We especially enjoyed presenting a case study of the World Bank-facilitated Red Sea–Dead Sea Water Conveyance Study Program between Jordan, Israel and the Palestinian Authority, on which Tony had worked on as an independent expert nominated by the Palestinians. During this period, he was actively championing the importance of farmers and their critical role in land and water management.

Tony never forgot his north-eastern English roots in Newcastle-on-Tyne. As a dedicated member of the Labour Party, he supported its approach to social equity with enthusiasm and, during elections, would comment that their 'message was being heard' by the voters. He also had great admiration for the employee-owned John Lewis Partnership which, he believed, provided a model of sharing work and benefits.

In reflection, what distinguished Tony from his peers – and maybe his most enduring legacy – is the integration of his exceptional intellectual depth, insight and creativity with his fullness as a human being – mixing knowledge with kindness, insight with action, and creativity with compassion, resulting in outcomes to truly make peoples' lives better. In this way, Tony will always be here.

Disclosure statement

No potential conflict of interest was reported by the author.

OPEN ACCESS

Scalable Water Balances from Earth Observations (SWEO): results from 50 years of remote sensing in hydrology

Tim Hessels, Jeffrey C. Davids and Wim Bastiaanssen

ABSTRACT

Strategic planning of water management at the river-basin scale requires (1) measurement and accounting of individual hydrological processes, (2) quantification of water resources, and (3) their optimal allocation. Scalable Water Balances from Earth Observations (SWEO) is an open-access parameterization enabling automated reporting of water footprints and Sustainable Development Goal (SDG) indicators. We present its systematic arrangement and input datasets, and demonstrate its accuracy by independent riverflow measurements. We also review some achievements in remote sensing for hydrology during the last 50 years in quantifying hydrological and water management processes, flows, fluxes and changes in storage from various independent sources; and append mathematical formulations.

Introduction

Tony Allan observed, 'the way that we consume water is detached' (Allan, 1997). He pointed out that the greater part of our water resources is consumed by the food supply chain. The World Economic Forum (2014) listed water scarcity as a global risk affecting business, society and the environment; Mekonnen and Hoekstra (2016) estimated that two-thirds of the global population (4.0 billion people) live under conditions of severe water scarcity at least one month of the year. The amount of water used to produce the goods, commodities and services that we rely on has been formally captured by the *water footprint* concept, and teams led by Allan and Hoekstra have used this concept to quantify water systems and water use at the product, field and river-basin scales (Hoekstra, 2013). Here, we focus on the water footprint per river basin.

Scarcity prompts users to develop accounting systems for water resources (e.g., Bastiaanssen, 2009; Karimi et al., 2013; Molden, 1997). Financial accounts present an overview of inputs, outputs and resulting balances that provide strategic insights: the same applies to water accounting (Figure 1). However, even more so than financial accounting, accurately measuring inflows, outflows and changes in storage in water

Supplementary data for this article can be accessed at https://doi.org/10.1080/02508060.2022.2117896

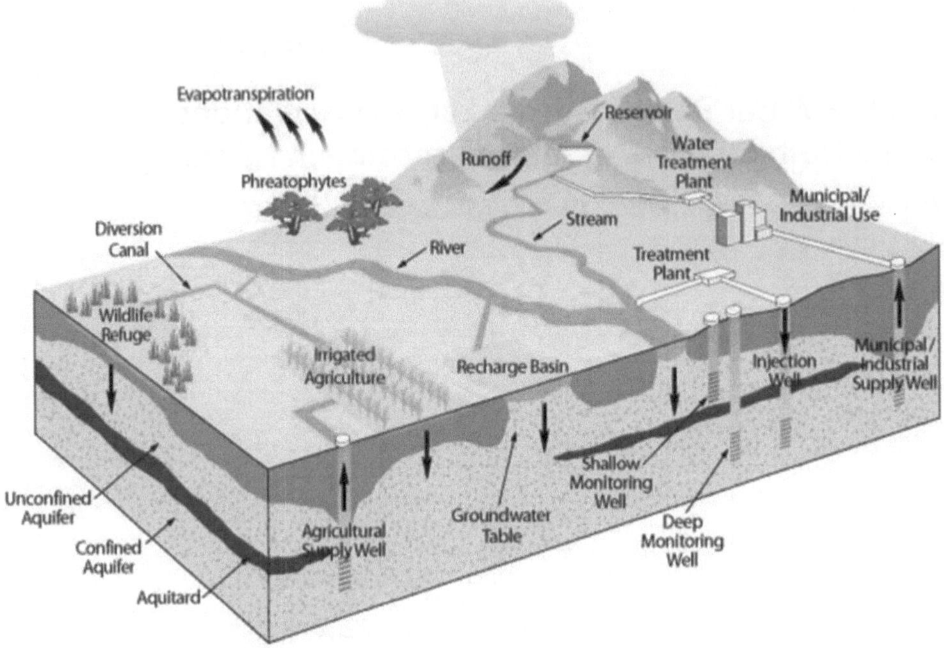

Figure 1. Physical processes in a river basin that need to be quantitatively described by water authorities for dealing with water scarcity and communications with society. Source: https://www.alevelgeography.com/drainage-basin-hydrological-system/.

systems is fraught with difficulty (Davids et al., 2019a). For instance, it is essential to know the difference between gross and net water withdrawals, but, in many cases, it is not directly measurable – even *blue water* in streams, lakes and reservoirs is hard to quantify (Davids et al., 2019b). Weekly or monthly water resources budgets would be helpful in support of fundamental decisions about water allocation, but the determination of all in- and outflows is hard to do and, therefore, not often done. So, it is the exception and not the rule to have reasonably complete water balances available for decision-makers. New tools and streamlined processes for water accounting are sorely needed.

Followers of water footprints use CropWat and AquaCrop models (Steduto et al., 2009) to relate water use to food production (e.g., Chukalla et al., 2015). While these models are robust, they are typically applied at the field scale; for the water footprint of river basins, more comprehensive methods are needed. Feng et al. (2021) have reviewed the analytical tools used for water footprint analysis, including some remote sensing methods that have emerged in recent years (e.g., Madugundu et al., 2018). Shilpakar et al. (2011) estimated monthly river flow in the East Rapti basin, Nepal, from remotely sensed monthly precipitation minus evapotranspiration (P-ET) data; Bastiaanssen et al. (2014) applied P-ET pixel information to determine river flow and water withdrawals in the Nile basin; and, similarly, Poortinga et al. (2017) estimated flow in the Red River, Vietnam. Moreira et al. (2019) went a step further by merging P, ET and storage changes (ΔS) from the independent GRACE gravity mission to estimate terrestrial water balances in South

America. Here, we report progress in preparing spatially distributed water balances using open-access Earth observations of P, ET and ΔS – in particular, the methods used in Scalable Water Balances from Earth Observations (SWEO) which enable the computation of consistent water balances at the river-basin scale.

50 years of remote sensing of hydrological processes (1970–2020)

Adequate water accounting systems require big datasets that are not easily available. SWEO uses globally standardized, open-access datasets: (1) to avoid dependence on unpredictable data-sharing attitudes from third parties; and (2) to alleviate any appearance of bias and potential data-sharing limitations associated with data ownership. Therefore, the results and achievements from water accounting can be shared with all relevant stakeholders. SWEO integrates Earth-observation data from the following individual hydrological processes into a vertical soil water balance for each pixel (currently either 250 or 1000 m): rainfall, snowmelt, evapotranspiration, soil moisture, water levels, surface-water change and groundwater-storage change. These data are currently available from various Earth-observation data archives (Table 1). These databases are based on recurrent satellite measurements, with time steps varying from daily to monthly. In addition, there are several data layers that, by nature, are relatively stable but essential for determining physical process, such as soil hydraulic conductivity that controls, for example, the partitioning of rainfall into surface runoff and infiltration.

Table 1. Open-access databases leveraged by Scalable Water Balances from Earth Observations.

	Spatially variable–temporally variable			Spatial variable–temporally constant		
Variable	Source	Spatial (m)	Temporal (days)	Variable	Source	Spatial (m)
Precipitation	CHIRPS	5000	1	Terrain height, accumulated pixels, drainage direction	SRTM or HydroShed	90, 250 or 1000
Rainfall days	CHIRPS	1000	1	Land cover	GlobCover and MCD12Q1	300 or 500
ET0 reference	FEWS	10,000	1	Population density	GPW V4	1000
ET actual	WaPOR, SSEBop	30, 250, 1000	1	Livestock density	FAO GWL 2005	5000
Air temperature	GLDAS, MERRA or GEOS	25,000 or 50,000	0.125 or 0.25	Sand content	SoilGrids	250
LAI (MOD15A2)	MODIS	500	8	Clay content	SoilGrids	250
fPAR (MOD15A2)	MODIS	500	8	Soil organic carbon content	SoilGrids	250
Snow cover (MOD10A2)	MODIS	500	8	Bulk density	SoilGrids	250
ΔS Ground Water Storage	GRACE	300,000 –400,000	30	Saturated hydraulic conductivity (K_{sat})	SoilGrids	250
Water body area	JRC	30	–	Specific yield (μ)	SoilGrids	250
				Piezometric heads at start	PCR-GLOBWB	9000
				Reservoirs and dams (optional)	GRAND	Points

Precipitation (P) and snow cover

There are several rainfall products based on radar satellites, notably Tropical Rainfall Measurement Mission (TRMM) and Global Precipitation Mission (GPM). Radar emits a burst of energy that is scattered if the energy strikes an object (raindrop, snowflake, hailstone); the density of rainfall droplets and the size of these droplets affects the scattering. In combination with radar measurements, thermal radiometers onboard geostationary satellites and radar satellites measure cloud temperatures that reflect water and ice concentrations. Remotely sensed precipitation products represent larger areas than a gauge on the ground, which has advantages and disadvantages. A larger representative area is great for determining catchment and river-basin rainfall volumes, but it is hard to validate because, by definition, there is a spatial-scale mismatch with *in situ* measurements. Satellite remote sensing products achieve global coverage but may suffer from random errors and bias (Koutsouris et al., 2016) arising from the indirect linkage between the observed parameters and precipitation and, also, imperfect algorithms.

SWEO uses the calibrated Climate Hazard Group InfraRed Precipitation Stations (CHIRPS) database for monthly rainfall and number of rainy days, making use of radar and geostationary satellites, ground stations and terrain information (Funk et al., 2015). The grid size is 5 km and data are available continuously since 1981, which is attractive for generating time series (e.g., Paca et al., 2020). CHIRPS has been widely tested and is well accepted (Dinku et al., 2018; Hessels, 2015; Hsu et al., 2021; Nawaz et al., 2021).

Precipitation includes rainfall and snowfall. Snowfall extends the time lag between precipitation and runoff and infiltration, so needs to be quantified as a separate process. MODIS images detect snow cover and its changes (https://modis.gsfc.nasa.gov/data/dataprod/mod10.php). SWEO uses eight-day snow cover maps to detect the timing of snowmelt, which is an essential input into monthly river flow determinations, especially in cold or mountainous regions where snowfall comprises much of total precipitation. The amount of snowmelt is computed from changes in snow cover and ambient air temperatures during this period (standard equations are provided in Appendix A in the supplemental data online). SWEO has delay factors built in for the location and timing of snowmelt and the arrival of meltwater in the nearest stream. In this way, snow cover information can be used to infer which parts of CHIRPS-based precipitation can be attributed to snowfall and which to rainfall. For further information on remote sensing of snow cover products, see Hall et al. (2002), Parajka and Blöschl (2006) and Rittger et al. (2013).

Evapotranspiration (ET) (actual)

ET represents consumptive use of water. It is the second largest term of the water balance, returning about two-thirds of precipitation to the atmosphere, so it is crucial for water footprint analysis but hard to measure without advanced equipment. It is commonly assumed that weather stations measure ET. Not so. Weather stations measure the state conditions of the atmosphere which can be used to assess a reference ET of grass or alfalfa. The United Nations Food and Agriculture Organization (FAO) has adopted procedures to compute reference ET (ET_0) from a strictly defined hypothetical surface (Allen et al., 1998), but these values may deviate substantially from actual ET; several ET algorithms have been developed but only a few operational spatial databases are related to actual ET fluxes. SWEO uses ET for the water balance, and ET_0 for computing the aridity index, which is used (among other processes) to develop Budyko curves for partitioning *green* and *blue* ET (Budyko, 1974; Simons et al., 2020a).

The first scientific publications using thermal infrared satellite measurements to express evaporative cooling date from the 1960s–70s (Gates, 1964; Jackson et al., 1977; Menenti, 1980; Rosema & Fiselier, 1990; Soer, 1977). Since then, many algorithms have been developed and tested against field measurements, amongst them the Surface Energy Balance Algorithm for Land – SEBAL (Bastiaanssen & Roebeling, 1993), which has been widely adopted and that led to other energy-balance models such as the Simple Surface Energy Balance Operational model (SSEBop) (Savoca et al., 2013; Senay et al., 2014), originally developed to monitor drought and crop development for Africa but now applied globally; METRIC (Allen et al., 2007); ETWatch (Wu et al., 2008); and SAFER (Teixeira et al., 2015). Another operational ET product, GlodET (www.glodet.nebraska.edu) from the Water for Food Institute, is based on the ALEXI model (Anderson et al. 2007a, 2007b).

The latest versions of SEBAL are implemented on Google Earth Engine, so they are becoming more easily available for automated implementation (Jaafar & Mourad, 2021; Laipelt et al., 2021). In principle, 30 m SEBAL outputs could be used for high-resolution applications of SWEO, but this has yet to be tested. SWEO is agnostic in terms of ET data and can use any raster-based ET dataset.

Soil water

Soil water controls various agronomic and ecological processes. Wet soils in arid climates suggest non-precipitation sources of water in the form of irrigation, floods or shallow groundwater. Determination of soil water by remote sensing usually makes use of thermal images (often in combination with vegetation cover) or microwave imagery. Passive microwave sensors measure the very weak natural emissions of the Earth surface after making corrections for vegetation (Jackson et al., 2011; Kerr et al., 2012; Mo et al., 1982; Owe & Van de Griend, 1998; Schmugge et al., 1986). At best, near-surface soil water values can be acquired for very large pixels (12.5–50 km) with sparse vegetation. The Soil Moisture and Ocean Salinity (SMOS) satellite measures radiation emitted in the microwave L-band (1.4 GHz). The Soil Moisture Active Passive (SMAP) satellite was to combine passive and active signals (Entekhabi et al., 2010), but the active radar system is not functioning; instead, radar-based soil moisture retrieval is based on Sentinel 3 data (Ojha et al., 2021). Lekshmi et al. (2014) review the retrieval of remotely sensed soil water.

Because of the poor spatial resolution of microwave techniques and their restriction to the near-surface layer, SWEO estimates soil water from thermal infrared measurements following the principles of a relative term of ET (e.g., Hain et al., 2011) or the evaporative fraction of the surface energy balance (e.g., Bastiaanssen et al., 1997; Scott et al., 2003). Canopy temperature reflects the access of vegetation to soil water in the entire root zone, thereby probing deeper into the soil than microwave technology. Earlier soil water detection work based on thermal infrared remote sensing (e.g., Carlson, 2007; Carlson et al., 1994; Gillies & Carlson, 1995) is based on the trapezoid between surface temperature and vegetation cover. Pang et al. (2020) compiled a good review of methods to assess soil water using thermal measurements.

Groundwater storage changes

Groundwater resources can be of immense value to ecosystems, and for domestic and industrial use. Recharge resulting from the infiltration of precipitation is the main source. Groundwater storage changes (ΔS_{gw}) are a result of differences between groundwater inflows (predominantly recharge at the river basin scale) and outflows (withdrawals, baseflow to streams, ET from groundwater-dependent ecosystems, and inter-basin groundwater flows). Due to geological variability, these underground processes are hard to quantify without calibrated local groundwater models, so GRACE satellite data are invaluable as an independent measure of changes in water storage. GRACE measures inter-satellite range changes and their derivatives between two co-planar satellites in low-altitude, polar orbits 220 km apart, achieving global coverage every 30 days; the orbits of the two independently orbiting systems are perturbed by the Earth's gravitational field, leading to inter-satellite range variations. Each system carries a Global Positioning System (GPS) receiver of geodetic quality and high-accuracy accelerometers to enable accurate orbit determination, spatial registration of gravity data and the estimation of gravity field models. Primarily, gravity fluctuations on land reflect storage changes of water, snow and ice. The value of GRACE data has been demonstrated in many hydrological studies (e.g., Jiang et al., 2014; Long et al., 2014; Rodell & Famiglietti, 2002; Sriwongsitanon et al., 2020). SWEO uses GRACE data from the Colorado Center for Astrodynamics Research (https://ccar.colorado.edu/grace/gsfc.html) to account for changes in groundwater storage.

Surface water storage changes

Lakes and reservoirs can store runoff to meet subsequent crop water demands. Changes in surface areas of open water bodies can be detected easily and accurately from spectral water indices (e.g., normalized difference water index – NDWI) and from radar backscatter signals. Changes of water bodies have recently been summarized into water occurrence maps (https://developers.google.com/earth-engine/datasets/catalog/JRC_GSW1_3_GlobalSurfaceWater); and attempts are being made to determine the width (perhaps, in future, flow) of large rivers (Yang et al., 2019).

SWEO relates the relationship between open-water surface areas (A) and water elevations (L) with water volumes using information from a pre-inundation Digital Elevation Model (Figure 2). First, water occurrence data are used to identify lakes and reservoirs. Next, changes in area can be related to changes in water level (Duan & Bastiaanssen, 2013). Unique A(L) relationships for each reservoir can be mathematically integrated to arrive at volume V changes. The relationship between the volume of water storage and the surface area (i.e., V = f(A) relationship) for each lake or reservoir is determined by SWEO and used to assess monthly surface-water storage changes (ΔS_{sw}). As an independent validation, water-level measurements can also be taken from satellite-borne altimeter data (e.g., https://dahiti.dgfi.tum.de/en/map/). For various

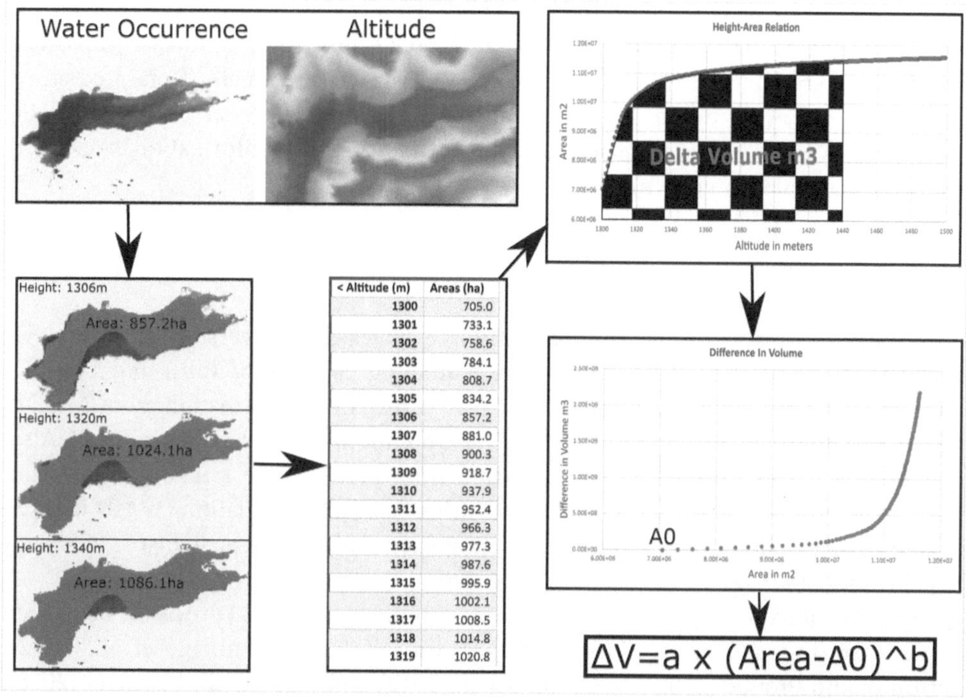

Figure 2. Remotely sensed determination of the relationship between water body surface area, water levels and water volume.

reasons, it is essential to make a distinction between natural water bodies that delay the natural runoff and reservoirs that provide a mechanism to store surface water for a longer period (months, years); SWEO uses the database of dams and reservoirs (GRAND) to differentiate between natural lakes and operational reservoirs (Lehner et al., 2011).

Land use

Land use influences hydrology and water management. For instance, forests intercept and consume a lot of water and can draw on shallow groundwater (McCulloch & Robinson, 1993), whereas urban areas have a completely opposite hydrological response. And hydrological processes affect land uses: for instance, cropland with sufficient but not abundant rainfall can be seasonally flooded and transformed into wetlands. SWEO uses land-use information to quantify and characterize the various services and benefits arising from consumptive use: bush and natural landscapes provide biodiversity and have reduced greenhouse gas emissions; irrigated agriculture provides jobs and food security, and also provides significant ecosystem services such as microclimate cooling, reduction of erosion and carbon capture.

SWEO uses customized land-use land cover (LULC) classes but, when dealing with global databases, it is often unavoidable to also include more generic land cover classes. At a minimum (pending the availability of more detailed LULC information), SWEO expresses water usage by means of the following classes retrieved from GlobCover (Arino et al., 2007) and MODIS (MCD12Q1): (1) bare land, (2) bushland, (3) forest, (4) pasture, (5) irrigated cereals, (6) irrigated non-cereals, (7) irrigated perennials, (8) rainfed crops, (9) other nature, (10) urban, (11) snow and ice, (12) natural water bodies, and (13) reservoirs (Table 1). When agencies have locally refined LULC information, this information can be incorporated into SWEO.

LAI and fPAR

In addition to ecosystem services, vegetation regulates the conversion of P into ET, enhances the infiltration capacity and water-holding capacity of soils, and provides food, stockfeed and fibre. SWEO uses leaf area index (LAI) information from MODIS to compute the partitioning of ET into E (evaporation) and T (transpiration). Also, interception and infiltration can be estimated from LAI and the fractional vegetation cover related to LAI. The fraction of photosynthetically active radiation (fPAR) is taken from MODIS to compute dry matter production from radiation using light-use efficiency (LUE) (Monteith, 1977).

Further reviews of the remote sensing to determine hydrological processes have been prepared by Engman and Gurney (1991), Schultz (1988), Schmugge et al. (2002), Pietroniro and Prowse (2002), Neale and Cosh (2012) and Frappart and Bourrel (2018).

SWEO parameterization

Hydrological input data are downloaded from independent satellite data archives. The risk is that each term has its own uncertainty and, by adding all terms together, an unrealistic water balance arises that is neither closed nor consistent: SWEO has a particular sequence of steps and procedures to avoid this. Each term is evaluated, bias corrected (when feasible), calibrated and tuned for conservation of mass. SWEO computes a two-layer vertical soil water balance for every pixel with a monthly time step. There is an upper soil water balance that represents the root zone with a fixed thickness of 0.5 m. Below is an unconfined shallow aquifer that represents a groundwater bucket with lateral exchanges. All flows between the two layers are assumed to be vertical. The upper soil water balance for the root zone is defined as:

$$P + Q_{sw} + Q_{gw} + q^{\uparrow} = ET + R_P + R_Q + q^{\downarrow} + \Delta S_{\theta} + \Delta S_{snow}$$

where P is precipitation, Q_{SW} is the surface water supply, Q_{GW} is the groundwater supply, $q^{\uparrow}$ is the capillary rise from shallow water tables, ET is the actual evapotranspiration, R_P is the surface runoff due to precipitation, R_Q is the surface runoff from water supply, $q^{\downarrow}$ is the percolation, ΔS_{θ} is the change of soil moisture storage, and ΔS_{snow} describes the storage changes of snow. Figure 3 presents these major processes where the Δ sign reveals incremental values related to Q_{sw} and Q_{gw}. These blue arrows originate both from natural (e.g., seepage, flood) and anthropogenic (e.g., pumping) water supplies.

Due to the importance of consumptive use, actual evapotranspiration ET is considered in more detail. One breakdown is to separate ET into the physically different processes E + T + I, where E is the evaporation from soil and water surfaces, T is the evaporation

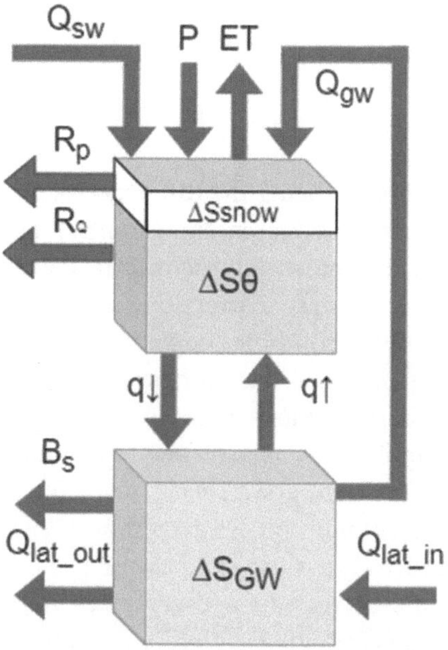

Figure 3. Vertical soil water balance applied to any pixel.

through stomata and I is the evaporation from interception; another is into the source of water, namely ET from rainfall (ET_{green}) and non-rainfall sources (ET_{blue}).

Runoff from precipitation (R_p) denotes snowmelt and overland flow. There is also runoff R_Q from the water supply (Q_{sw} + Q_{gw}) that is specified a priori for different types of land use such as the leakage from a water utility or the tail water of an irrigation furrow. The groundwater bucket receives percolation water which can be considered as recharge of the shallow aquifer. The water balance of the shallow aquifer is defined as:

$$q^{\downarrow}+Q_{lat\_in}=Q_{gw}+B_S+Q_{lat\_out}+q^{\uparrow}+\Delta S_{GW}$$

where $q^{\downarrow}$ is the recharge from leaking fields, Q_{lat} is the lateral groundwater flow, Q_{gw} is the groundwater abstraction by crops and groundwater dependent ecosystems, B_S is the base flow towards the main river for provision of dry season flow, $q^{\uparrow}$ is capillary rise, and ΔS_{GW} is the changes of groundwater storage. Recharge $q^{\downarrow}$ occurs from snow-covered land being saturated with water, forests, rivers, lakes and from irrigated land. SWEO considers recharge to be a three-month moving average of percolation water leaving the root zone because it takes time before moisture touches the zone of saturation where pores and fractures of the ground are saturated. Soil water needs to be above field capacity for generating percolation, albeit drier soils can also convey water but at a lower rate. Appendix A in the supplemental data online provides the mathematical formulation of percolation and other flows and fluxes presented in Figure 3.

Two types of bias corrections are included in SWEO. First, P is compared with individual rain gauges; bias factors can be derived to explain the difference with observations, and these factors can be inserted to modify monthly CHIRPS data. Next is the verification of the total mass balance of the river basin. Without any knowledge on internal water (re-)distribution in the basin, it can be stated in general that a bulk water balance always applies:

$$\Sigma Q = \Sigma P - \Sigma ET - \Sigma \Delta S$$

where ΣQ is the net basin outflow, being the difference between transboundary inflow (i.e., aqueduct, aquifer) and outflow (i.e., sea, sink, aquifer). Because ΣP is bias corrected by gauges and ΣET is the second most important term that far exceeds ΣS in absolute numbers, SWEO introduces a bias correction for ΣET to make the balance. A prerequisite for applying this second bias correction is that ΣQ is measured. In the absence of outflow measurements, at best the longer term ΣQ term can be approximated from global hydrological models or other sources of information. The bias correction for ΣET will, for simplicity, be applied to any pixel and to any month. This is realistic as the equation for net radiation can have a systematic deviation that applies to any pixel. A constraint is built in to prevent monthly ET exceeding 250 mm/month. Hence, ΣP and ΣET at the river basin scale are always congruent with ΣS and ΣQ.

With P and ET being bias corrected for every pixel, information on P-ET can be deployed to obtain first insights into the location of net water-generating areas (P > ET) and net water-consuming areas (P < ET). Figure 4 is an example from Afghanistan showing mountains and water divides with P-ET > 500 mm/yr simultaneously with irrigated and wetland areas with P-ET ~ –1000 mm/yr. Water must move from positive to negative P-ET values, and where river topography cannot explain this, interflow and groundwater flows must take care of this lateral transport. Hence, the vertical pixel columns are hydrologically connected through streams and aquifers.

Surface runoff ($R_P + R_Q$) is routed from each pixel to the nearby stream using topographical information. The same occurs with baseflow B_S that is, by definition, an extra source of water to the river system. However, B_S cannot drain all excess

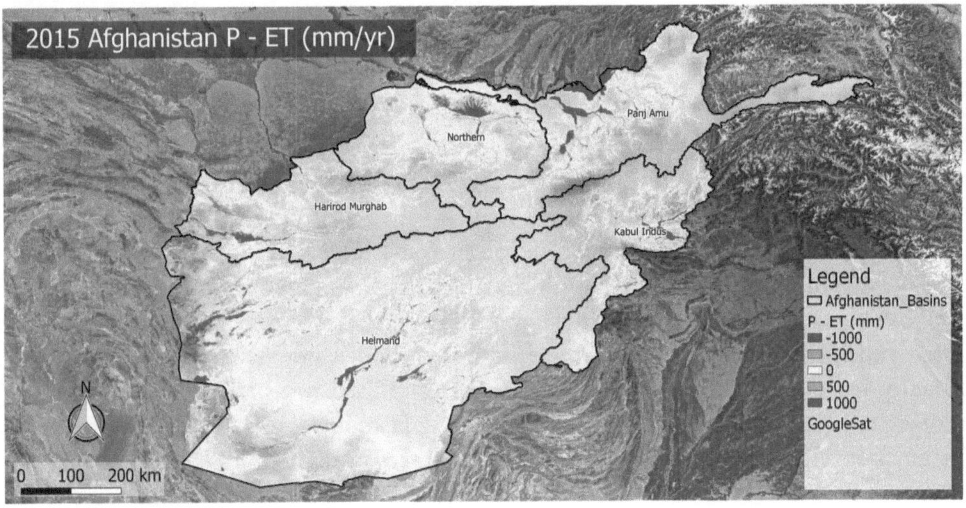

Figure 4. Annual scale values of P – ET in Afghanistan during 2015. Note: Positive values represent net water-generating areas; and negative values are net water consumers (wetlands, irrigation systems).

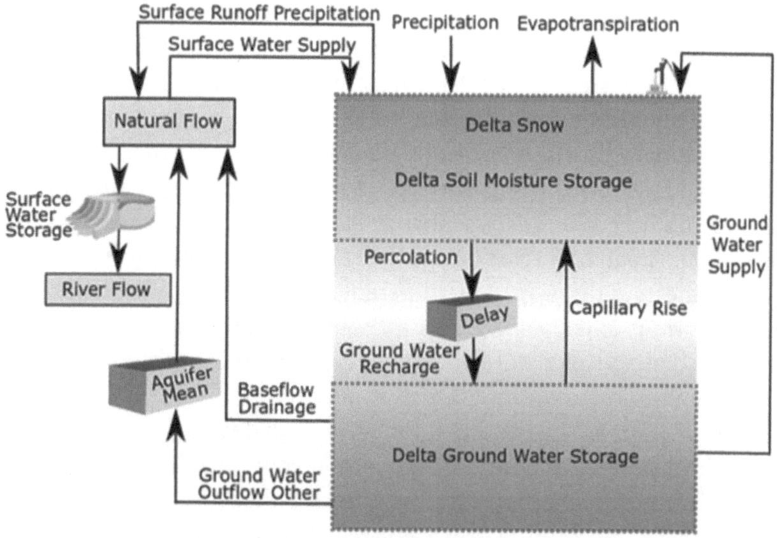

Figure 5. Schematic coupling of various components of Scalable Water Balances from Earth Observations in the regional basin context.

groundwater to rivers because it depends on hydraulic heads and hydraulic conductivity (see Appendix A in the supplemental data online). There is a maximum capacity to B_S that cannot be exceeded.

With the vertical soil and groundwater bucket for every pixel being determined, the water resources of aggregated areas such as natural catchments, sub-basins, irrigation schemes or administrative units can be quantified. Figure 5 shows the schematic coupling of the vertical pixel based balance and the regional basin including also horizontal components. In the absence of detailed, available groundwater models, SWEO considers lateral groundwater movements Q_{lat} in the regional context. The Q_{lat} sign of a pixel can be positive or negative depending on other terms of the groundwater bucket. Because groundwater storage changes ΔS_{GW} are measured by large GRACE pixels and, thus defined, a lateral groundwater process is required to obtain an equilibrium between all smaller pixels, a mixed-cell approach of all pixel-based groundwater buckets is applied to check whether there is an equilibrium. If the balance is neutral, then Q_{lat} from SWEO is considered as an assessment of local lateral groundwater movement without any geohydrological basis. If not in balance, B_S is enhanced to a certain maximum by syphoning groundwater to streams. If still not in balance, the separation of Q_{sw} and Q_{gw} will be adjusted until Q_{lat} at the regional scale (aquifer level) is zero. This is another important calibration process to get realistic groundwater balances. A key assumption in this process is that recharge is reasonably good (and has been internally calibrated from green water pixels; see the next section).

SWEO computes river flow for every 250 m reach by accumulating surface runoff (R_p and R_Q) and baseflow B_S from the nearest pixels; in general, though not in drylands, river flow increases from the upstream to the downstream ends of the basin. Lakes and reservoirs create a storage-and-release system that is included in the flow accumulation: ΔS_{sw} are computed from the V(A) relationship that varies with each surface-water body. The complexity of this system can be defined by the user by defining the minimum size of a water body that should be considered in SWEO.

Surface water is withdrawn in transit by various agro-ecosystems. These withdrawals diminish flow and large flood plains and wetlands or irrigation systems can reduce river flow to virtually nothing. So, it is essential to have a number of checkpoints for actual flow; in case of a mismatch of river flow, corrections on the ratio of surface water (Q_{sw}) versus groundwater supply (Q_{gw}) for different land-use classes are introduced. The assessments of various sinks and sources make it feasible to reconstruct a hydrograph at any reach. In the case of natural catchments with small withdrawals and supplies, corrections on infiltration and runoff are introduced.

The calculation of $q^{\downarrow}$, R_p, Q_{sw} and Q_{gw} for every pixel depends on P, ET and soil water θ. To simplify the process, first $q^{\downarrow}$ and R_p are computed for pixels where water supply can be safely excluded, hence Q_{sw} and Q_{gw} are zero for this special group of pixels. This occurs when P > ET: they are also known as *green water pixels*. The notion of *green water* was introduced by Falkenmark and Rockström (2006, p. 130) to better understand agro-ecosystems that need no water supply other than 'water from precipitation that is stored in the root zone of the soil and evaporated, transpired or incorporated by plants'. This definition is followed in SWEO. For green water pixels, $q^{\downarrow}$ can be approximated as P-ET – R_p – ΔS. ET_{green} is a natural process that is hard to change (only by land-use changes). In SWEO, a first estimation of ET_{green} is computed from the Budyko curve that prescribes the breakdown of gross rainfall into runoff and ET_{green} using the aridity index. This will give a first separation between rainfed and irrigated pixels required to calculate $q^{\downarrow}$. A more precise ET_{green} value can be computed from a separated soil water balance that does not have any other water source then P (Dogrul et al., 2017). Therefore, one soil water balance in SWEO is only for green water determination, and the same pixel has a second soil water balance for all inflows and outflows.

The counterpart ET_{blue} relates to water supply from surface water systems (streams, rivers, lakes, reservoirs, wetlands, lagoons) and groundwater systems (unsaturated zone, unconfined and confined aquifers). Floods, irrigation, seepage, interflow, capillary rise and deep-rooting plants all supply water to vegetation that evaporates into the atmosphere. The water footprint network defines *blue water* as 'water that has been sourced from surface or groundwater resources and is either evaporated, incorporated into a product or taken from one body of water and returned to another, or returned at a different time'. SWEO defines ET_{blue} as the incremental consumptive use due to water supply from a piped utility, hydrants, irrigation canal, river inflow, flood, interflow, capillary rise, perched water table, groundwater pumping or direct withdrawal from deep rooting systems. (Note: ET_{blue} is computed as ET – ET_{green}, so any error in ET_{green} is translated into ET_{blue}.) The isolation of computing percolation $q^{\downarrow}$ and surface runoff R_p from ET_{green} pixels, followed by determination of surface water Q_{sw} and groundwater supplies Q_{gw} is a practical method to solve terms of the water balance that cannot be derived from Earth observations otherwise.

SWEO case studies

Some national-scale hydrology models may serve in national water resources plans. However, hardly any river basins have their own models, and water resources planners have to rely on open-access data. Most hydrological models focus on P and river flow. They are not suitable to describing complex irrigation, wetland, bushland and forest hydrological processes, being the major users of renewable water resources. Against this background, the FAO and TUDelft have been involved in setting up SWEO studies for different basins (Table 2).

The success of merging different hydrology terms into a consistent, calibrated water balance can be checked with independent field measurements; river discharge at key locations is suitable information for such validation. Note that river flow is not a standard SWEO input parameter but infiltration, baseflow, groundwater withdrawals and surface water withdrawals are tuned to match the longer term flow at strategic locations. Figures 6 and 7 demonstrate the accuracy attainable following the standard SWEO internal calibrations.

Figure 6 shows the predicted inflow into Shardara reservoir, a key reservoir in the transboundary Syr Darya River system. Inflow at this point is of paramount importance

Table 2. Scalable Water Balances from Earth Observations case studies completed. In most cases the hydrological period 2003–19 was analysed.

Countries/region	River basin
Pakistan	Indus
Afghanistan	All basins
Central Asia	Syr Darya
Myanmar	Irrawaddy
Palestine–Israel	Jordan basin and mountain aquifer
Chile	Seven selected basins

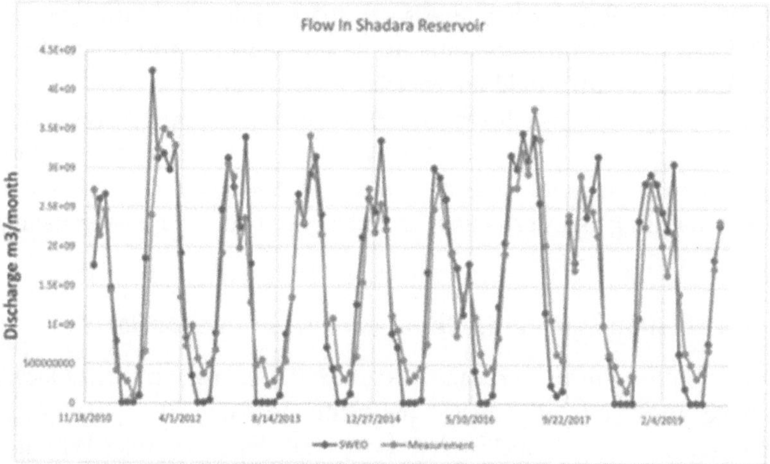

Figure 6. Estimated monthly inflow into the Shardara reservoir based on Scalable Water Balances from Earth Observations water accounting and discharge measurements by KhazHydroMet (Kazakhstan).

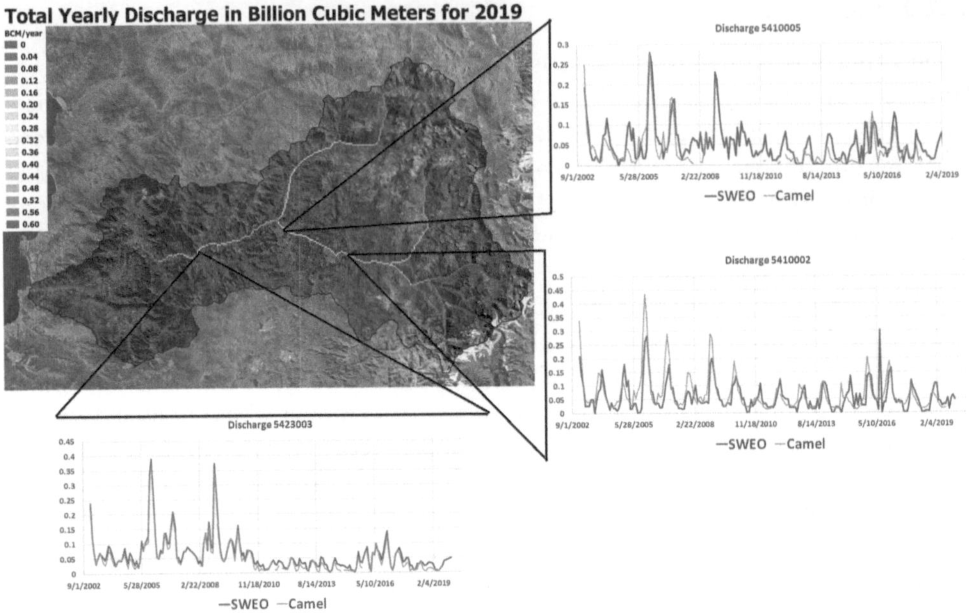

Figure 7. Estimated and measured river flow at three key discharge stations with longer term records from the Aconcagua River.

for Kazakhstan, and depends on excess water from Kyrgyzstan and other countries. The graphs show good agreement of monthly flows; the only difference is a systematically higher inflow measured during the low-flow season, which suggests that baseflow B_S should be increased or groundwater abstractions decreased to obtain a more exact match.

The second case covers three strategic discharge stations on the Aconcagua River (Chile). In this case, low flow is predicted satisfactorily, so baseflow was apparently good. Months with peak flow are determined well, and this confirms that spatio-temporal P-ET data and surface runoff R_p are accurate. Also, the partitioning into surface runoff and the withdrawals for irrigation and other water use sectors must be rather accurate, otherwise the flow at particular points would not match.

These two examples demonstrate that river flow can be rather well reproduced indirectly from satellite measurements, and that SWEO can be used for various allocation analyses including withdrawals, supplies, consumptive use, non-consumptive use and recycling. This is the basis for using the information for evaluating good water management practices.

SWEO implementation in water footprint analysis of river basins

Society's water footprint must be reduced (Allan, 2003; Chapagain et al., 2006). This will require courageous – that is, unpopular – interventions that will have to be supported by reliable water accounts: How much water do we really have available? Who is currently getting the water? Is the water used efficiently? Is it over-exploited? Can we import food grown elsewhere? Perry's (2013) ABCDE(F) framework on water resources management is germane: the first step is the determination of the water accounts (being the capital A)

Table 3. Information produced by Scalable Water Balances from Earth Observations for monthly water analysis.

Hydrology	Water management	Water resources indicators	Efficiency
P	External renewable water	Water utilization: withdrawal/available water	Water productivity
ET	Internal renewable water	Water scarcity: supply – demand	Water footprint
$q^{\downarrow}$	Water demand	Water stress: $ET_{pot} - ET_{act}$	Irrigation efficiency
$q^{\uparrow}$	Non-utilized outflow	Recycling factor: Σwithdrawal/Σsupply	Beneficial fraction
R_P	Q_{sw} per land use	Changes SW storage	Consumed fraction
R_Q	Q_{gw} per land use	Changes GW storage	
B_S	E per land use	Changes irrigated area	
Q_{sw}	T per land use	Changes wetland area	
Q_{gw}	ET_{green} per land use	Groundwater dependency	
Q_{lat}	ET_{blue} per land use		
River flow	Consumptive use, domestic		
Groundwater flow	Consumptive use, industry		
	Consumptive use, livestock		
	Consumptive use, hydropower		

that are necessary for defining longer term policies, rules, operations and laws. Withdrawals should be analytically related to caps on consumptive use, with an explicit recognition of consumptive use by ET_{green} and ET_{blue} (Yan et al., 2020).

Decisions require information on the longer term storage changes of lakes, reservoirs and aquifers, the volume of renewable water resources, natural and anthropogenic water supplies, return flows, recycling, reuse, and virtual water trade. Water-use efficiency, water productivity, water footprint, irrigation efficiency and beneficial fraction all help to evaluate whether water is used efficiently, and standard indicators are suggested by AquaStat and the Sustainable Development Goals (SDGs), both promoted by the United Nations. SWEO computes these performance indicators from water balances, vegetation photosynthesis and land use (Table 3).

Conclusions

Changing precipitation patterns and increasing demands for water have created countless situations where most, if not all, available water resources are utilized and demands can no longer be met. Such water-scarce situations demand a rapid water-accounting system that quantifies hydrological and water management processes and presents them in a clear, consistent way. The accounts must include withdrawals, supplies, consumptive use and non-consumptive use whereby flows are returned into the river systems. Simultaneously, benefits such as dry matter production and the efficient use of scarce water resources should be described, including water footprints, water productivity, and efficiency and recycling coefficients. However, no standard model can produce all this information in an automated, standardized, rapid and affordable manner.

This paper discusses a breakthrough in water accounting – SWEO – that leverages spatially and temporally explicit information from open-access Earth observation databases and integrates these independent estimates of flows, fluxes and storage changes into a consistent water balance. This parameterization includes bias factors and internal calibration coefficients to ensure closure of the water budget; *in situ* measurements of precipitation and streamflow at one or more locations, especially the river basin outflow (in the case of streamflow), are necessary. However, this is a much-reduced data demand compared with classical methods for preparing monthly water balances and river basin profiles, and vital for water accounting in data-scarce regions such as Afghanistan or Myanmar. SWEO's philosophy is to exploit recurrent Earth observation data to the greatest extent possible to provide accurate, timely, spatially discrete information in an automated manner using a standardized open-source model developed in Python. Over the last five years, six pilot studies with SWEO have been completed in collaboration with enthusiastic end-users. In order to broaden SWEO's accessibility and usefulness, and to limit data bandwidth and computing requirements, the idea of implementing SWEO on Google Earth Engine has been suggested and is currently under consideration.

This paper summarizes: (1) SWEO's essential open-access Earth observation data; (2) the primary mathematical framework; (3) related computational steps to achieve consistency through bias-corrections and calibration of typical breakdown; and (4) the partitioning of parameters (e.g., ET into E and T). Further, we validated SWEO with 17 years of river flow data in the Syr Darya and Aconcagua rivers. SWEO's formalized data sources and workflow and its flexibility that enables customized tuning for particular conditions makes this new model applicable to a range of environmental, physiographical and anthropological conditions. Its outputs are standardized and include several SDG and AquaStat indicators. Finally, critical areas of development in future versions of SWEO include: (1) recycling and reuse of non-consumptive use (Simon et al., 2020b); (2) downscaling of the groundwater storage (because GRACE satellite pixels are very large; Miro & Famiglietti, 2018; Vishwakarma et al., 2021); and (3) inclusion of locally refined land-use information to improve reporting of water-related benefits and services.

Disclosure statement

No potential conflict of interest was reported by the authors.

ORCID

Tim Hessels http://orcid.org/0000-0002-7264-7220
Jeffrey C. Davids http://orcid.org/0000-0003-4343-5410

References

Allan, J. A. (1997). *Middle East water: Local and global issues*. SOAS Water Issues Group, University of London.

Allan, J. A. (2003). Virtual water – The water, food, and trade nexus. Useful concept or misleading metaphor? *Water International*, *28*(1), 106–113. https://doi.org/10.1080/02508060.2003.9724812

Allen, R. G., Pereira, L. S., Raes, D. S., & Smith, M. (1998). *Crop evapotranspiration. Guidelines for computing crop water requirements* [Irrigation and Drainage Paper 56]. FAO

Allen, R. G., Tasumi, M., Morse, A., Trezza, R., Wright, J. L., Bastiaanssen, W., Kramber, W., Lorite, I., & Robison, C. W. (2007). Satellite-based energy balance for mapping evapotranspiration with internalized calibration (Metric) – Applications. *Journal of Irrigation and Drainage Engineering*, *133*(4), 395–406. https://doi.org/10.1061/(ASCE)0733-9437(2007)133:4(395)

Anderson, M. C., Norman, J. M., Mecikalski, J. R., Otkin, J. A., & Kustas, W. P. (2007a). A climatological study of evapotranspiration and moisture stress across the continental United States based on thermal remote sensing: 1. Model formulation. *Journal of Geophysical Research: Atmospheres, 112*(D10). https://doi.org/10.1029/2006JD007506

Anderson, M. C., Norman, J. M., Mecikalski, J. R., Otkin, J. A., & Kustas, W. P. (2007b). A climatological study of evapotranspiration and moisture stress across the continental United States based on thermal remote sensing: 2. Surface moisture climatology. *Journal of Geophysical Research: Atmospheres, 112*(D11). https://doi.org/10.1029/2006JD007507

Arino, O., Gross, D., Ranera, F., Leroy, M., Bicheron, P., Brockman, C., Defourny, P., Vancutsem, C., Achard, F., & Durieux, L. (2007). GlobCover: ESA service for global land cover from MERIS. *2007 IEEE international geoscience and remote sensing symposium*, Piscataway, NJ: IEEE

Bastiaanssen, W. G. M. (2009). *Water accountants: The new generation of water management auditors.* Inaugural Address, Delft University of Technology.

Bastiaanssen, W. G. M., Karim, P., Rebelo, L.-M., Duan, Z., Senay, G., Muthuwatte, L., & Smakhtin, V. (2014). Earth-observation based assessment of the water production and water consumption of Nile Basin agro-ecosystems. *Remote Sensing, 6*(11), 10306–10334. https://doi.org/10.3390/rs61110306

Bastiaanssen, W. G. M., Pelgrum, H., Droogers, P., de Bruin, H. A. R., & Menenti, M. (1997). Area-average estimates of evaporation, wetness indicators and top soil moisture during two golden days in EFEDA. *Agricultural and Forest Meteorology, 87*(2–3), 119–137. https://doi.org/10.1016/S0168-1923(97)00020-8

Bastiaanssen, W. G. M., & Roebeling, R. A. (1993). *Analysis of land-surface exchange processes in two agricultural regions in Spain using thematic mapper simulator data* [IAHS Publication 407-416]. International Association of Hydrological Sciences.

Budyko, M. I. (1974). *Climate and life.* Academic press.

Carlson, T. (2007). An overview of the triangle method for estimating surface evapotranspiration and soil moisture from satellite imagery. *Sensors, 7*(8), 1612–1629. https://doi.org/10.3390/s7081612

Carlson, T. N., Gillies, R. G., & Perry, E. (1994). A method to make use of thermal infrared temperature and NDVI measurements to infer surface soil water content and fractional vegetation cover. *Remote Sensing Reviews, 9*(1–2), 161–173. https://doi.org/10.1080/02757259409532220

Chapagain, A. K., Hoekstra, A. Y., Savernije, H. H. G., & Gautum, R. (2006). The water footprint of cotton consumption: An assessment of the impact of worldwide consumption of cotton products on the water resources in the cotton producing countries. *Ecological Economics, 60*(1), 186–203. https://doi.org/10.1016/j.ecolecon.2005.11.027

Chukalla, A. D., Krol, M. S., & Hoekstra, A. Y. (2015). Green and blue water footprint reduction in irrigated agriculture: Effect of irrigation techniques, irrigation strategies and mulching. *Hydrology and Earth System Sciences, 19*(12), 4877–4891. https://doi.org/10.5194/hess-19-4877-2015

Davids, J. C., Davkota, N., Pandey, A., Prajapati, R., Ertis, B. A., Rutten, M. M., Lyon, S. W., Bogaard, T. A., & van de Giesen, N. (2019a). Soda-bottle science – Citizen science monsoon precipitation monitoring in Nepal. *Frontiers in Earth Sciences, Hydrosphere, 7*, 46. https://doi.org/10.3389/feart.2019.00046

Davids, J. C., Rutten, M. M., Pandey, A., Devkota, N., van Oyen, W. D., Prajapati, R., & van de Giesen, N. (2019b). Citizen science flow – An assessment of simple streamflow measurement methods. *Hydrology and Earth System Sciences, 23*(2), 1045–1065. https://doi.org/10.5194/hess-23-1045-2019

Dinku, T., Funk, C., Peterson, P., Maidment, R., Tadesse, T., Gadain, H., & Ceccato, P. (2018). Validation of the CHIRPS satellite rainfall estimates over Eastern Africa. *Quarterly Journal of the Royal Meteorological Society, 144*(S1), 292–312. https://doi.org/10.1002/qj.3244

Dogrul, E. C., Kadir, T. N., & Brush, C. F. (2017). *IWFM demand calculator, revision 63. Theoretical documentation and user's manual.* California Dept Water Resources.

Duan, Z., & Bastiaanssen, W. G. M. (2013). Estimating water volume variations in lakes and reservoirs from four operational satellite altimetry databases and satellite imagery data. *Remote Sensing of Environment, 134*, 403–416. https://doi.org/10.1016/j.rse.2013.03.010

Engman, E. T., & Gurney, R. J. (1991). *Remote sensing in hydrology*. Chapman and Hall.

Entekhabi, D., Njoku, E. G., O'Neill, P. E., Kellogg, K. H., Crow, W. T., Edelstein, W. N., Entin, J. K., Goodman, S. D., Jackson, T. J., Johnson, J., Kimball, J., Piepmeier, J. R., Koster, R. D., Martin, N., McDonald, K. C., Moghaddam, M., Moran, S., Reichle, R., Shi, J. C., . . . Van Zyl, J. (2010). The soil moisture active passive (SMAP) mission. *Proceedings of the IEEE, 98*(5), 704–716. https://doi.org/10.1109/JPROC.2010.2043918

Falkenmark, M., & Rockström, J. (2006). The new blue and green water paradigm: Breaking new ground for water resources planning and management. *Journal of Water Resources Planning and Management ASCE, 132*, 3. https://www.eqb.state.mn.us/sites/default/files/documents/Falkenamark_20493345.pdf

Feng, B., Zhou, L., Xie, D., Mao, Y., Gao, J., Xie, P., & Wu, P. (2021). A quantitative review of water footprint accounting and simulation for crop production based on publications during 2002–2018. *Ecological Indicators, 120*, 106962. https://doi.org/10.1016/j.ecolind.2020.106962

Frappart, F., & Bourrel, L. (2018). *The use of remote sensing in hydrology*. MDPI Books.

Funk, C., Peterson, P., Landsfeld, M., Pedreros, D., Verdin, J., Shukla, S., Husak, G., Rowland, J., Harrison, L., Hoell, A., & Michaelsen, J. (2015). The climate hazards infrared precipitation with stations – a new environmental record for monitoring extremes. *Scientific Data, 2*(1), 1–21. https://doi.org/10.1038/sdata.2015.66

Gates, D. M. (1964). Leaf temperature and transpiration 1. *Agronomy Journal, 56*(3), 273–277. https://doi.org/10.2134/agronj1964.00021962005600030007x

Gillies, R. R., & Carlson, T. N. (1995). Thermal remote sensing of surface soil water content with partial vegetation cover for incorporation into climate models. *Journal of Applied Meteorology and Climatology, 34*(4), 745–756. https://doi.org/10.1175/1520-0450(1995)034<0745:TRSOSS>2.0.CO;2

Hain, C. R., Crow, W. T., Mecikalski, J. R., Anderson, M. C., & Holmes, T. (2011). An intercomparison of available soil moisture estimates from thermal infrared and passive microwave remote sensing and land surface modeling. *Journal of Geophysical Research: Atmospheres, 116*(D15). https://doi.org/10.1029/2011JD015633

Hall, D. K., Riggs, G. A., Salomonson, V. V., DiGirolamo, N. E., & Bayr, K. J. (2002). MODIS snow-cover products. *Remote Sensing of Environment, 83*(1–2), 181–194. https://doi.org/10.1016/S0034-4257(02)00095-0

Hessels, T. M. (2015). *Comparison and validation of several open access remotely sensed rainfall products for the Nile Basin* [MSc thesis]. TU Delft.

Hoekstra, A. Y. (2013). *The water footprint of modern consumer society*. Earthscan.

Hsu, J., Huang, W.-R., Liu, P.-Y., & Li, X. (2021). Validation of CHIRPS precipitation estimates over Taiwan at multiple timescales. *Remote Sensing, 13*(2), 254. https://doi.org/10.3390/rs13020254

Jaafar, H., & Mourad, R. (2021). GYMEE: A global field-scale crop yield and ET mapper in Google Earth engine based on Landsat, weather, and soil data. *Remote Sensing, 13*(4), 773. https://doi.org/10.3390/rs13040773

Jackson, T. J., Bindish, R., Cosh, M. H., Zhao, T., Starks, P. J., Bosch, D. D., Seyfried, M., Moran, M. S., Goodrich, D. C., Kerr, Y. H., & Leroux, D. (2011). Validation of soil moisture and ocean salinity (SMOS) soil moisture over watershed networks in the US. *IEEE Transactions on Geoscience and Remote Sensing, 50*(5), 1530–1543. https://doi.org/10.1109/TGRS.2011.2168533

Jackson, R. D., Reginato, R. J., & Idso, S. B. (1977). Wheat canopy temperature: A practical tool for evaluating water requirements. *Water Resources Research, 13*(3), 651–656. https://doi.org/10.1029/WR013i003p00651

Jiang, D., Wang, J., Huang, Y., Zhou, K., Ding, X., & Fu, J. (2014). The review of GRACE data applications in terrestrial hydrology monitoring. *Advances in Meteorology*, 725131. https://doi.org/10.1155/2014/725131

Karimi, P., Bastiaanssen, W. G. M., & Molden, D. (2013). Water accounting plus (WA+) – A water accounting procedure for complex river basins based on satellite measurements. *Hydrology and Earth System Sciences, 17*(7), 2459–2472. https://doi.org/10.5194/hess-17-2459-2013

Kerr, Y. H., Waldteufel, P., Richaume, P., Wigneron, J. P., Ferrazzoli, P., Mahmoodi, A., Al Bitar, A., Cabot, F., Gruhier, C., Juglea, S. E., Leroux, D., Mialon, A., & Delwart, S. (2012). The SMOS soil moisture retrieval algorithm. *IEEE Transactions on Geoscience and Remote Sensing, 50*(5), 1384–1403. https://doi.org/10.1109/TGRS.2012.2184548

Koutsouris, A. J., Chen, D., & Lyon, S. W. (2016). Comparing global precipitation data sets in Eastern Africa: A case study of Kilombero Valley, Tanzania. *International Journal of Climatology, 36*(4), 2000–2014. https://doi.org/10.1002/joc.4476

Laipelt, L., Kayser, R. H. B., Ruhoff, A. S., Ruhoff, A., Bastiaanssen, W., Erickson, T. A., & Melton, F. (2021). Long-term monitoring of evapotranspiration using the SEBAL algorithm and Google Earth engine cloud computing. *ISPRS Journal of Photogrammetry and Remote Sensing, 178*, 81–96. https://doi.org/10.1016/j.isprsjprs.2021.05.018

Lehner, B., Lierman, C. R., Revenga, C., Vörösmarty, C., Fekete, B., Crouzet, P., Döll, P., Endejan, M., Frenken, K., Magome, J., Nilsson, C., Robertson, J. C., Rödel, R., Sindorf, N., & Wisser, D. (2011). High-resolution mapping of the world's reservoirs and dams for sustainable river-flow management. *Frontiers in Ecology and the Environment, 9*(9), 494–502. https://doi.org/10.1890/100125

Lekshmi, S., Singh, D. N., & Baghini, M. S. (2014). A critical review of soil moisture measurement. *Measurement, 54*, 92–105. https://doi.org/10.1016/j.measurement.2014.04.007

Long, D., Longuevergne, L., & Scanlon, B. R. F. (2014). Uncertainty in evapotranspiration from land surface modeling, remote sensing, and GRACE satellites. *Water Resources Research, 50*(2), 1131–1151. https://doi.org/10.1002/2013WR014581

Madugundu, R., Al-Gaadi, K. A., Toler, E., Hassaballa, A. A., & Kayad, A. G. (2018). Utilization of Landsat-8 data for the estimation of carrot and maize crop water footprint under the arid climate of Saudi Arabia. *PLOS ONE, 13*(2), e0192830. https://doi.org/10.1371/journal.pone.0192830

McCulloch, J. S., & Robinson, M. (1993). History of forest hydrology. *Journal of Hydrology, 150*(2–4), 189–216. https://doi.org/10.1016/0022-1694(93)90111-L

Mekonnen, M. M., & Hoekstra, A. Y. (2016). Four billion people facing severe water scarcity. *Science Advances, 2*(2), e1500323. https://doi.org/10.1126/sciadv.1500323

Menenti, M. (1980). *Defining relationships between surface characteristics and actual evaporation rate. Meeting of working group 2 of the TELLUS project*. Joint Research Centre of the European Communities.

Miro, M. E., & Famiglietti, J. S. (2018). Downscaling GRACE remote sensing datasets to high-resolution groundwater storage change maps of California's Central Valley. *Remote Sensing, 10* (1), 143. https://doi.org/10.3390/rs10010143

Mo, T., Choudhury, J., Schmugge, T. J., Wang, J. R., & Jackson, T. J. (1982). A model for microwave emission from vegetation-covered fields. *Journal of Geophysical Research: Oceans, 87*(C13), 11229–11237. https://doi.org/10.1029/JC087iC13p11229

Molden, D. (1997). *Accounting for water use and productivity*. International Water Management Institute.

Monteith, J. L. (1977). Climate and the efficiency of crop production in Britain. *Philosophical Transactions of the Royal Society of London. B, Biological Sciences, 281*(980), 277–294. https://doi.org/10.1098/rstb.1977.0140

Moreira, A. A., Ruhoff, A. L., Roberti, D. R., Souza, V. D. A., da Rocha, H. R., & Paiva, R. C. D. D. (2019). Assessment of terrestrial water balance using remote sensing data in South America. *Journal of Hydrology, 575*, 131–147. https://doi.org/10.1016/j.jhydrol.2019.05.021

Nawaz, M., Farooq, I. M., & Irfan, M. (2021). Validation of CHIRPS satellite-based precipitation dataset over Pakistan. *Atmospheric Research, 248*, 105289. https://doi.org/10.1016/j.atmosres.2020.105289

Neale, C. M., & Cosh, M. H. (2012). *Remote sensing and hydrology*. IAHS Press.

Ojha, N., Merlin, O., Suere, C., & Escorihuela, M. J. (2021). Extending the spatio-temporal applicability of DISPATCH soil moisture downscaling algorithm: A study case using SMAP, MODIS and Sentinel-3 data. *Frontiers in Environmental Science, 9*, 40. https://doi.org/10.3389/fenvs.2021.555216

Owe, M., & Van de Griend, A. A. (1998). Comparison of soil moisture penetration depths for several bare soils at two microwave frequencies and implications for remote sensing. *Water Resources Research, 34*(9), 2319–2327. https://doi.org/10.1029/98WR01469

Paca, V., Espinoza-Davales, G. E., Moreira, D. M., & Comair, G. (2020). Variability of trends in precipitation across the Amazon River Basin determined from the CHIRPS precipitation product and from station records. *Water, 12*(5), 1244. https://doi.org/10.3390/w12051244

Pang, Z., Qin, X., Jiang, W. S., Fu, J., Yang, K., Lu, J., Qu, W., Li, L., & Li, X. (2020). The review of soil moisture multi-scale verification methods. *ISPRS Annals of Photogrammetry, Remote Sensing and Spatial Information Sciences, 5*(3), 395–399. https://doi.org/10.5194/isprs-annals-V-3-2020-395-2020

Parajka, J., & Blöschl, G. (2006). Validation of MODIS snow cover images over Austria. *Hydrology and Earth System Sciences, 10*(5), 679–689. https://doi.org/10.5194/hess-10-679-2006

Perry, C. (2013). ABCDE+ F: A framework for thinking about water resources management. *Water International, 38*(1), 95–107. https://doi.org/10.1080/02508060.2013.754618

Pietroniro, A., & Prowse, T. D. (2002). Applications of remote sensing in hydrology. *Hydrological Processes, 16*(8), 1537–1541. https://doi.org/10.1002/hyp.1018

Poortinga, A., Bastiaanssen, W. G. M., Simons, G., Saah, D., Senay, G., Fenn, M., Bean, B., & Kadyszewski, J. (2017). A self-calibrating runoff and streamflow remote sensing model for ungauged basins using open-access Earth observation data. *Remote Sensing, 9*(1), 86. https://doi.org/10.3390/rs9010086

Rittger, K., Painter, T. M., & Dozier, J. (2013). Assessment of methods for mapping snow cover from MODIS. *Advances in Water Resources, 51*, 367–380. https://doi.org/10.1016/j.advwatres.2012.03.002

Rodell, M., & Famiglietti, J. S. (2002). The potential for satellite-based monitoring of groundwater storage changes using GRACE: The high plains aquifer, Central US. *Journal of Hydrology, 263*(1–4), 245–256. https://doi.org/10.1016/S0022-1694(02)00060-4

Rosema, A., & Fiselier, J. (1990). Meteosat-based evapotranspiration and thermal inertia mapping for monitoring transgression in the Lake Chad region and Niger Delta. *International Journal of Remote Sensing, 11*(5), 741–752. https://doi.org/10.1080/01431169008955054

Savoca, M. E., Senay, G. B., Maupin, M. A., Kenny, J. F., & Perry, C. A. (2013). *Actual evapotranspiration modeling using the operational simplified surface energy balance (SSEBop) approach*. US Dept Interior, US Geological Survey.

Schmugge, T. J., Kustas, W. P., Ritchie, W. P., Jackson, T. J., & Rango, A. (2002). Remote sensing in hydrology. *Advances in Water Resources, 25*(8–12), 1367–1385. https://doi.org/10.1016/S0309-1708(02)00065-9

Schmugge, T. J., O'Neill, P. E., & Wang, J. (1986). Passive microwave soil moisture research. *IEEE Transactions on Geoscience and Remote Sensing GE, 24*(1), 12–22. https://doi.org/10.1109/TGRS.1986.289584

Schultz, G. A. (1988). Remote sensing in hydrology. *Journal of Hydrology, 100*(1–3), 239–265. https://doi.org/10.1016/0022-1694(88)90187-4

Scott, C. A., Bastiaanssen, W. G. M., & Ahmed, M. (2003). Mapping root zone soil moisture using remotely sensed optical imagery. *Journal of Irrigation and Drainage Engineering, 129*(5), 326–335. https://doi.org/10.1061/(ASCE)0733-9437(2003)129:5(326)

Senay, G. B., Gouda, P. H., Kagone, S., Bohms, S., Howell, T., Friedrichs, M., Marek, T., & Verdin, J. (2014). Evaluating the SSEBop approach for evapotranspiration mapping with Landsat data using lysimetric observations in the semi-arid Texas high plains. *Hydrology and Earth System Sciences Discussions, 11*(1), 723–756. https://doi.org/10.5194/hessd-11-723-2014

Shilpakar, R. L., Bastiaanssen, W. G. M., & Molden, D. (2011). A remote sensing-based approach for water accounting in the East Rapti River Basin, Nepal. *Himalayan Journal of Sciences, 7*(9), 15–30. https://doi.org/10.3126/hjs.v7i9.5785

Simon, G. W. H., Droogers, P., Contreras, S., Sieber, J., & Bastiaanssen, W. (2020b). Virtual tracers to detect sources of water and track water reuse across a river basin. *Water, 12*(8), 2315. https://doi.org/10.3390/w12082315

Simons, G., Bastiaanssen, W. G. M., Cheema, M. J. M., Ahmad, B., & Immerzeel, W. W. (2020a). A novel method to quantify consumed fractions and non-consumptive use of irrigation water: Application to the Indus basin irrigation system of Pakistan. *Agricultural Water Management, 236*, 106174. https://doi.org/10.1016/j.agwat.2020.106174

Soer, G. (1977). *The TERGRA model: A mathematical model for the simulation of the daily behaviour of crop surface temperature and actual evapotranspiration.* Instituut voor Culturtechniek en Waterhuishouding, The Netherlands.

Sriwongsitanon, N., Suwawong, T., Thianpopinug, S., Williams, J., Jia, L., & Bastiaanssen, W. (2020). Validation of seven global remotely sensed ET products across Thailand using water balance measurements and land use classifications. *Journal of Hydrology: Regional Studies, 30*, 100709. https://doi.org/10.1016/j.ejrh.2020.100709

Steduto, P., Raes, D., Hsiao, T. C, Fereres, E., Heng, L. K., Howell, T. A., Evett, S. R., Rojas-Lara, B. A., Farahani, H. J., & Izzi, G. (2009). Concepts and applications of AquaCrop: The FAO crop water productivity model. In W. Cao, J. W. White, E. Wang (Eds.), *Crop modelling and decision support* (pp. 177–191). Springer.

Teixeira, A. H. D. C., Padovani, C. R., Andrade, R. G., Leivas, J., Victoria, D., & Galdino, S. (2015). Use of MODIS images to quantify the radiation and energy balances in the Brazilian pantanal. *Remote Sensing, 7*(11), 14597–14619. https://doi.org/10.3390/rs71114597

Vishwakarma, B. D., Zhang, J., & Sneeuw, N. (2021). Downscaling GRACE total water storage change using partial least squares regression. *Scientific Data, 8*(1), 1–13. https://doi.org/10.1038/s41597-021-00862-6

World Economic Forum. (2014). *Global risks 2014.* www.weforum.org/risks

Wu, B.-F., Xiang, J., Yan, N., & Du, X. (2008). ETWatch for monitoring regional evapotranspiration with remote sensing. *Advances in Water Science, 19*(5), 671–678. http://skxjz.nhri.cn/en/article/id/490

Yang, X., Pavelsky, T. M., Allen, G. H., & Donchyts, G. (2019). RivWidthCloud: An automated Google Earth engine algorithm for river width extraction from remotely sensed imagery. *IEEE Geoscience and Remote Sensing Letters, 17*(2), 217–221. https://doi.org/10.1109/LGRS.2019.2920225

Yan, N., Wu, B., & Zhu, W. (2020). Assessment of agricultural water productivity in arid China. *Water, 12*(4), 1161. https://doi.org/10.3390/w12041161

Professor Tony Allan and Libya

Salem Maiar

Professor Allan's introduction to Libya

Professor Tony Allan's association with Libya went back to the mid-1960s when his old Durham friend, Keith McLachlan, persuaded him to apply for a fellowship at SOAS (the University of London's School of Oriental and African Studies) to work on Libya instead of going to work for British Aluminium. This uncharacteristically sudden decision to switch careers, as described by Tony himself, was the beginning of more than half a century of interdisciplinary teaching and research. From the outset, he was involved in organizing the Libyan University–London University Joint Research Project that lasted three years and gave him a grounding in the allocation and management of water resources in Libya and, subsequently, in the Middle East and North Africa (MENA) region.

Several visits to the country enabled him to complete, in 1971, his PhD on 'Changes in the Economic Use of Land and Water in the Vicinity of Tripoli'. He made many friends with academics, staff and students at the University of Libya and he joined Professor Richard Goodchild as a founding member of the Society for Libyan Studies, which fostered academic links between British and Libyan scholars. Doctorate accomplished, Tony began lecturing at SOAS and, with Keith McLachlan and Edith Penrose, he wrote *Libya: Agriculture and Economic Development* (1973), followed by *Libya: The Experience of Oil* (1981) and *Libya Since Independence, Economic and Political Development* (1982).

His insight and recommendations

Tony was in a prime position to observe Libya's metamorphosis from kingdom to republic and the technological and social experiments of the new, young leadership that assumed power in September 1969. Not least of these experiments was the full-throated financial commitment to the piping of non-renewable groundwater from the Nubian Sandstone Aquifer beneath the Sahara to the northern coastal belt. The Great Man-made River Project has been a social success in that bad water services have been replaced by reliable ones, but these technical and social achievements should be measured against the economics of the project and the extent to which it prevented Libyan industry and society from building up its own capabilities. Inevitably, the construction was accomplished by international consortia. The all-out commitment to this single project prevented Libya from gaining experience in alternative technologies and has blinded water users and managers to the real source of water security.

Tony was always concerned whether any government of Libya would grasp the nature of the political economy of water security – and invest accordingly: he insisted to his Libyan colleagues that water endowments do not determine food security and that water security itself lies beyond the water sector. He organized many workshops and seminars, in both Libya and London, attempting to convey to the decision-makers in Libya that the country can easily meet the *small water* needed for municipal and industrial (jobs) water – both at present and in the future – but it cannot access the enormous volume of water needed to meet its food requirements.

Libya's management of its water resources over the past four decades has been characterized by weak understanding of water endowments, the potential of green and blue water, or of the environmental services provided by water resources. It has been especially weak in marshalling the skills of investment, contract management and, particularly, the supervision of these inputs. Libya has not established institutions capable of understanding how what Tony called its *big water* security has been achieved. The role of international trade and food imports has not been grasped by its policymakers or by the Libyan people. Nor does Libya have the operational capacity to maintain efficient water supply and sewerage services, which we might describe as the *small water* element of water security.

In the period 2003–05, Libya's relations with the international community improved and, in a further attempt to advocate his vision for the country, Tony brought SOAS along with Oxford's Water Research Centre (headed by Professor Mike Edmunds, a pioneer in arid zone hydrology and champion of international collaboration) to draw up proposals, in which I was involved, for integrated inputs by Oxford and London universities to universities and other institutions in the Libyan water sector, including, *inter alia*, the General Water Authority, the Great Man-made River Project, the University of Libya, the Environment Agency and the Agricultural Research Centre. Unfortunately, as oil prices rose, the Libyan leadership lost interest in this initiative.

I also arranged for Tony and his friend Brian Chatterton, a former South Australian Minister of Agriculture and dryland farming pioneer, to visit Libya in 2009 in the course of which they gave lectures at Tripoli, Benghazi and Omar Al-Mukhtar universities and demonstrated the unexploited potential of *green water*. Tony concluded, and continued to argue, that Libya can continue to enjoy food and water security, but it could do this more cheaply, and with greater consideration for the environmental services water provides, if it changed its approach to managing and investing in water resources. The behaviour of food consumers is a big element in food and water security; current patterns of food consumption impact both environmental and human health. Necessary changes would require very difficult political, social and economic reforms that could only be implemented over a decade or more. Given the current circumstances in Libya, the main deficit is in institutional capacity. Much is to be gained by addressing the policy, institutional, managerial and operational issues of water and food security in Libya both internally and from a critical external perspective.

Tony Allan's love for Libya was deeply rooted. He was very concerned with the development of the natural and human resources of the country and he won, and kept, everyone's respect. His death is a great loss to Libya's academic tradition and to Libyan links, not only to the Society for Libyan Studies but also to British and Libyan universities at large. It is hard to imagine a future without being able to get in touch with him to access his unique expertise and the amiable conversation that would take place.

Disclosure statement

No potential conflict of interest was reported by the author.

Of intellectual friendship in *fin-de-siècle* London

Chibli Mallat

Intellectual friendship

To Tony Allan I owe joy in intellectual friendship. When I arrived in SOAS (the University of London's School of Oriental and African Studies) in October 1986, Noel Coulson, whose guidance I was seeking, had just died; there was no substitute in SOAS or elsewhere in the world. I had to fend for myself, first, by entering an immense and understudied field, then by questioning it so that it might come of intellectual age as a respectable discipline. Tony led the way for me with the most rigorous intellectual friendship possible. He and his lifelong partner Mary O'Shea were uniquely supportive.

Many of you will remember the isolation of the PhD candidate in that exacting journey; quite a few give up in the face of a hostile or careless faculty. At one point, I was seriously minded to give up or transfer. It is thanks to Tony, in Geography, and to John Wansbrough and Robin Ostle, in the Near East Department, that I stayed on at SOAS as a graduate student. It is also to them that I owe my first tenured position.

Tony was there with his subtle collegiality, always preventing petty jealousies from prevailing. Academic mediocrity is part of the field but, fortunately, so is academic elegance. To members of the younger generation lured by the demanding career of teaching and research, count your blessings if you have a Tony Allan on your side. It was more than the pleasure of regular breakfasts at Guy's around the corner, or the occasional dinner. Nor was it just academic friendship, that rarefied mentorship where the more mature scholar offers qualitative attention and support to younger colleagues whose potential he believes in and supplies with the right dose of emotional engagement. Academic mentorship, academic disciple-building, comes naturally to any dedicated teacher with an element of power and with a projection of status and hierarchy: Tony's empathy was altogether different. While we both sought multidisciplinarity, my focus on law prevented me for being his disciple, nor did he ever project himself as a mentor.

His was a special type of academic friendship, developing into a broader intellectual friendship over a full decade and remaining – to the end – with no hierarchy, no power, just a constant, subtle, intellectual engagement that went way beyond academe. Together, we entered the minefield of daily life in the Middle East, sometime prodding an Iraqi opposition in disarray; sometime engaging David Gore-Booth, a dear and influential colleague at the Foreign Commonwealth Office (FCO) with Palestine; sometime spreading academic wings with Edward Mortimer at the *Financial Times* on such unusual topics as Israeli high-technology or proto-federalism in the late Ottoman Empire; always opening up the SOAS platform to the human rights of the Middle East citizen – his freedom of speech, her right to gender equality, his right to water, her right to good governance.

Early on, Tony opened the Centre of Middle Eastern Studies for me, a mere graduate student, to convene an international conference on Islamic finance. The SOAS name and his support worked magic. Not a single expert from around the world declined the invitation of the graduate student in London. Like most of our later conferences, the international day was carried out to near perfection, on a shoestring. This was when the Apple 'Mac' had just arrived. Looking at it in antiquarian shops now, the screen appears direly small. But the little box produced wonders. Between Tony's personal Mac, which he made available to his most active graduate students, and the one that postgrads shared on the fourth floor, the Islamic Finance proceedings were ready on the day of the conference in the form of a book with a proper introduction and a consequential table of contents (Mallat, 1988). I still think this was a first, fit for the *Guinness Book of Records*.

Academic conferences remain just a blip if they leave no written trace. Another lasting lesson from the Islamic Finance meeting was that papers must be ready beforehand as an advanced draft. If the papers that would be chapters in the book are not ready, if the rational division of sessions and contributions does not match a coherent table of contents, if the authors are not told upon invitation to think of their talks as publications, if timely reminders are not sent with general directions for length and style, the chances for a book are slim. Others of Tony's lessons have stood me in good stead: do not participate in a conference if your contribution has not been put to paper – writing up forces your creative power and prevents wallowing in repetition; do not give the same lecture twice if you can avoid it – the lecture repeated is a scholar's tragedy. There was a further germane lesson, a practical one, that I learnt *a contrario*: for your equanimity, do not convene a conference *and* write a paper for it, for the administrative task elbows out the scholarship needed to produce it in time.

I still regret convening a couple of conferences at SOAS that did not result in publication. In 1990, Yemen had just united, and I organized with the Middle East Centre a multidisciplinary conference to better understand a country that was always the poor cousin of Middle Eastern scholarship. It featured some brilliant talks, but no publication; all that remains are just a few pleasant memories. Around 1992, I sought to open up SOAS's scholarship to the lively Arab scene in London with a two-day meeting held in Arabic that featured a formidable group of Arab thinkers. Again, I had not prepared the speakers adequately to contribute finished or advanced papers. The mere names of those who attended stand out in memory in a bitter yet creative decade in *fin-de-siècle* London, but nothing remains but the names of some prominent Arab writers and a few scattered papers.

These fleeting conferences were the exceptions. Most of the events organized with Tony turned into solid books, and many chapters are still read and cited. As the world scholar on water – and on scarce Middle East water – he naturally drew me to his interests. There was the Nile book which he put together with a specialist colleague (Allan & Howell, 1994), to which I contributed a chapter (Mallat, 1994). I also co-wrote the introduction with him (1995a), and contributed a chapter (1995b) for a state-of-the art book resulting from a big conference on Water in the Middle East which we convened in

November 1992 (Allan & Mallat, 1995). He was amused to find out, some years later, that it had been translated into Arabic and published in Damascus without us being approached or informed by the translators or the publisher, which was none other than the Syrian Ministry of Culture.

Conferences, I also learnt from Tony, could endanger one's career. On the plus side, they are necessary to understand topics that one cannot address alone, for they help bring together a pool of experts. Because I alone could not tackle the intricacies of Islamic finance (Mallat, 1988), of Islamic family law (Mallat & Connors, 1990), of public law in Islam (Mallat, 1993), or of commercial law in the Middle East (Lewis Ruttley & Mallat, 1995), their respective grounding conferences and lecture series helped bring together the best and the brightest to exchange ideas as they mapped the field in ways inaccessible to a single scholar.

Ambition

Yet, conference proceedings and edited books are not enough. The intellectual marker in an academic career is set by a book, or by one or more purposefully seminal articles. Our respective professional trajectories matched this acute conscience of the ultimately personal nature of scholarship. In a joint, silent understanding with Tony, each of us sought to compose a marker. It was obvious, in his case, that it would be on water, understood differently in the convergence of several disciplines. It was obvious, in my case, that it would be in Islamic law understood differently from the traditional juxtaposition of some Qur'anic verses and *hadith*, or from a focus on family law where traditional rules survived in contemporary statutes.

It took us almost the length of our careers to write it. His *The Middle East Water Question – Hydropolitics and the Global Economy* (Allan, 2001) appeared in 2001. My *Introduction to Middle Eastern Law* was published in 2007. Both, I believe, marked our respective fields with a scientific break, Bachelard's *coupure*, Kuhn's paradigm shift. No scholar writing today on water can ignore its *virtual* reality and the structural consequences that this concept has since imposed on the whole field, a concept which Tony built slowly and ever more comprehensively with careful scientific testing over three decades. I like to think that no legal scholar can ignore Middle Eastern law as a 'family of law' or 'legal tradition', a discrete world legal corpus like the common law, civil law or Roman law, informed by court decisions both in their rediscovery in the classical age of law registers and in the way they are decided by the contemporary judge in a mixture of religious and secular traditions.

Tony Allan and I kept in touch. The last time was when he was feted at the American University of Beirut for his leadership in water studies. There were several guests who were joining him as a world-renowned figure in the field. My expertise in water law had lost its edge over time, so that we did not have much of a technical discussion. There was, however, floating in the air, a muted mutual appreciation of the professional trajectory we covered, each in his own field, animated by intellectual friendship over three decades. Here we were, celebrating it again a last time over the Mediterranean sunset.

Disclosure statement

No potential conflict of interest was reported by the author.

References

Allan, J. A. (2001). *The Middle East water question: Hydropolitics and the global economy.* IB Tauris.
Allan, J. A., & Howell, P. (eds). (1994). *The Nile: Sharing a scarce resource.* Cambridge University Press.
Allan, J. A., & Mallat, C. (eds). (1995). *Water in the Middle East: Legal, political and commercial implications.* IB Tauris (Arabic translation, Damascus).
Lewis Ruttley, H., & Mallat, C. (1995). *Commercial Law in the Middle East.* Kluwer Law International.
Mallat, C. (editor). (1988). *Islamic law and finance.* Graham and Trotman.
Mallat C. (editor). (1993). *Islam and public law: Classical and contemporary studies.* Graham and Trotman.
Mallat, C. (1994). Law and the Nile river: Emerging international rules and the *shari'a.* In J. A. Allan & P. Howell (Eds.), *1994 The Nile: Sharing a scarce resource* (pp. 365–384). Cambridge University Press.
Mallat C. (1995a). Introduction (with J.A. Allan). In J. A. Allan & C. Mallat (eds.), *Water in the Middle East: Legal, political and commercial implications* (pp. 1-18). IB Tauris.
Mallat C. (1995b). The quest for water use principles: Reflections on shari'a and custom in the Middle East. In J.A. Allan & C. Mallat (eds.), *Water in the Middle East: Legal, political and commercial implications* (pp.127-137). IB Tauris.
Mallat, C. (2007). *Introduction to Middle Eastern law.* Oxford University Press.
Mallat, C., & Connors, J. (eds). (1990). *Islamic family law.* Graham and Trotman.

Recollections of a peacemaker

Munther J. Haddadin

Entry into peace-making

My professional career began in structural engineering research in the USA and after that in related fields in Jordan. I then spent a decade and a half in Jordan working on planning and implementation of integrated social and economic development in the Jordan Valley. I then entered academia teaching at American universities, publishing professional papers and worked for several years as an international consultant on water issues, culminating in three years peace-making between Jordan and Israel and as a Jordanian cabinet minister and, afterwards, running Jordan's phosphate mining company and, separately, managing urban development projects.

I came to know Professor Tony Allan in September 1991 when Jordan was mobilizing capabilities to engage in the Middle East Peace Process that started the following month in Madrid. Our meeting was in the Faculty lunch room at SOAS (the University of London's School of Oriental and African Studies). We met again later that year, and our friendship lasted until his life on Earth ended in 2021. After our initial exchange of views, Dr Allan was invited for consultation by His Royal Highness Crown Prince Hassan Bin Talal, who was assigned the difficult task of directing the Jordan Peace Delegation. HRH asked him, in my presence and the presence of the Head of the Delegation, Dr Abdul Salam Majali (later Prime Minister), if he would assist in Jordan's peace negotiations. Dr Allan answered: 'Your Royal Highness, with Dr Haddadin on board with you, you don't need any further assistance.'

I was in charge of negotiations over water, energy and the environment in the Bilateral Negotiations Conference that convened in Washington, and was head of the Jordanian delegation to the Water Resources Working Group in the Multilateral Negotiations Conference that commenced its meetings in Moscow on 26 January 1992, and met later in several other capitals – deliberating ways and means to boost peace, once it was arrived at, by implementing projects to develop water resources and cooperation among former adversaries. Dr Allan was a frequent visitor to the Middle East and would always stop by and see me, when we would converse over dinner, and discuss progress in multilateral negotiations and how to boost the stock of water resources available to countries in the Levant.

Shadow water

One day in 1996, Dr Allan stopped in Amman on his way to a conference in Cairo, carrying with him the notion of *virtual water*. We had long discussed the water needs of countries in the region and the role that trade in commodities played in satisfying such needs. Finally, he came up with this term to describe that role. He asked me: 'What do you think of this term, virtual water, to indicate the role that trade in food commodities plays in meeting the needs of countries for water?' I responded: 'My mother tongue is Arabic, not English. Although the term signifies the objective in your mother tongue, it doesn't quite do the same translated into Arabic.' I tried hard to come up with a term for the Arabic literature, and he waited, but the urgency the issue demanded did not help me engage in a creative-thinking mode! Tony was to leave for that conference in Cairo the next day and he duly introduced virtual water.

Thereafter, we saw each other often, either in London or in Amman. When I retired from government service as minister in 1998, I had time to think and write about water-resources issues, one of which was the issue of *green water*. I also had the chance to conduct research and publish, especially when I taught Water Resources and Environment at Oregon State University in 2001 and 2003. I finally came up with a term that would answer Dr Allan's question about a term, other than *virtual water*, which would reflect the intended meaning when translated into Arabic: that new term was *shadow water*, which translates as مياه الظلال.

A key role of imported commodities is to save the importing country the water that would be needed to produce the imports, food or otherwise. That saved quantity would be the shadow of the quantity used in the exporting county to produce the commodity in question, but the shadow of the actual water used is not necessarily equal to the amount actually used in its production. In other words, a ton of wheat grown in the US state of Iowa would require, say, 800 cubic metres of water, but in semi-arid Jordan it takes about 1200 cubic metres of water to produce a ton of wheat. The latter figure is the shadow of the first, but the two are not equal. Introducing the *shadow water* concept provided an analytical method to quantify the role of water saved by importing commodities, something the term virtual water does not. Tony was very pleased with this new term and encouraged me to publish it – which I did.

More interaction with Dr Allan took place over the Peace Project: Integrated Development of the Jordan Rift Valley, which had the Dead Sea–Red Sea linkage as its backbone. Dr Allan was one of three experts chosen by the World Bank to assess the technical, financial and environmental soundness of the project.

Over the years I was contacted individually by some of Tony's students, seeking assistance in their research on water resources in semi-arid areas; he had previously asked me if he might suggest to his students that they contact me and, of course, I agreed.

As a human being, Dr Allan was distinguished by integrity, candour and decency. As a professional, he was superb and almost unparalleled as an expert and a leader in his field, and he contributed greatly to the way we think about and treat water resources. My wish is that time will bring another to the profession who will be close to Tony's qualities and competence. May God have mercy on his soul.

Disclosure statement

No potential conflict of interest was reported by the authors.

Sanctioned discourse and the power of hegemonic imaginings

Charles Tripp

A conversation long interrupted

There is a fitting personal aspect to this commentary as it allows me to take up once more the threads of a conversation with Tony Allan from some 25 years ago, when we were colleagues at SOAS. The phrase 'sanctioned discourse' came up in that conversation, but the setting – standing outside the lift at SOAS – probably did not allow for greater elaboration at the time. As Tony very generously acknowledges, he was particularly struck by the concept and used it to good effect in his own fields of scholarship (Allan, 2002, p. xvi). It is for me now to expand upon what I meant by the term and to suggest how it might be used most effectively.

Sanctioned is one of those curious words that has two, in some respects opposing, senses or at least connotations. Most commonly it means that which is authorized or allowed. But it can also refer to the imposition of penalties should people act or speak in ways that are specifically disallowed. In this latter sense the verb is more rarely used than the noun *sanction* to indicate the penalty itself. Nevertheless, it is worth bearing these two meanings in mind when considering the significance of the term in relation to the construction and sustaining of a particular discourse. The first meaning could be said to be the permissive aspect that enables and empowers a particular discourse; the second acts as the prohibitive feature that keeps a discourse within permissible boundaries. Both, therefore, act upon the terms of the discourse to help shape it and to lend it its imaginative power.

In this sense, all discourse is sanctioned. That is, it is associated with the mechanisms, attitudes, norms and usages, but also the institutions and practices that help to define that a discourse is about X rather than about Y or Z. The sanctioning of a discourse is, therefore, about the authorization of the terms appropriate to a topic that help to delineate the discourse itself (and, by the same token, that prevent it from straying into taboo or prohibited zones). Sanctioning gives shape to the topography of a discourse, both its internal features but also its boundaries, marking it off from any other distinct discourse.

This is a merely definitional observation. More interesting and more significant for understanding the intimate relations between power and discourse, the role of discourse in the constitution of knowledge and the relationship of knowledge to power, are the questions that then follow. For instance, who or what sanctions (allows) a particular discourse, validates its terms, its logical structure, its limits and its internal forms, as well as the words and/or performances that may be associated with it? How do these constructs work internally and how do they lend it agency with an impact on the world? What effects and consequences does such discourse produce and what are the relations of power that are thereby enhanced or, indeed, degraded by the prevalence of a particular discourse?

In seeking to answer these questions, the more interesting implications of a *sanctioned discourse* become apparent. They can serve as the diagnostic tools that will enable the investigation of the power of a particular discourse in a specific domain, as well as the broader relations of social power in which it and the domain itself are set. From this one can understand better the nature and the limits of the authority that sanctions a particular discourse. This may be related to specific situations and moments in history, but also to the general shape and logic of discourse as a form of human exchange, of knowledge extension, and of domination. Thus, it provides a programme for the dissection of the vocabularies, idioms, and performances of those who act and articulate them within sanctioned arenas – and for the scrutiny of what has been included and excluded to sustain discursive hegemony. It furthermore opens the way for an examination of the other sociological and political features that determine questions of access, the bearing and behaviour of those allowed to participate – in short, the rules, spoken and unspoken, written and unwritten, that govern the boundaries and the forms of sanctioned discursive practice (Van Dijk, 1992).

At the same time, an understanding of the structures and dynamics of power that both sustain and are created by a particular discourse also draws attention to the forces that may contest it – their nature, forms and the sources from which they spring. Some of these may be internal to a particular discourse, as new vocabularies emerge to challenge the dominance of the sanctioned language of a domain for a variety of reasons, possibly connected to the growing inadequacy of the given to maintain a convincing explanation or narrative of the world. Some of these may be due to events that may change power relations, throwing into question or displacing the authorities whose sanction has hitherto sustained the discourse. Such features have been prominent in recent history – both in the Rhodes Must Fall campaigns, and more sharply still in the Black Lives Matter movement (Lowery, 2017; Chantiluke et al., 2018). These very sanctions are now seen no longer to have the power to police the boundaries between the acceptable and the taboo. This may be the result of long-developing processes of social change, intellectual development, political resistance and normative revision, even if historically they may seem to have been precipitated by a set of dramatic events. They may not have an immediate effect but they will work upon existing discourses, the fields in which they operate and the sanctions that keep them in place. This can lay the foundations for new discourses, backed by the newly imagined authority of a significantly altered dispensation (Morreira et al., 2021).

In his *History of Sexuality* (1986–90) Michel Foucault provides a detailed examination of the ways in which discourses – and their sanctions in both senses of the word – emerge, proliferate and help to constitute the subject, often with unforeseen and subversive consequences. He is careful to point out that this process of discourse formation can be distinct from sovereign power, that is, the power of the state. Such discourses may intersect with state power at certain moments and in certain guises, as in the criminalization of sexual acts or of forms of sexuality which bring their own punitive sanctions. But what interests Foucault is the nature of the authority behind the classification systems that distinguish the normal from the perverse, the constellations of power revealed by these distinctions and the ways in which they become internalized, as well as giving rise to several specialized discourses, each with their own vocabularies, their own authorities and their own sanctions. In the domain of sexuality and sexual behaviour the ruling on what is and what is not aberrant, and thus requiring 'treatment' of varying kinds, depended on the discursive focus, becoming central historically to a variety of disciplines: psychiatry, medicine, ethical systems (the Church but also secular norms), the state and the law (Foucault, 1981).

In this process, the constraints on and the sanctioning of what is allowed are key elements in the constitution of a discourse and in its power over social imagination. More generally, as Foucault states:

> Truth is a thing of this world: it is produced only by virtue of multiple forms of constraint. And it induces regular effects of power. Each society has its regime of truth, its general politics of truth: that is, the types of discourse which it accepts and makes function as true; the mechanisms and instances which enable one to distinguish true and false statements, the means by which each is sanctioned; the techniques and procedures accorded value in the acquisition of truth; the status of those who are charged with saying what counts as true. (Foucault, 1980, p. 131)

At the core of these processes are questions regarding the rules to be followed, what they prescribe, why people follow them, and who is authorized to set them and to supervise them. Pierre Bourdieu took this further, examining the constituent elements of social power, but also focusing on the ways in which distinct and specialized discourses took shape and began to exercise disciplinary power over specific fields. He dissected the ways in which language was used not simply to communicate with but also to exercise power over others, directly but also indirectly by the framing of a discourse and the setting of its terms. Ever alert to the inequalities and hierarchical organization of social relations, he focused on the ways in which language can be used and inflected to keep people in the place accorded to them in the social hierarchy or indeed to deny them any voice at all. He applied this to the analysis of the formation of certain professions and disciplines (with a particularly sharp and critical focus on academia), bringing out the ways in which language and behaviour, and those who would use it, are sanctioned, in both senses of the term (Bourdieu, 1991).

As Bourdieu argues, these processes help to reinforce discourses that themselves sustain distinct elites and secure their symbolic, cultural and economic capital, the possession of which has historically been a key part of the establishment and enhancement of their authority (Bourdieu, 1986, pp. 241–258). It is the accumulation of cultural and symbolic capital – closely linked to and enabled by class position, that is, the possession of economic capital – that gives authority to those who possess it. This lends weight to their sanctioning of the particular languages, performances and practices that sustain and reinforce specific discourses, as well as the maintenance of the institutional structures that enable discursive reproduction.

For Bourdieu and others, most notably – although in a rather different context – Antonio Gramsci (Zeitoun et al. 2022, in this issue) the power of hegemonic discourse was due not simply to the material power of those who were able to make the rules and to authorize what could be said and not said. It was also because the forms of censorship thereby established were often invisible to those immersed in the discourses that they took for granted and who were therefore unaware of the implicit forms of inequality and domination that sustained a particular order. For Gramsci this was particularly poignant for the predicament of the southern peasantry in Italy whose objective situation at the bottom of the class pyramid was dire, but who seemed unable to do otherwise than to accept the authority of those whose discursive strategies were geared to convincing the peasantry that this is the way the world should be (Gramsci, 1988, pp. 178–185, 325–343). Cumulatively, this meant that the 'common sense' of the ruling class became constitutive of the 'common sense' of the subalterns, limiting thereby their capacity to imagine the world otherwise.

Bourdieu also took up this theme of the dominant *common sense*: the power of a discourse that precludes certain significant questions. Although he was interested in the social and historical canvas on which class conflict and political contestation were played out, he was very much aware that the same dynamics could be applied to the understanding of disciplinary fields, to particular subjects, and to the structure of institutions and the ways of life and systems of validation that they sustained (Bourdieu, 1984). For him, elements of the unthinking acceptance of norms, rules and procedures helped to shape, if not wholly to determine, their outlook:

> structural censorship is exercised through the medium of the sanctions of the field [... the subject] is, in a way, censored once and for all, through the forms of perception and expression that he has internalized and which impose their form on his expressions. (Bourdieu, 1991, p. 138)

Nevertheless, powerful as particular discourses might be and hegemonic in their effects, all those who have examined the ways in which sanctioned discourses work have also given much thought to the circumstances under which they lose their power – the moment when their magic wears off. The reasons and causes of disenchantment have varied greatly, depending on the focus and the analytical preoccupations of the author, but underneath it lies the understanding that, as with all human endeavour, all discourses are potentially malleable and subject to change. As some of the analysts of the fallout of the global economic crisis of 2008–09 have rather ruefully pointed out, even a development of this magnitude was not sufficient to shake belief in the dominant discourse of global capitalism in its neo-liberal incarnation. It did lead to some changes in language, some recasting of norms and some re-engineering of the mechanisms of profit, but it did not fundamentally change power relations or indeed the authority of the discourse that had blinded many to the adverse consequences of globalization (Farnsworth & Irving, 2018). Such resilience is visible in any number of other fields. And yet, as Galileo (1564–1642), a key actor in the Copernican Revolution that upended the hitherto sanctioned discourse of Western astronomy, is alleged to have put it: *E pur si muove* (And yet it moves). In this case, he and others in the 16th and 17th centuries introduced a new positivist and empiricist discourse that challenged and eventually made irrelevant the hitherto accepted authority of Aristotle and of the Roman Catholic Church. In time, of course, this discourse created its own authorities and institutions that sanctioned a range of discourses linked to the diverse projects of scientific discovery and helped to shape the world that we now inhabit.

Disclosure statement

No potential conflict of interest was reported by the author.

ORCID

Charles Tripp http://orcid.org/0000-0002-8670-4521

References

Allan, J. A. (2002). *The middle east water question: Hydropolitics and the global economy*. IB Tauris.
Bourdieu, P. (1984). *Distinction: A social critique of the judgement of taste*. (tr. R. Nice). Routledge.
Bourdieu, P. (1991). *Language and symbolic power*. (tr. G. Raymond & M. Adamson). Polity Press.
Bourdieu, P. (1986). The forms of capital. In J. G. Richardson (Ed.), *Handbook of theory and research for the sociology of education* (pp. 241–258). Greenwood Press.
Chantiluke, R., Kwoba, B., & Nkopo, A. (Eds.). (2018). *Rhodes must fall: The struggle to decolonise the racist heart of Empire*. Zed Books.
Farnsworth, K., & Irving, Z. (2018). Austerity: Neoliberal dreams come true? *Critical Social Policy*, *38*(3), 461–481. https://doi.org/10.1177/0261018318762451
Foucault, M. (1981). *The history of sexuality*, Vol 1: *An introduction* (tr. R. Hurley). Penguin.
Foucault, M. (1980). *Power/Knowledge: Selected interviews and other writings 1972–1977*. (C. Gordon, Ed.) Pantheon.
Gramsci, A. (1988). *A Gramsci reader – Selected writings 1916–1935*. (D. Forgacs, Ed.) Lawrence and Wishart.
Lowery, W. (2017). *'They can't kill us all': The story of Black Lives Matter*. Penguin.
Morreira, S., Luckett, K., Kumalo, S. H., & Ramgotra, M. (Eds.). (2021). *Decolonising curricula and pedagogy in higher education – Bringing decolonial theory into contact with teaching practice*. Routledge.
Van Dijk, T. A. (1992). Discourse and the denial of racism. *Discourse and Society*, *3*(1), 87–118. https://doi.org/10.1177/0957926592003001005
Zeitoun, M., Cascão, A. E., Daoudy, M., Greco, F., Mirumachi, N., & Warner, J. (2022). Power plus: Tony Allan's contributions to understanding transboundary water arrangements. *Water International*, *47*(6), 1001–1015. https://doi.org/10.1080/02508060.2022.2125234

Water wars, conflict and cooperation – how the virtual water concept helped change the discourse

Anders Jägerskog and Jan Lundqvist

Entrenched ideas and virtual water

Professor Tony Allan used to remind his students that for new ideas or scientific concepts to take hold – no matter how right they are – usually takes decades and, sometimes, longer (Allan, 2001, 2011). Building on the work of Kuhn (1962) showing how slowly scientific paradigms change, and inspired by French philosophers such as Michel Foucault and Pierre Bourdieu, he highlighted how prevailing discourses are entrenched and, therefore, *sanctioned* through a process in which power – in all its forms – underpins the basic tenets of the discourse.

Over the years, Tony focused on how politics and differences in economic power shape the actual usage of water – and discourse on how water is used – in one particular area: that is as water embedded in agricultural produce and exported to societies in other geographical areas (Allan, 2001). He changed the way the world thinks about water and global trade through highlighting the role of this embedded water in his *virtual water* concept, for which he was awarded the Stockholm Water Prize in 2008. This did not happen overnight but led to the concept of the *water footprint*[1] and significantly affected the global water debate.

Tony cared for the next generation. He was often surrounded by young people, and he liked it. He visited the World Water Week in Stockholm 18 times and had a great impact on it, running between sessions, ignoring pains in his legs. Justice and fairness were important dimensions in his thinking. He captured the absurdity of situations where, in water-scarce regions, water is exported in the form of agricultural produce, often to countries that are relatively rich in water as well as in purchasing and bargaining power. Thus, trade in agricultural commodities is often trade in water. Contrary to conventional notions of trade, it sometimes implies a negative comparative advantage but, in other cases, trade is *the* opportunity to improve their food security. Tony had a perceptive social sensitivity on these issues and often emphasized that farmers are the de facto water managers, but they do not dictate trade and prices over the goods and services they produce. In this commentary, we posit that the concept of virtual water also changed the way in which we view conflict and cooperation.

As a young PhD Student at Linköping University in the late 1990s, my (Anders') focus was on the water conflicts in relation to Israel and the Palestinian Authority as well as between Israel and Jordan. I had begun building upon what was, at the time, the prevailing discourse – namely that the next war in the Middle East would be about water. Water scarcity was viewed from a Malthusian perspective and highlighted as a source of conflict and even war (e.g., Cooley, 1984; Falkenmark, 1986; Homer-Dixon, 1994; Shiklomanov, 1990; Starr, 1991). It was assumed that the already challenging situation in the Middle East and North Africa (MENA) would get even worse and, therefore, increase the risk of conflict as population growth continued and countries developed – thus increasing their water needs. Coupled with statements from politicians and representatives of the international system along the same lines, it seemed reasonable at the time for a student interested in research on conflicts to focus on water and how it would lead to conflict. The topic of my PhD focused on the water negotiations as part of the broader peace negotiations between Israel and the Palestine Liberation Organization (PLO) and Israel and Jordan, respectively (Jägerskog, 2003).

Upon the good advice of my PhD supervisor and co-author of this article, Professor Jan Lundqvist, I spent part of my PhD research time as a visiting researcher at the School of Oriental and African Studies (SOAS) at the University of London with Professor Allan. As my research and the academic debate on water conflict and cooperation developed, it became apparent that some of the early prophecies of *water wars* had been overly deterministic. Later in the debate, and as a result of significant undertakings establishing a quantitative understanding of water conflict and cooperation pioneered by Professor Aaron Wolf and colleagues at Oregon State University, it has been shown that conflicts over transboundary water (be it groundwater or surface water in rivers and lakes) tend to lead to cooperative outcomes rather than, as first assumed, to deeper conflict and wars (Transboundary Freshwater Dispute Database, Oregon State University).[2]

However, in addition to the research by Aaron Wolf and his team (Wolf, 1995; Wolf et al., 2005), Tony made a significant contribution in explaining *why* local conflicts over water do not lead to war. Through analyses of the global trade system and the embedded water (virtual water) that is traded, he highlighted its ameliorating role in regions where water scarcity has been most acute, such as the MENA region. At the same time, he recognized that for poorer countries with a large part of the population directly or indirectly dependent on agriculture, issues around distributive justice come to the fore. The virtual water trade has enabled a country such as Egypt – which basically ran out of water to produce the foodstuffs it needed in the 1970s – to compensate through the increased import of virtual water (Allan, 2001). Not only has Tony's contribution been important from a global trade perspective but also, arguably, from a peace and conflict perspective in that it enabled analysts to deepen their understanding of water's relation to peace and security. The early research and analyses largely ignored how imported virtual water compensated for the water shortage countries were facing and, so, decreased the risk of conflict (Jägerskog et al., 2014) so the research community overplayed the risk of conflict. While perhaps not intended as a contribution to the study of war and peace, virtual water trade brought nuance to the debate, importantly complementing the research of groups such as the Oregon School that emphasized the important role of *interdependence* between states that share a river, lake or groundwater resource.

From an international relations perspective, the basic theoretical perspective in the MENA region has been realism, drawing on the school of thought stemming from Macchiavelli and Hobbes in which the security of the state is the primary objective and everything is seen as a competition – emphasizing the zero-sum nature of any engagement. From a water perspective, this encourages a focus almost exclusively on conflict and, thereby, misses the many cooperative events on water in the region highlighted by the work of Wolf and also, recently, by Borgomeo et al. (2021). The idealist or functionalist approach (Mitrany, 1975; Wendt, 1992; Keohane, 1984; Haftendorn, 2000 taking states as rational actors that can recognize the importance of interdependence and mutuality) sees cooperative events as an outcome of rational behaviour in which, in the case of water, hydrological interdependence is emphasized. However, both of these perspectives miss the broader political economy perspective that Tony provides through the focus on virtual water that significantly decreases the need to see any interaction through a zero-sum perspective (Allan, 2001; and see Waterbury, 2022, in this issue).

As with discourse shifts in societies, it is clear that they also shift, though slowly, in the sphere of water relations in the MENA region (Jägerskog, 2002a, 2002b). Tony Allan was key not only in reshaping the discourse on water scarcity and trade in water-intensive foodstuffs but also in moving the discourse on to appreciate that cooperative outcomes – albeit not always between equally powerful partners as highlighted by Jägerskog (2003) and Zeitoun and Warner (2006) – were more likely than conflictual.

Notes

1. For example, see https://waterfootprint.org/en/.
2. See https://transboundarywaters.science.oregonstate.edu/content/transboundary-freshwater-dispute-database (accessed on 29 November 2021).

Disclosure statement

No potential conflict of interest was reported by the authors. The views expressed by Anders Jägerskog are his own and do not necessarily reflect the views of the World Bank.

References

Allan, J. A. (2001). *The Middle East water question: Hydropolitics and the global economy.* IB Tauris.

Allan, J. A. (2011). *Virtual water: Tackling the threat to our planet's most precious resource.* IB Tauris.

Borgomeo, E., Jägerskog, A., Zaveri, E., Russ, J., & Damania, R. (2021). *Ebb and flow, Volume 2: Water in the shadow of conflict in the Middle East and North Africa.* World Bank.

Cooley, J. K. (1984). The war over water. *Foreign Policy, 54*(54), 3–26. https://doi.org/10.2307/1148352

Falkenmark, M. (1986). Freshwater: Time for a modified approach. *Ambio, 15*(4), 192–200.

Haftendorn, H. (2000). Water and international conflict. *Third World Quarterly, 21*(1), 51–68. https://doi.org/10.1080/01436590013224

Homer-Dixon, T. F. (1994). Environmental scarcities and violent conflict: Evidence from cases. *International Security, 19*(1), 5–40. https://doi.org/10.2307/2539147

Jägerskog, A. 2002a *Hydrosolidarity as seen from a political perspective – Overcoming sanctioned discourse obstacles.* Balancing human security and ecological security interests in a catchment: Towards upstream/downstream hydrosolidarity. SIWI Proceedings 2002, Stockholm, Stockholm International Water Institute.

Jägerskog, A. (2002b). *The sanctioned discourse – A crucial factor for understanding water policy in the Jordan River Basin.* SOAS Occasional Paper 41. SOAS, University of London.

Jägerskog, A. (2003). *Why states cooperate over shared water: The water negotiations in the Jordan River Basin.* PhD dissertation, Linköping University, Linköping Studies in Arts and Science

Jägerskog, A., Swain, A., Öjendal, J. (2014). Water security – Origins and foundations. In A. Jägerskog, A. Swain, & J. Öjendal (Eds.), *Water security, A four volume set of SAGE major works* (pp. xxiii-xl). SAGE Publications.

Keohane, R. (1984). *After hegemony.* Princeton University Press.

Kuhn, T. (1962). *The structure of scientific revolutions.* Chicago University Press.

Mitrany, D. (1975). *The functional theory of politics.* St Martin's Press.

Shiklomanov, I. A. (1990). Global water resources. *Nature and Resources, 26*(3), 34–43.

Starr, J. R. (1991). Water wars. *Foreign Policy, 82*(82), 17–36. https://doi.org/10.2307/1148639

Waterbury, J. (2022). From zero-sum to variable-sum on the Nile. *Water International, 47*(6), 909–911. https://doi.org/10.1080/02508060.2022.2118366

Wendt, A. (1992). Anarchy is what states make of it: The social construction of power politics. *International organization, 46*(2), 391–425. http://www.jstor.org/stable/2706858

Wolf, A. T. (1995). *Hydropolitics along the Jordan River: Scarce water and its impact on the Arab-Israeli conflict.* United Nations University Press.

Wolf, A. T., Kramer, A., Carius, A., & Dabelko, G. D. (2005). Managing water conflict and cooperation. In Worldwatch Institute (ed.), *State of the World 2005: Redefining global security.* (pp. 80–99). WorldWatch Institute.

Zeitoun, M., & Warner, J. (2006). Hydro-hegemony. A framework for analysis of transboundary water conflicts. *Water Policy, 8*(5), 435–460. https://doi.org/10.2166/wp.2006.054

How virtual water saved the Middle East from water wars

Greg Shapland

Little more than 30 years ago, the notion that war between Middle Eastern states over the water resources they shared was highly likely – if not actually inevitable – began to gain currency in some writing about the region. The authors were able to cite statements by figures of international standing, including President Sadat of Egypt, in support of this contention (Starr, 1991). The idea benefitted, in terms of its ability to grab attention, from a dramatic focus on armed conflict rather than on less exciting ideas such as conflict resolution, cross-border cooperation and the global trade in food. (And, after all, had Middle Eastern states not shown a distinct proclivity to go to war over other matters, so why not water?) The *water wars* thesis also benefitted from the power of alliteration.

Proponents of the probability of armed conflict over water were right to point out that rapidly rising populations in Middle Eastern countries would need to be fed. They were also right to assert that an attempt to grow sufficient food within the region through irrigated agriculture would require quantities of water that were simply not available. They were wrong, however, to conclude that this would lead to an ever-more-intense competition for water that would eventually lead the states in the region to go to war in a Darwinian struggle to feed their citizens. Indeed, it is a matter of empirical observation that wars over water have not occurred.

The most important reason for this failure of states in the region to live up to their bellicose reputation was their pragmatic recognition that it was cheaper and less risky to buy food from abroad than to go to war to protect 'their' water resources. Middle Eastern regimes could instead reduce the volume of water available to their farmers in order to maintain the supply to more economically profitable sectors (industry, commerce, tourism, etc.) or more politically significant groups (urban consumers). That meant an increase in imports of food, mainly wheat and other grains, which was not a problem as long as the global market could provide that food and governments in the region had the forex to pay for it.

Middle Eastern leaders did not trumpet this policy, which remained, as Tony Allan put it, 'economically invisible and politically silent' (Allan, 2004). Indeed, they preferred to give the impression that they would be able to provide food for all from within the boundaries of the national territory. This they did mainly for internal political reasons: authoritarian regimes wanted to convince their citizens that even if they were deprived of most freedoms, they would at least be fed and would not have to depend on foreign sources for their food.

Although water wars have not occurred, there continue to be disputes over transboundary water resources. There is probably more than one reason for this. Foremost among them, perhaps, is the reluctance of downstream states to accept a reduction in the volume of water (and in its quality) that they formerly received. This might not be worth fighting a war over but it could still be worth contesting if negotiations or pressure from allies (or both) could mitigate the damage. Another powerful reason may be the need for regimes to demonstrate that they are defending the national interest. As a partial corollary, they may also find it useful to deflect attention from failings at home by constructing an external threat. The accompanying nationalist rhetoric is usually uncompromising and, sometimes, downright belligerent. This may help to attract the interest and support of allies but undoubtedly inhibits the making of concessions when the contending states do engage in negotiations. Egypt and Ethiopia offer a classic case: decades of hostile rhetoric have made it extremely difficult for the two regimes to grasp the win–win solution that is available to them (Brichieri-Colombi 2022, in this issue). Such disputes are not likely to go as far as armed conflict. In situations of international tension, however, one should never exclude the possibility of escalation or miscalculation, which may occur as a result of factors that have little or no connection to the declared subject of the dispute. Even if armed conflict does not occur, the absence of cooperation impedes the economic advancement of the states concerned.

There is much here that diplomacy by outside powers could do to help. To have a chance of success, diplomatic initiatives must respect national sensibilities. Moreover, they are likely to be most effective when there is a high degree of international coordination. The Trump administration's foray into the hydropolitics of the Nile in early 2020 was, in both respects, an object lesson in how not to do it. Other outside powers, including some European states, have done valuable groundwork in terms of technical assistance and cooperation between countries sharing water resources (so has the World Bank). However, they have eschewed involvement in the international politics of Middle East water.

For its part, the UK has been reluctant to engage in hydro-diplomacy in the Middle East at all. It would be pleasing to think that this is because it has embraced the concept of the invisible hand of virtual water and therefore sees no need to do so. However, the experience of working for several decades in the Foreign and Commonwealth Office (FCO) suggests to the author that the explanation lies elsewhere. One reason is undoubtedly the multiplicity of other, more immediately pressing issues, in the region, such as terrorism and internal strife of various kinds. Another has been the tendency among FCO ministers and officials to regard Middle East water as a development question rather than a political one. Meanwhile, the Department for International Development (DfID) generally chose to refrain from engaging with the subject because the core of its mission was poverty reduction – so ruling out Middle Eastern countries that fall mainly into the middle- or higher income categories. When DfID did involve itself in water in the region – in the Occupied Palestine Territories – it had the discouraging experience of seeing expensive infrastructure falling prey to military action and discontinued this activity. However, this detached stance looks set to change now that DfID has been merged with the FCO as the Foreign, Commonwealth and Development Office (FCDO).[1]

In future, there could well be a stronger case for more sustained and energetic engagement in Middle Eastern water politics on the part of outside powers, including the UK. The virtual water system has served the region very well to date in terms of maintaining food security for most people in the region and hence averting inter-state conflict. However, the continued smooth functioning of the system cannot be taken for granted. Indeed, Russia's invasion of Ukraine has demonstrated its vulnerability by disrupting the export of around 30% of the wheat normally reaching the world market via the Black Sea; other foodstuffs are affected too.[2]

Middle Eastern oil and gas exporters will be able to afford the higher prices of food on the global market that this disruption has caused. And they will provide forex to their allies – notably Egypt – so that they can do so too. In the not-too-distant future, sufficient security may be restored to the Black Sea region to allow the resumption of exports, which will reduce global food prices. But longer term challenges will remain. For one thing, the growth of the population of the planet will test the capacity of its farmers to provide enough food for the world in general: in 2050, they will have to provide for a world population of 9.7 billion people, 2 billion more than today (Roser et al., 2019).

Moreover, the impact of climate change could threaten the supply of food on the global market if crops in more than one major grain-growing region of the world are affected simultaneously. And higher demand for water in these regions may mean that its export as virtual water becomes ever more problematic. Given that these regions are to be found in different climatic zones, it seems improbable that all or most of them would experience drought at the same time. It is nonetheless possible that drought could affect crops in one region while intense rainfall events and floods affect them (or damage transport infrastructure) in others. If such a scenario materializes, Middle Eastern grain importers will face greatly increased prices.

Past form suggests that Middle Eastern regimes will, when faced with higher grain prices, make the necessary forex available to buy it: their survival could be in jeopardy if they did not. But doing so may be a good deal more difficult in future than it has been to date, as revenues flowing into the region decline as a function of the transition to renewable energy in countries that import hydrocarbons from the Middle East. Even those states in the region that do not receive income from oil and gas exports benefit indirectly through 'side payments or the employment of their nationals by the major oil and gas exporting countries' (Springborg, 2020, p. 26). The funds available to all states in the region could be significantly reduced.

This nightmare scenario may not come to pass. It is, however, sufficiently credible for it to merit the attention of the US State Department, the FCDO and other foreign ministries. Whether the pressure of other concerns will allow them the time and space to do so is another matter.

Notes

1. *Prime Minister Boris Johnson announces merger of Department for International Development and Foreign Office.* No. 10 Downing Street press release, 16 June. https://www.gov.uk/government/news/prime-minister-announces-merger-of-department-for-international-development-and-foreign-office.

2. 'Ukraine war: More countries will "feel the burn" as food and energy price rises fuel hunger, warns WFP'. https://www.wfp.org/stories/ukraine-war-more-countries-will-feel-burn-food-and-energy-price-rises-fuel-hunger-warns-wfp.

Disclosure statement

No potential conflict of interest was reported by the author.

References

Allan, J. A. (2004). Why no Middle East water wars: Global solutions to local deficits? Ch. 1.11, and F. Zereini & W. Jaeschke (Eds.), *Water in the Middle East and in North Africa* (pp. 111-125). Springer.

Brichieri-Colombi, S. (2022). Egypt's water balancing act. *Water International*, *47*(6), 912–933. https://doi.org/10.1080/02508060.2022.2118372

Roser, M., Ritchie, H., & Ortiz-Ospina, E. (2019) World Population Growth (first published in 2013; most recent substantial revision in May 2019). *Our World in Data* website. https://ourworldindata.org/world-population-growth. Last visited 31 October 2021

Springborg, R. (2020). *Political economies of the Middle East and North Africa*. Polity Press.

Starr, J. (1991). Water wars. *Foreign Policy*, *82*(82), 17. https://doi.org/10.2307/1148639

From zero-sum to variable-sum on the Nile

John Waterbury

In the early 1970s my work as an academic correspondent for the American Universities Field Staff in Egypt focused on Egypt's system of agricultural production in light of the country's rapid population growth. There was, I confess, a hint of neo-Malthusianism in this which at the time, and even today, seemed warranted. The Aswan High Dam was just coming into full operation, and the Sudan was not fully using its allotment of 18.5 bcm under the 1959 agreement between Egypt and the Sudan on the full utilization of the Nile River, but I anticipated a day when water supply would be the major constraint on Egypt's agricultural expansion. It had ambitious plans for horizontal expansion of its cultivated area and equally ambitious plans for vertical expansion through multi-cropping existing cultivated surfaces. Would the water be there? To answer that question, I needed to better understand the Sudan's plans for agricultural expansion. (Note: the Sudan had not identified significant petroleum deposits at this time.)

Agriculture and livestock appeared to be the country's comparative advantage.

In 1975 I made the first of a number of trips to the Sudan to learn first-hand about the country's plans. It soon became clear that they were quite ambitious, leading in almost any scenario to demand far exceeding Sudan's 1959 allotment. I realized that the Sudan and Egypt were locked in a zero-sum game that meant that the gains in water supply of one would come at the expense of diminished supply for the other. Zero-sum situations bear the seeds of (violent) conflict (Waterbury, 1976a, 1976b, 1976c).

I also realized that over time Egypt would become a major net importer of food and of other agricultural products while the Sudan would become a net exporter. I versed myself in relative efficiencies (crop for drop ratios) that indicated that the Sudan might enjoy advantages in agricultural production as compared with Egypt. To take but one example, the relatively vast area between the Blue and White Niles in Sudan had only been minimally exploited through the Gezira-Managil cotton scheme. Otherwise, there were millions of feddans[1] of deep alluvial soils that had never been farmed and that were open to gravity-flow irrigation because the course of the Blue Nile lies at a higher elevation than that of the White Nile. To make a long story short, I concluded that Egypt should cede some of its water to the Sudan and import water-intensive crops such as cane sugar and rice from the Sudan. The Sudan could become a welcoming environment for Egyptian investors. This was, after all, the period of Anwar Sadat's 'Open Door' policy (*al infitah*) (Waterbury, 1979, 1986).

This was about two decades before Tony Allan began to develop his notion of virtual water (Allan, 1993). Had I had it in my tool kit I would have seen how a zero-sum game could have variable-sum solutions. What I envisaged was a kind of double whammy for Egypt. It would see its share of the Nile (55.5 bcm as per the 1959 agreement) diminish all the while paying the Sudan for the virtual water embedded in the Sudan's expanded agricultural production. In retrospect, the way out for Egypt would have been to invest in Sudanese agriculture and then import virtual water from anywhere in the world that made economic sense. As it was, Sudan could not realize its enormous agricultural potential without jeopardizing Egypt's economy, or so it seemed in the pre-virtual water era.

It is important to recall the international context of all this. In October 1973, Egypt, Syria and Israel engaged in a brief war (the Ramadan/Yom Kippur war) that saw Egypt register some early victories which were undone by a large-scale Israeli counter-offensive. King Faisal of Saudi Arabia was convinced that only massive intervention by the United States in support of Israel reversed the tide of battle. He then authorized a boycott of petroleum sales on the part of the Arab members of OPEC to the United States. Oil prices increased vertiginously. There were lightly veiled threats from the United States that it might use food exports as a threat to agriculture import-dependent Arab oil exporters. Those same Arab states, led by the Kuwait-based Arab Fund for Economic and Social Development (AFESD), began to look to the Sudan as a swing producer of agricultural exports (including livestock) that could meet the demand of the petroleum exporters. Egypt at the time was seeking direct foreign investment, not providing it, so the expansion of Sudanese agriculture was not within Egypt's control.

I will not go into what actually transpired in the ensuing eight or nine years. Suffice it to say, ramping up Sudanese production in the midst of economic crisis and the reigniting of the civil war in the south meant that Egypt was given a reprieve, one that it managed quite successfully. It greatly increased the efficiency with which it used water in agriculture while successfully diversifying its sources of imported virtual water. It became the world's largest importer of wheat. Those years in the mid-1970s saw the birth in real time of what we now call the water–energy–food nexus.

We may see a re-enactment of the situation of the 1970s. The Sudan has lost its southern provinces to the Republic of South Sudan and along with it most of its petroleum deposits. So the Sudan is moving briskly and painfully to a post-oil future. Aggressive development of its agricultural sector, along with downstream processing of agricultural produce, will lead the Sudan to once again focus on developing millions of feddans of virgin land between the two Niles. In this respect, Sudan becomes the natural ally of Ethiopia from whom it will import power. The Grand Ethiopian Renaissance Dam and other projected undertakings on the Blue Nile and its tributaries can help meet the Sudan's needs for hydropower and managed flows of the river(s) for agricultural expansion. Once again, wealthy Arab investors may see an opportunity to enhance the food security of the oil exporters. Egypt is, as in the 1970s, offered an opportunity to profit, on balance, from a new variable-sum game.

Disclosure statement

No potential conflict of interest was reported by the author.

References

Allan, J. A. (1993). Fortunately there are substitutes for water, otherwise our hydro-political futures would be impossible. In *Priorities for water resources allocation and management* (pp. 13–26). Overseas Development Administration.

Waterbury, J. (1976a). The Sudan in quest of a surplus, Part I: Dreams and realities, American Universities field staff. *Northeast Africa Series, XXI*(8).

Waterbury, J. (1976b). The Sudan in quest of a surplus, Part II: Domestic and regional politics: American Universities field staff. *Northeast Africa Series, XXI*(9).

Waterbury, J. (1976c). The Sudan in quest of a surplus, Part III: Capital packages and regional prospects. *American Universities Field Staff, Northeast Africa Series, XXI*(10).

Waterbury, J. (1979). *The hydropolitics of the Nile Valley*. Syracuse University Press.

Waterbury, J. (1986). Soudan: Priorité au Nil Bleu ou au Nil Blanc? *Bulletin du CEDEJ, 19*(1), 71–109.

Egypt's water balancing act

Stephen Brichieri-Colombi

ABSTRACT

Egypt's 2007 National Water Resources Plan projected that water supply and demand would still be in balance by 2017. However, in 2010, the National Water Resources Plan – Coordination Project found that agricultural demands were far higher than previously estimated and demand already exceed the officially available resources. The first part of this paper deals with the estimation of the Nile water balance in 2010; the second discusses measures, including actions in the wider economy, that would allow Egypt to manage the situation and cohabit with its co-riparians as they implement their own plans to use Nile waters.

Introduction

In 2005, Egypt's Ministry of Water Resources and Irrigation published the National Water Resources Plan (NWRP) for the period 1997–2017, prepared with the assistance of Dutch-funded consultants (MWRI, 2005). Five years later, MWRI initiated the NWRP Cooperation Project (NWRP-CP, 2011) to help implement the plan with the assistance of other Dutch-funded consultants.[1] There were major differences in the balance of water supply and demand in Egypt as estimated in the NWRP and as reviewed in NWRP-CP Technical Report 57 (TR57).[2] Whereas the NWRP indicated that by 2017 Egypt would still be within its claimed share of 55.5 billion cubic metres (bcm)/year of Nile flow, TR57 indicated that it was already using far more than this. With resources already stretched, Egypt was vulnerable to possible reductions in Nile flows. Although the NWRP identifies many interventions that might improve water-use efficiency and marginally increase available water resources, these would have little impact on the balance between supply and demand. More radical changes are required to make the economy less dependent on the Nile.

Background

The preface to the 1997–2017 NWRP states:

> The main objective of the NWRP is to describe how Egypt will safeguard its water resources in the future, both with respect to quantity and quality and how it will use these resources in the best way from a socio-economic point of view.

Table 1. National Water Resources Plan (NWRP) selected outcomes.

NWRP (p. 4.32)	1997	2017	% Change
Population (millions)	59.3	83.1	140%
Urbanization	44%	40%	91%
Gross domestic product (GDP) (billions LE)[a]	246	789	321%
Irrigation area (millions of *feddan*[b])	7.985	11.026	138%
Crop intensity	2.1	1.5	71%
Cultivated area (millions of *feddan*)	16.769	16.539	99%
Cultivated area (millions of ha)	7.043	6.946	0.414
Net production value (LE/*feddan*)	2812	2075	74%
Net production value (LE/ha)	1181	872	74%
Net production value (LE/m^3)	0.64	0.66	103%
Export/import value	100	86	86%
Fisheries production	100%	86%	86%
Population in desert areas	3%	23%	767%
Employment in agriculture (millions/year)	5.01	6.24	125%
Employment in industry (millions/year)	2.18	4.19	192%
Average income farmers (LE/year)	5362	4629	86%
Drinking water supply coverage	97.3%	100%	103%
Sanitation coverage	28%	60%	214%
Cereal self-sufficiency	73%	53%	72%
Outflow to sinks from the Nile system (bcm)	16.3	17.6	108%
Water availability (m^3/*feddan*/year)	4495	3285	73%
Water availability (m^3/ha/year)	1888	1380	73%

Notes: [a]In 2005, US$1 = Egyptian pounds (LE).
[b]1 *feddan* = 1.03 acres, 4168 m^2, 4.168 ha.

The plan analysed the water situation in 1997 and projected outcomes in 2017 under two scenarios: the reference case if no additional measures were taken, and the second if the proposed strategy – Facing the Challenge (FtC) – were implemented (Table 1). As several of these outcomes were unwelcome, particularly to farmers who account for 25–30% of the workforce, the NWRP-CP questioned whether the NWRP should be fully implemented.

The water balance for the reference case and FtC case shows the same Nile resources, with FtC showing a reduction in outflows to the sea, largely offset by increase in flows lost to the desert, with a small net gain of 1.0 bcm to all sectors except fisheries (summarized in Table 2). To better understand the situation at the start of the implementation of the plan, the NWRP-CP TR57 made a new water balance estimate for 2010 (Table 2). By contrast, the water balance estimate produced by TR57 suggests that Egypt was already using far more water than the annual quota of 55.5 bcm agreed with Sudan in 1959.

Table 2. Comparative estimates of resource and demands in 2017 (bcm).

	NWRP	TR57	Increase
Demands			
Municipal	1.6	2.8	1.2
Industrial	0.9	1.4	0.5
Agricultural	38.7	44.8	6.1
Fisheries, fallow and watercourses	3.3	3.3	0.0
Agriculture drainage to desert	2.6	2.2	−0.4
Pumped drainage to sea	9.7	16.5	6.8
Total	56.8	71.0	14.2
Resources			
Treaty Aswan High Dam releases and rainfall	56.8	57.5	0.7
Deficit on treaty	0.0	13.5	13.5
Aswan High Dam flood control releases		6.5	

The main reason for the difference is that the NWRP was prepared in response to the political imperatives of the time, which included affirmation that the quota of Nile flows agreed with Sudan was adequate to implement various aspects of the Horizontal Expansion Programme (HEP). This is made clear in NRWP references to the three mega-projects: Toshka, El-Salam and the West Delta Project: 'It is all but certain that sufficient water would be available for these [horizontal] expansions to become viable.' However, this statement is qualified elsewhere:

> More *feddans* will share the same water, so the supply per *feddan* would drop. From an economic point of view, it is more beneficial to give the best lands full irrigation, rather than to spread it far and wide. It is therefore a highly political decision to what extent horizontal expansion should be pursued. (NWRP Measure 95)

> Present plans till 2017 will increase the stress on the system and reduce the water availability per *feddan*. The measures proposed [...] are not sufficient to bring this water availability per *feddan* at the old level and, hence, farmers will be able to grow less per *feddan*. ... Considering the increasing stress on the system and the future reduction in water availability per *feddan* it is proposed to limit the horizontal expansion and make it dependent on the availability of additional water sources. (NWRP Measure 112)

A comparison between the crop consumptive use (CCU), as calculated from the water balance and from calculations based on crop water requirements over the 14-year period from 1985 to 1998, shows that the water being applied to crops was less than optimal (NWRP, 2005, p. 91 – box); crops were already under stress by up to 5% from 1994 onwards. The concerns of the NWRP-CP were entirely justified and the above opening statement clearly demonstrates the political pressure that was being applied. HEP was meant to attract foreign investors and produce commodities for export, rather than for the local market, and generate few jobs per unit of water consumed. TR57 was unequivocal: it recommended complete suspension of all HEP projects and a comprehensive public review of the entire policy.

Modelling the water balance

Under TR57, the water balance was modelled in two different ways: the first was a static model that simply summed estimates of the inflows, net abstraction by the four sectors (municipal, industrial, agricultural and fisheries) and discharges; and the second was a dynamic model using a suite of programmes developed under the NWRP. Both models conclude that water use in Egypt greatly exceeds the supply available under the 1959 treaty.

Static model

Section 2 of TR57 explains in detail how these estimates were made. The differences in municipal and industrial demands arise largely because the NWRP assumes all wastewater returns to the Nile, whereas TR57 identifies the cities that return their wastewater to the desert or the sea and adjusts returns accordingly. However, the major difference is in the estimate of agricultural consumption: in the case of TR57, this is based on estimates of CCU of each crop in Upper, Middle and Lower Egypt; in the case of the NWRP it is arrived at as a closure term to balance total demands and resources.

TR57 examined the overall inflows and outflows from the Nile system, and the three main recycling paths in Egypt: the Nile aquifer, the drains that discharge into the Nile upstream of the Delta barrage, and the pumping stations which are mostly in the delta. The total recycling capacity from these sources is 13.6 bcm/year, rising to 19.8 bcm/year by 2017. Water that was being recycled once in 1997 was being recycled twice by 2017 but the corresponding increase in the salinity of the recycled water sets a limit on recycling as a solution to Egypt's water problems.

Over the period 1987–2017, normalized Nile inflows[3] averaged 91.4 bcm, well above long-term average flows (Figure 1), and Sudan has been taking less than its 1959 treaty quota of 18.5 bcm, so Aswan High Dam (AHD) releases have been greater than planned (Figure 2). In exceptional years (1998–2001, 2007), such releases were necessary to safeguard the integrity of the dam due to the limited capacity of the Toshka spillway.

Coastal barrages on the Rosetta and Damietta branches of the Nile release a minimum environmental flow of around 0.2 bcm/year to the Mediterranean. In addition, numerous pump stations in Lower Egypt lift water at an average salinity of around 2500 ppm from the drainage canals, where the water level is below sea level, into the coastal lakes connected to the sea. They remove some 40 million tonnes of salt each year that would otherwise build-up in the soil. Data for the period 1997–2010 from the Drainage Research Institute are shown in Figure 3.

The trend line ignores the years in which volumes were increased because of exceptional AHD releases. The increasing volume pumped is consistent with a steady increase in irrigated area in the delta and the increasing quantity of rice grown there. The estimated outflow through these stations amounts to 15.5 bcm in 2010. The record shows clearly that excess water released from the AHD is used to supply Egypt's needs, rather than simply being passed directly to the sea.

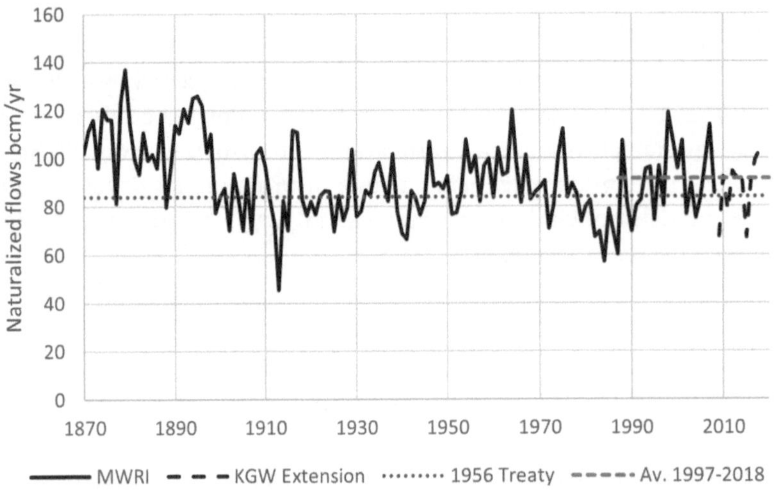

Figure 1. Naturalized flows.

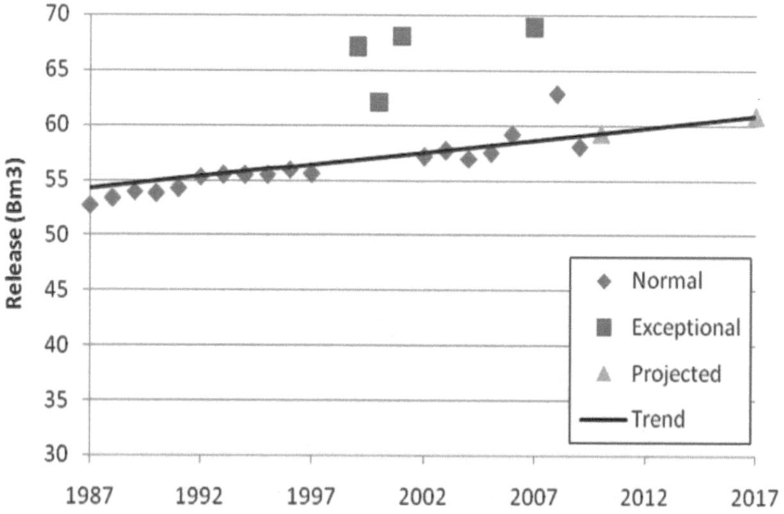

Figure 2. Aswan High Dam releases.

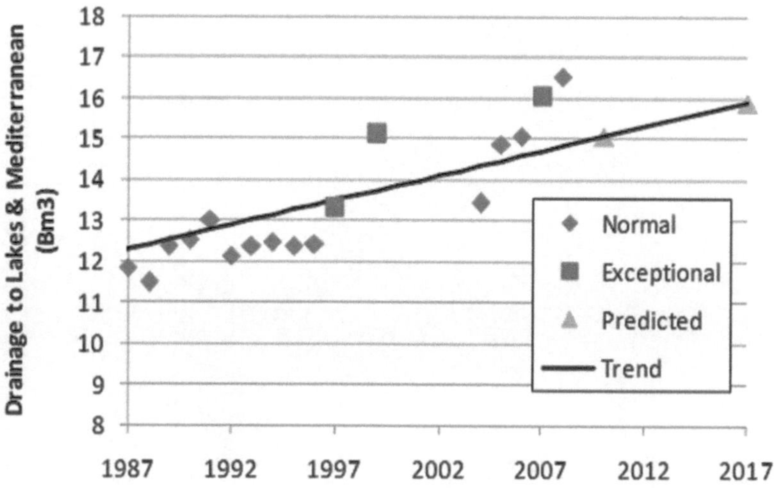

Figure 3. Drainage flows to lakes and the Mediterranean.

Dynamic model

The NWRP developed a decision support system that consists of a hydrodynamic model (RIBASIM) to simulate flows in the Nile system between the AHD and the delta barrage, and another (SIWARE) to model flows in the delta. A further module (DELWAQ) computes water quality, including salinity, at points in the system. The full complexity of the system is shown in Figure 4.

In addition, an optimization model (ASME) runs in two modes: a simulation mode that uses the actual cropping patterns; and an optimization mode that identifies the

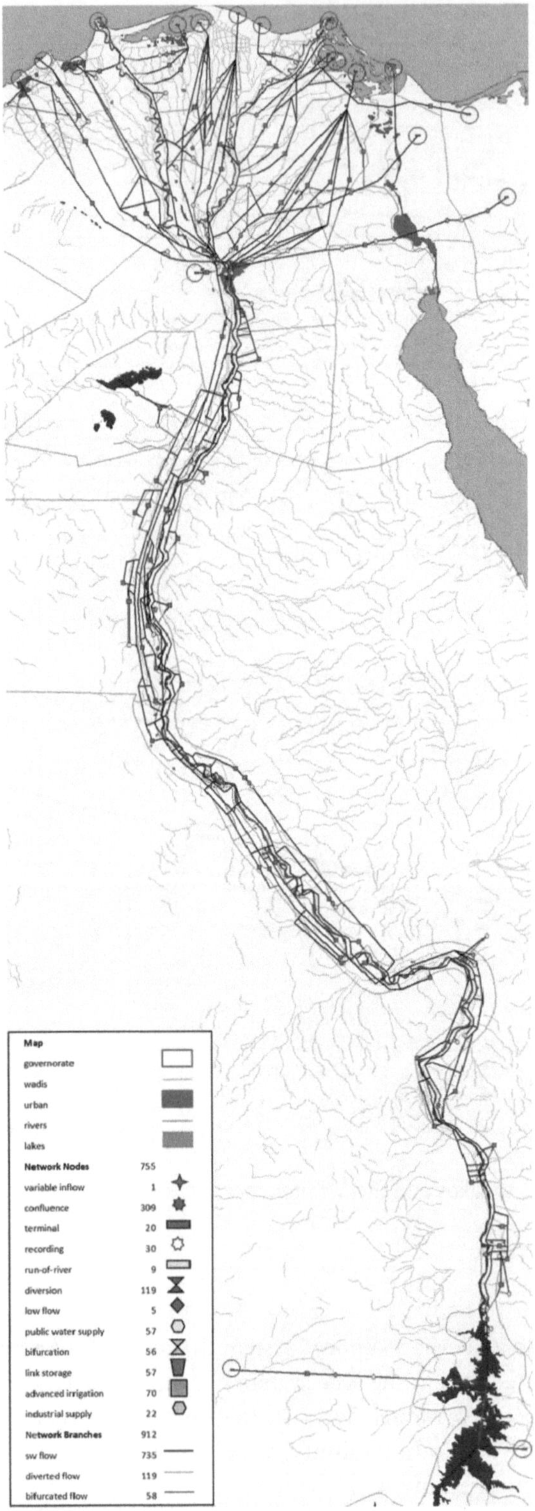

Figure 4. Water distribution system.

cropping patterns which would achieve specified objectives subject to the constraints imposed by the distribution system and other factors such as water quality and sub-optimal irrigation supply.

The decision-support system also includes a geospatial database, a module to calculate the annual evaporation from each crop, a set of constraints, an output database and an optimization routine to maximize one or other of a set of objectives. The geospatial database includes data on:

- Areas of different crops grown in each governorate in different seasons by relating cropping information collected by the Ministry of Agriculture and Land Resources to the command areas of branch canals and canals of higher order (Azim & Viala, 2011).
- Crop factors such as yield, rooting depth, salt tolerance, input costs and sale values for up to 50 main crops grown in different seasons.
- Climate factors such as rainfall and potential evapotranspiration in Upper, Middle, and Lower Egypt.
- Soil factors such as moisture-holding capacity and percolation.
- Crop consumptive use (CCU): the volume of water evaporated by individual crops, calculated from formulae embracing climate and soil factors, and date of planting (FAO, 1977, 1997). Estimates of CCU for 38 different crops grown in Lower, Middle and Upper Egypt were used in all the water-balance models; the gross crop water requirement calculated by adding losses and dividing by the system efficiency.
- Outputs such as water evaporated, transpired and lost; values and tonnages of primary and secondary products; labour and production costs. The tonnages may reflect reductions due to lack of water at critical growth periods or excessive salinity.
- A set of constraints, such as the physical area available, places limits on the options.

The model consists of 756 nodes representing features of the system (irrigation areas, pump stations, etc.) and 913 links that connect them (surface water, diverted and bifurcated flow). There is a single supply point – the inflow to Lake Nasser – and 20 end nodes which represent sinks, mainly in the Mediterranean and coastal lakes; the only sink between Aswan and Cairo is Lake Fayoum. For a given set of crops in each canal command area, the program computes the water balance in each irrigation area. However, in the calibration runs, percolation was set to zero so there is no allowance for water draining into deeper aquifers; the model is therefore forced to show that the system is highly efficient although it is difficult to believe that there is no leakage from the system covering the width of the Nile Valley over a length of 1000 km.

The model can simulate the flow in each waterway (Nile reaches, canals and drains), and estimate the effects of drought and salinity over thirty 10-day periods a year for various scenarios. The NWRP-CP report on the model (TR40) compared simulated flows at several points with those actually observed, which showed the performance was good in some places and far less good in others.

The ASME model was run using the cropped areas for 2006/07 from the Ministry of Agriculture and Land Resources' estimates of municipal and industrial use, the actual

Table 3. ASME model water balance and impacts for the Nile system, 2006/07.

Mode	Opt	Sim	Opt	Sim
Inflow (bcm)	55.5	61.5	61.5	75.0
Outflow (bcm)				
Evaporation from water surfaces	3	3	3	3
Municipal	1.6	1.6	1.6	1.6
Industrial	0.4	0.4	0.4	0.4
Fishponds	1.7	1.7	1.7	1.7
Fallow	0.3	0.2	0.3	0.2
Net crop consumptive use (CCU) from surface water	33.8	36	37.4	42.7
Drainage outflow to sinks	4.0	4.1	5.5	4.8
Drainage outflow to sea	10.6	14.3	11.5	15.8
Environmental flow to sea	0.2	0.2	0.2	0.2
Additional Nile flow to sea	0	0	0	4.7
Total	55.5	61.5	61.5	75.0
Surplus at market prices (billions LE)	775.0	757.2	775.4	761.1
Surplus (% of optimal at 55.5 bcm)	100.0%	97.7%	100.0%	98.2%
Impacts				
Irrigated area surface irrigation (millions of *feddans*)	8.65	8.65	8.65	8.65
Cropped area surface irrigation (millions of fedddans)	12.91	14.77	13.77	14.77
ET maximum/ET actual (%)	97.5	87.6	99.0	100.0
Values at farm gate prices				
Returns to water (LE/m^3)	3.7	3.2	3.3	2.9
Marginal value of water (LE/m^3)	0.13	1.97	0.01	0
Gross agricultural production value (billons LE)	154.0	143.0	155.6	155.9
Net agricultural production value (billons LE)	123.8	114.1	125.1	124.6
Net agricultural production value (LE/fed)	14,310	13,188	14,458	14,398

Notes: Effective rainfall of 0.6 bcm throughout.
Sim, simulation mode; Opt, optimization mode.

AHD release, and the outflow as documented by the Drainage Research Institute. The results of four runs are shown in Table 3.

The first run simulated the actual situation and found it to be infeasible because the model constrains the allocation to agriculture to a minimum of 85% of demand (below which crop yields fall so far that they would not be grown). This indicates that in 2006/07 the cropped area had already expanded beyond the point where farmers could get the water they needed to satisfy crop water requirements. Other runs show the CCU is less than the water required by freely transpiring crops, even with an inflow of 61.5 bcm. The fact that farmers did manage to produce crops in that year, despite the apparently infeasible conditions, suggests that much more water was being released into the system than shown in the model; the full agricultural demand (a CCU of 42.7 bcm) was found by running the model with an artificially high inflow of 75 bcm. Constraints on agricultural exports at the time meant that there is very little advantage to both producers and consumers in Egypt however much the inflow exceeds 55.5 bcm. The marginal value of water of 0.13 LE/m^3 in optimum mode with an inflow of 55.5 bcm corresponds to US$23 million/bcm at the rate of exchange in 2007 of 5.64 LE to the dollar, or about US$28 million/bcm in 2021 dollars.

Other models

Other modellers have come to similar conclusions about the effectiveness of changing cropping patterns. Researchers from three Egyptian research institutes

Table 4. Water saving by changing crop rotations (thousands m³/ha).

Zone	Region	Present	Proposed	Reduction	Saving (%)	Area (millions of ha)	Average saving (%)
Old lands	Lower	45.9	42.7	3.3	7%	1.70	8.0%
Old lands	Middle	50.3	46.3	4.0	8%		
Old lands	Upper	53.4	48.7	4.6	9%		
Calcareous land	Lower	48.0	38.5	9.5	20%	0.04	19.8%
Sandy soil	Lower	36.4	36.2	0.2	0%	0.80	0.7%
Sandy soil	Middle	39.3	39.1	0.2	1%		
Sandy soil	Upper	42.5	42.0	0.5	1%		
Salt-affected land	Lower	57.8	47.5	10.3	18%	0.90	17.8%
Sugarcane areas	Upper	40.7	29.5	11.2	28%	0.03	27.6%
Total/average						3.47	9.1%

Note: Areas are the author's estimates as they were not shown in the study.

computed the value of irrigation water consumed by each crop in each governorate and derived an *optimum* cropping pattern that maximized these returns (El-Gafy et al., 2014). They assessed the water requirements of 22 summer crops, 21 winter crops and 6 *nili* (July–October) crops grown in 22 governorates. The value of water with the actual cropping pattern in 2009 ranged from 2.25 LE/m^3 in Giza to 0.67 LE/m^3 in Aswan, with an average of 0.98 LE/m^3. The optimum cropping pattern in the exemplar governorate, Behiera, eliminated sugarcane, reduced tomato and onion by 21–22%, and rice by 10%; with vegetables and sugar beet increased by up to 30%. However, this pattern brought about a water saving of only 2%, an increase in total net value of 0.4% and a 2% increase in average economic value to 1.01 LE/m^3. Additionally, Table 4 summarizes the overall effect of modelled changes of crop rotations in nine different areas computed by researchers from the Egyptian Agricultural Research Centre (Ouda & Zohry, 2015).

Half the overall reduction of 9.1%, is due to the elimination of sugarcane, which is an obvious target given the high productivity of sugar beet in new lands. However, powerful lobbies would have to be overcome to close down sugarcane production.

Conclusions on water balance

All the modelling suggests that little is to be gained from optimizing cropping patterns, in terms of either water saving or profit. This is unsurprising – Egyptian farmers have been in business for five millennia, they know the best crops to grow and, since the agricultural reforms of 1986 and 1990, they have been free to choose what to grow, save conditions on the areas of cotton and sugarcane; but the modelling assumes, unrealistically, that farmers will grow whatever is in the national interest or follow instructions from higher authorities.

The water balances show that Egypt is already fully utilizing not only the 1959 treaty flow of 55.5 bcm but also the additional water released due to high Nile inflows. The dynamic model suggests the excess was 7.5 bcm in 2007, while the static model suggests the excess would have risen to 11.1 bcm in 2010 if all crops have been supplied with their water requirements, although it is probable that there were some shortages. Differences are inevitable but the only conclusion can be that Egypt is using far more than its 1959

Nile water quota. Egypt has been protected from the negative consequences of the Horizontal Expansion Program by increased Nile inflows but, even with this additional inflow, the ASME model shows the crops are using 12% less than the crop water requirements. The situation is unsustainable and severe readjustment will be required if Nile flows return to normal.

Managing water scarcity

Egypt has exceeded the limit of agricultural production with a Nile water supply of 55.5 bcm/year. Since water demands cannot be met from the available resources, which are themselves under threat, other means of managing water scarcity must be found. Some of these are discussed below, grouped broadly into agricultural, augmentation and myriad socio-economic measures that move well beyond the water sector per se. None will be a silver bullet, and all are subject to the uncertainties of climate change.

Agricultural measures

Measures to date include a change from furrow to overhead or trickle irrigation in desert areas. Small amounts of water might be reallocated from other sectors, and yields raised by introducing higher yielding cultivars, improving water quality in the delta by controlling discharges, and land and soil management practices. These are unlikely to make a major impact and the magnitude of the water deficit far exceeds any possible savings in the municipal and industrial sectors, so adjustments must come from the agricultural sector.

This will not be easy. Any reduction in supply to the Old Lands will reduce the area cultivated there and affect the lives of millions of people; 32% of the workforce in Egypt is employed in the sector, mostly in these lands. Although some of this area is being lost to urbanization, the corresponding reduction in water demands is far from enough to allow continued expansion into New Lands. Other ways must be found to manage the water deficit.

The priority now assigned to the agricultural sector is reflected in the Memorandum of Agreement signed on 27 October 2007 by the Minister of Water Resources and Irrigation and Minister of Agriculture, in which the MWRI agreed to provide whatever water requirements were needed according to the cropping pattern selected by farmers, 'to avoid any shortages of water delivery needed by different crops'. As there is no proviso to suggest that supply can be curtailed in times of drought or shortage,[4] this policy can only increase vulnerability to water scarcity.

The NWRP notes that the yields of almost all crops in the New Lands are lower than in the Old Lands: for the 11 crops for which comparisons were possible in 2008, yields on new lands are lower than those on old lands in Upper, Middle and Lower Egypt by 21%, 8% and 13%, respectively. Moreover, since water used on new lands is much less likely to return to the Nile, the productivity (t/m^3 of water) is also less and, although investors in the New Lands grow crops for export rather than domestic staples, this does not mean it is better for Egypt's national economy to allocate water to New Lands. The HEP does not change the amount of water consumed by crops: it merely diverts the benefits from farmers on old lands to those on new lands, and diverts drainage from the Mediterranean to the desert.

The ASME model examined the trade statistics and calculated how much water was involved in the production of imported and exported crops for the year 2009–10 (TR75, p. 46). Imports were equivalent to a consumptive use of 13.3 bcm (24.2 bcm gross), and exports 3.5 bcm (6.3 bcm gross), so the effect of importing food is to add almost 18 bcm (29%) to the amount of water in the budget. For comparison, Gad and Ali (2009, pp. 301–310) estimated that, in 2005, Egypt imported the equivalent of 12.8 bcm of virtual water, in the shape of 9 million tons of wheat and maize, and exported 1.3 bcm (11.5 bcm net). They argued for optimization of economic use of water – producing (and exporting) high-value crops and commodities with the scarce internal water resources and importing low-value commodities with high a virtual water content, such as rice.

Overall, it appears unlikely that Egypt will be able to significantly improve its water security through agricultural measures.

Augmentation measures

For millennia, Egypt was the regional hegemon, expecting the gods to assure the annual floods on the Nile, and seeking little from the upper riparians except slaves, ivory and elephants for its armies. Briefly, in the 13th century, there were concerns that Ethiopia might be able to influence the flow of the Nile, but those soon passed. In the 19th century, cotton became an important cash crop. Because of its planting date, it became necessary to store water for release prior to the annual flood and this led to the construction of dams of increasing height at Aswan to store the relatively silt-free water at the tail end of the flood season.

The comfortable view of the Nile as a reliable source was disturbed by the suggestion of a French railway engineer, Victor Prompt, that a dam could be built at Fashoda, a few kilometres downstream of Malakal, and operated in a way that would disrupt the cotton crop. Measures to assure the Nile flow for Egypt by occupation of, or treaty with, upper riparians or their colonial masters were soon initiated and were completed in 1934 when Belgium agreed that any water abstracted from the Kagera would be returned to the river before it crossed the border.

Meanwhile, planners set to work to find places for over-year storage before settling on a high dam at Aswan in exchange for a concession to allow increased abstraction by Sudan, and joint financing of the Jonglei canal project intended to increase Nile flows. Egyptian hegemony allowed them to ignore the protests of other riparians, but this hegemony has been weakened by the arrival of China as a source of development finance rivalling the Western institutions in which Egyptians had much influence, and the fungible nature of finance allows counties such as Ethiopia to raise funds for dam projects whether there is direct or indirect external investment. Despite the treaties, generally repudiated by the other riparians, new hydraulic works are being built, most notably in Ethiopia, where the Tekezze Dam with a live storage of 5.3 bcm was completed in 2009 and the Grand Ethiopian Renaissance Dam (GERD), with a live storage of 59 bcm, is nearly complete.

In 1909, Sir William Garstin first proposed a project to build a canal to bypass the Sudd and reduce evaporation losses there; similar proposals have been made for the Machar Marches in the Baro-Akobo catchment. The Jonglei I project would augment net

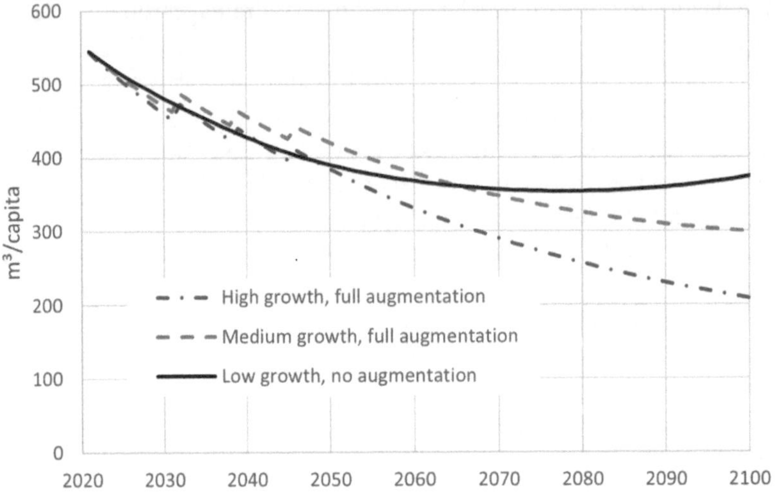

Figure 5. Augmented Nile flow per capita.

flows by some 3.8 bcm/year, Jonglei II by 3.4 bcm/year and the Machar Marshes canalization by 3.2 bcm/year (Howell & Allen, 1994, p. 215). Fierce local opposition brought about the abandonment of the 360 km-long Jonglei canal after two-thirds had been built and the Government of South Sudan made it clear in 2013 that 'any potential plans to develop Jonglei canal would be strongly opposed' (ENTRO, 2014, p. 35). This is unsurprising as South Sudan would suffer all the social and environmental costs without any direct benefits, but, in purely economic terms, it could be a good investment. The cost estimate for Jonglei I in 1972 was £32 million (about US$213 million in 2021).[5] If each project took seven years to construct with cost proportional to the net flow increase, operation and maintenance costs were 10% of capital costs, and benefits to Sudan and Egypt were US$28 million per bcm as calculated above, the internal rate of return would be 22%, less the social and environmental costs arising in South Sudan.

However, these flow-augmentation possibilities would make little difference in the face of population growth. Figure 5 shows the Nile flow per capita for high, medium and low population growth projections (United Nations [UN], 2020) if all three drainage projects were implemented in succession, starting in 2025. The augmentation schemes would be a poor substitute for population management policies and delay severe water scarcity by only a few years unless they were accompanied by low population growth.

A proposal to create a Nile–Congo link to pump 110 bcm/year from the Congo to the Nile was briefly considered in 2013 but rejected by MWRI on numerous grounds (Al-Youm, 2015). In the same year, it was proposed to make the Nile navigable from Lake Victoria to Alexandria at a cost of US$10–12 billion (Valadi, 2020) and, in 2016, the African Development Bank (AfDB) approved a loan to MWRI for prefeasibility studies that investigated six proposed routes, all of which included a canal past the Sudd. For safe navigation on the Nile, a canal 100 m wide with a depth more than 2.3 m is required (Abdel-Aziz, 2004), around half the conveyance capacity of the 5 m-deep Jonglei canal.[6] If it were built, the argument in favour of dredging it by another 1 m or so to equal the conveyance of the Jonglei Canal would be almost irresistible. Even so, augmentation projects are unlikely to be a solution to water scarcity in Egypt.

Adapting to upstream developments

In 1959, Egypt and Sudan agreed between them that they would share the entire flow of the Nile, with a quota of 55.5 bcm going to Egypt. As seen in the NWRP, water planning in Egypt is based on this flow. The other riparians have never accepted this agreement and wish to abstract water from the river for their own needs: such developments could affect Egypt's water security.

Although colonial-era treaties generally forbid such abstraction, Egypt has tolerated various projects that, in principle, contravene them: Tanzania has abstracted water from Lake Victoria for drinking water supplies around Mwanza; Kenya has constructed several dams around Kisumu; and Burundi, Rwanda and Tanzania are implementing the Rusumo Falls hydropower project on the Rwanda–Tanzania border; Ethiopia has built several dams, including Finchaa and its associated irrigation scheme, the Tekeze Dam; and, despite initial objections, the GERD on the Sudanese border is nearing completion - however, there is now a major disagreement about the duration of the filling of the GERD reservoir.

Egypt has appealed to international water law in its dealings with upper riparians, with little success, and none of the Nile riparians is a signatory to the 1997 UN Convention on the Non-Navigable Uses of International Watercourses, which took 24 years to draft. In essence, the convention balances the principles equitable sharing of a watercourse with an obligation not to cause significant harm to other users; but even the simplest analysis shows that all riparians would be entitled to a share, and so reduce the flows claimed by Egypt and Sudan (Brichieri-Colombi, 1966).

The role of the Sudd in the flow of the Nile makes it almost impossible for Egypt to demonstrate that it has suffered significant harm from water used upstream of Malakal, or indeed the construction of hydropower dams anywhere in the basin except where evaporation is significant, as at the Main Nile dam sites.

Whether Egypt could be accused of contributory negligence by irresponsibly expanding irrigation to the point where any abstraction by an upper riparian causes it harm is a legal point that has yet to be tested; but the fact that it continues to expand irrigation during prolonged discussions at fora such as the Nile Basin Initiative suggests that it is trying to establish facts on the ground to support its arguments about significant harm and, so, pre-empt upstream development.

While international treaties serve a purpose to codify agreements among or between riparians, there is little evidence that legal arguments pave the way to agreement, and recourse to international water law is unlikely to furnish a solution.

Given its diminishing hegemony, Egypt has three options in its dealings with upper riparians:

- *Cooperation*: all attempts to come to agreement over water sharing have foundered on Egypt's claim that the various treaties signed on its behalf by the British government bind other riparians not to use water from the Nile. In 2010, six countries agreed a Comprehensive Framework Agreement that was ratified by Ethiopia, Rwanda, Tanzania and Uganda, and signed by Kenya (2010) and Burundi (2011).

As neither of the latter two has yet ratified the agreement, it has not yet entered into force – and both Egypt and Sudan have made it clear they will not sign, as it does not respect their claims to acquired rights. In January 2021, Ethiopia's discussions with Egypt and Sudan over the filling of the GERD reservoir again broke down; in September, the UN Security Council again urged the two countries to resume talks under the auspices of the African Union. Whittington's (2022, in this issue) analysis makes clear that the problem is intractable because the parties to the dispute cannot agree a baseline from which to measure the impacts of the filling process.

The difficulty of cooperation is 'rhetorical self-entrapment' (Schimmelfennig, 2003, p. 221), a situation when parties find themselves constrained by their previous arguments, their tenets so entrenched that they cannot re-examine whether their theoretical underpinnings are still, or ever were, valid. The twin ideas that Egypt has acquired water rights on the Nile and that no other riparian except for Sudan can withdraw water from the river, and that Egypt depends for its very existence on the Nile, so dominate its thinking that it cannot envisage a situation in which it can set aside discussion about its rights and concerns – even if it would benefit from doing so. These arguments have been so often reiterated to the Egyptian population that it is now difficult to backtrack and there is little prospect of formal cooperation among the full set of riparians on issues that touch on water sharing.

- *Conflict*: threats of water war on the Nile gained credence when it was revealed that Egypt had discussions with Sudan about the provision of facilities to the Egyptian Air Force for a putative raid on the GERD. In May 2010, Wikileaks cited an Egyptian diplomatic source saying that: 'Sudanese president Umar al-Bashir has agreed to allow the Egyptians to build a small airbase in Kusti to accommodate Egyptian commandos who might be sent to Ethiopia to destroy water facilities on the Blue Nile. [...] It will be their option if everything else fails.'

  And in April 2021, Egypt and Sudan engaged in a five-day military exercise at Merowe involving commandos and air forces in what was widely viewed as a warning to Ethiopia about filling the dam (Sabry, 2021). Naturally, Ethiopia has installed military countermeasures and would be operating close to its own air-fields.

  The dam is a mass concrete structure that could resist anything but a very powerful bomb which, if successful in breaching the dam, could release a major flood.[7] On 14 August 2021, the level reached 571 m and the volume stored 4.4 bcm; if this were released over 24 hours, the average flow would be 51,000 m^3/s, three times the spillway capacity of Roseires dam, so it, the Sennar dam and Khartoum would all be at grave risk. The GERD is likely to be finished in the next few years; filling will necessarily continue during the coming flood seasons, and once the construction gap is closed, the storage volume would be equal to the average annual flow over the site and it would be possible to hold back the entire Blue Nile flow for a year, which would be a disaster for both Sudan and Egypt.

  Economic conflict is also possible but unlikely. Cutting trade links would not impact either economy significantly; denial of access to the Suez Canal would be in contradiction of Article 1 of the Rules of Navigation of the Suez Canal which states: 'Transit through the Suez Canal is allowed to vessels of all nations subject to compliance with the conditions stated in the present Rules of Navigation' (Suez Canal Authority, 2015), and would contradict its claim that the Suez Canal is Egypt's gift to the world; denial of use of airspace would be possible, but such actions can provoke unintended consequences and unexpected alliances.

Conflict with upper riparians can only jeopardise service industries, particularly tourism, and thus Egypt's economic growth and food security. As a way of resolving disputes, conflict is seldom productive, and unlikely to be used by Egypt to resolve issues on the Nile – it would be incompatible with Egypt's role as a peacemaker in the Middle East, a role that continues to assure it considerable status on the world stage.

- *Cohabitation*: the third option is cohabitation, in the sense used to describe the relationship between president and prime minister in France when the two are from different political parties; the parties recognize they have important differences but find a formula that allows dialogue without concession, and pragmatic solutions when needed. Studies of the GERD filling process show that, eventually, additional regulation capacity on the Nile can only be of benefit to all three countries. They also show the reservoir can be filled in five years without reducing AHD releases below 55.5 bcm/year when flows are at or above average over a 10-year period if, as at present, filling starts while the AHD is full (Wheeler et al., 2020). The analysis is subject to many variables that include possible actions by Sudan to extend irrigation or fill dams on the Main Nile,[8] and by Egypt to adjust storage at the AHD so that the combined storage never fell below 60 bcm on 31 July, the level at which water reductions are introduced. If this happened, a tripartite group would need to decide on alleviation measures.

There is little reason to believe that Ethiopia would take steps that caused significant adverse downstream impacts. If it did, Egypt and Sudan would have far more support at the UN and from Western allies in its arguments on the Nile than it does at present. The safest scenario for Egypt and Sudan is encourage Ethiopia to fill the dam as quickly as possible, even if there are short-term consequences, and take up power from GERD – which would make it in Ethiopia's interest to operate it as designed, filling each year from a low level at the start of the flood in August so that it is able to attenuate any unusually high inflows that might cause damage in Sudan. Feasibility studies for the construction of a 3000 km, 500 kV power interconnector from GERD to the Sudanese grid have been completed with a budget of US$1.8 billion. The project has not progressed since January 2017 (PIDA, 2021), but this kind of project would help resolve the issue. Moreover, Sudan, long seen as Africa's breadbasket, harvested 111,372 t of wheat from an area of 315,500 ha in 2020, double that recorded five years earlier (AfDB, 2021) and it may well become more efficient for Egypt to import wheat from Sudan than rely on domestic production.

The GERD will almost certainly be filled in the next few years and a pragmatic operational regime will emerge, even if not formally agreed. However, as Whittington (2022) notes, it sets a dangerous precedent for Egypt: while developments upstream of Malakal may have a negligible impact, any major new irrigation development in Ethiopia certainly would risk reducing inflows to Egypt.

Socio-economic measures

The objective of the NWRP was to safeguard its water resources and use them in the best way through the application of integrated water resources management (IWRM) principles, which are referred to 38 times in the NWRP report. However, even if this objective is achieved, it will not prevent water scarcity in Egypt or significantly reduce its vulnerability to changes in Nile flows.

An alternative approach is required that sets water in the national economy (WINE) using principles that focus on objectives of the state, rather than of managing water optimally (Brichieri-Colombi, 2009). WINE is an integrated approach that recognizes that water is but one of several resources available to meet the state objectives of providing security, increasing living standards, and ensuring food and energy supplies. It goes beyond the narrow confines of IWRM and the river basin and, in its cast of managers, water planners are a minority (Figure 6).

The issue for Egypt is not whether it has enough water to justify further investment in activities such as the HEP, which will always be vulnerable at times of reduced Nile inflows: the issue is food security. As shown by Allan (2011), Egypt has achieved food security by importing food rather than by developing local water resources to grow it at home. The Industrial Development Authority still needs to regulate the water supply and effluent treatment of the industrial sector to minimize water use and protect water quality in the lower reaches of the Nile – which will improve crop yields – but there is no economic advantage in limiting water allocations to the profitable industrial and services sectors to protect less-profitable investments in agriculture.

Food security has many dimensions, some associated with distribution rather than overall supply, but it cannot be solved unless the overall supply is adequate. At the national level, it can be measured by the net cost of importing food as a proportion of the gross national income. The Food and Agriculture Organization (FAO) tabulates the annual import and export values of 386 crops and livestock products, which are plotted for the period 1987–2019 in Figure 7. Over the 21-year period, the net cost of imports of all agricultural products has fallen from around 5% of gross national income to half that

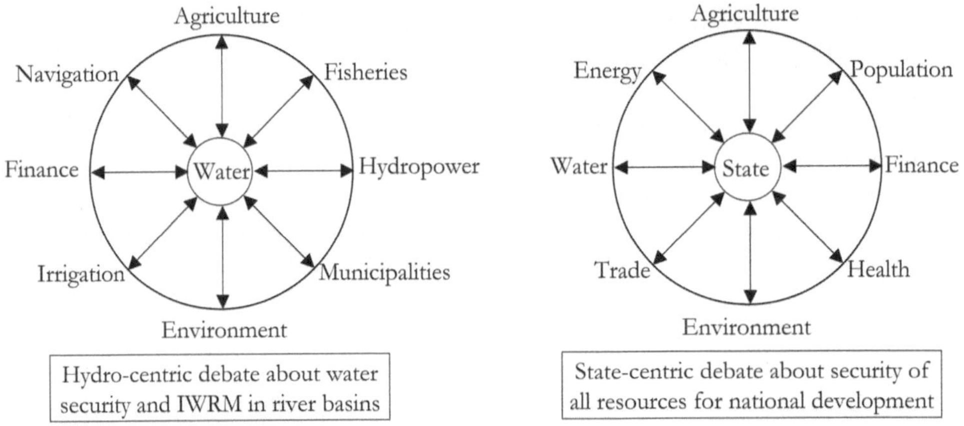

Figure 6. Water in the national economy (WINE).

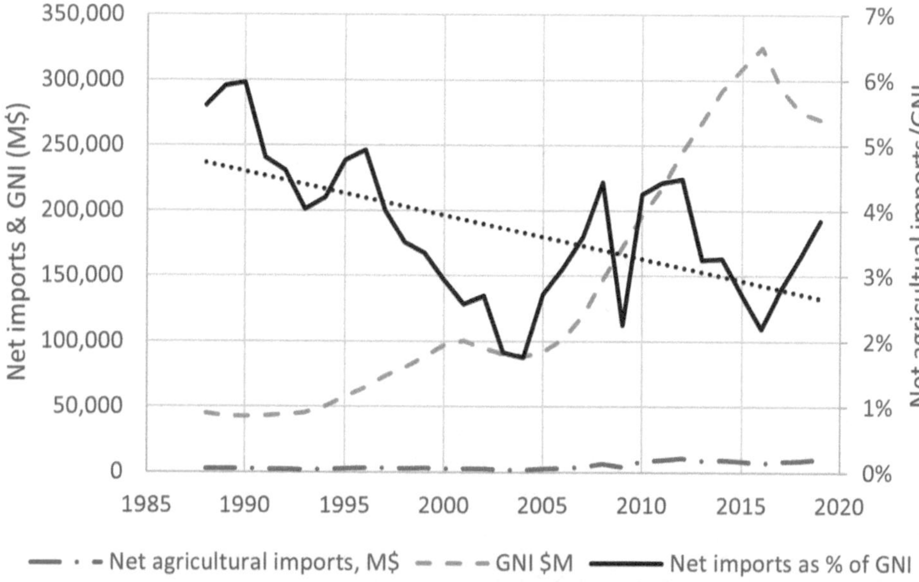

Figure 7. Trends in net costs of agricultural products and gross national income.

figure, which suggests that food security in Egypt is improving. Provided that the country refrains from extravagant agricultural projects and, instead, invests in non-agricultural sectors of the economy, the trend is likely to continue.

There are few indications that *per caput* food supply is an issue; dietary energy supplies were already well above the minimum levels in 2000 and show an increasing trend (Figure 8).

In the long term, Egypt can increase food security by managing population growth to reduce the total fertility rate (TFR) within 25 years from the present 2.88 children per woman to 2.00 in accordance with the UN low-fertility projection. Over the last 50 years, 36 out of 74 countries with similar TFRs have reduced their TFRs by at least this ratio.[9] Six counties in the region have achieved this outcome;[10] in Algeria and Tunisia, TFR has already fallen to 1.5. Figure 9 shows the fertility decline in these counties since their TFRs first fell to around 3.0 compared with the low-fertility projection for Egypt. If Egypt were to follow the low-fertility target and the TFR fell to 1.5 within 60 years, the population in 2080 would be 161 million and falling rather than 207 million and rising.

Population management is usually achieved by social and economic empowerment of women, largely through access to education, which also helps the development of the service and industrial sectors that need to expand if Egypt is to maintain its ability to import more food. Over the last 20 years, this transition has gone well: out of 45 countries that are making similar transitions, only 11 have reduced the relative importance of the agricultural sector at a faster rate. Figure 10 shows that if Egypt continues with the current exponentially declining trend, by 2050 agriculture will contribute less than 6% of the economy (World Bank, 2021). The trend in neighbouring counties indicates that this is achievable.

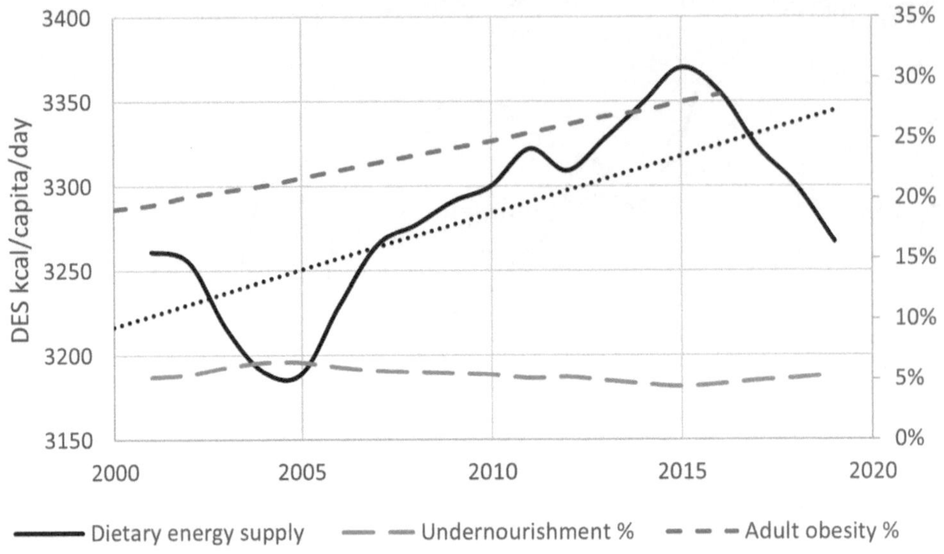

Figure 8. Food security indicators (FAO, 2020).

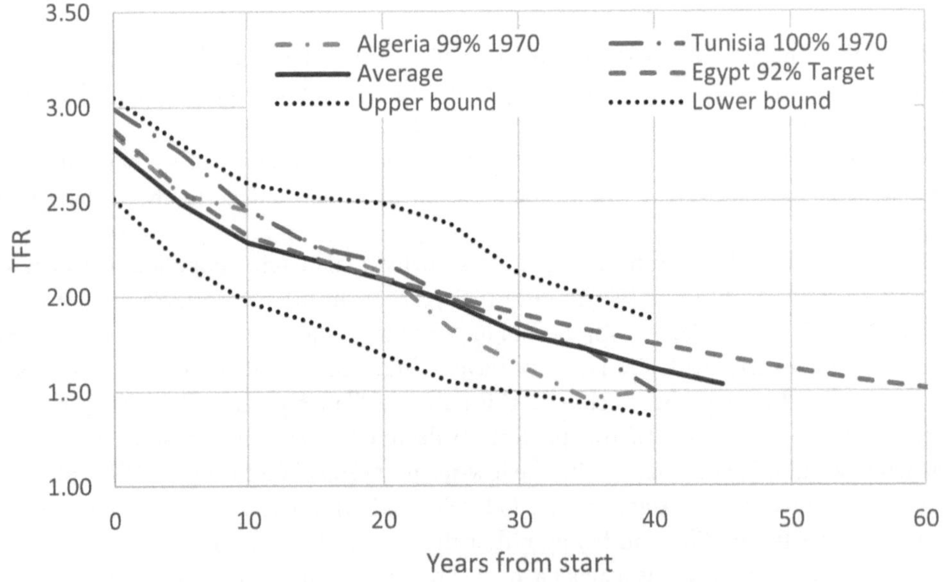

Figure 9. Total fertility rate (TFR) decline in selected countries.

Climate change

Egypt's food security may be at risk from climate change. Evaporation in the Sudd means that whatever happens in the East African riparian countries will have little effect on the flow reaching Sudan and Egypt. Studies of the Blue Nile catchment using three general circulation models agree that flows entering Sudan from Ethiopia are likely to increase significantly by mid-century (27% compared with the baseline 1988–2004), but differ

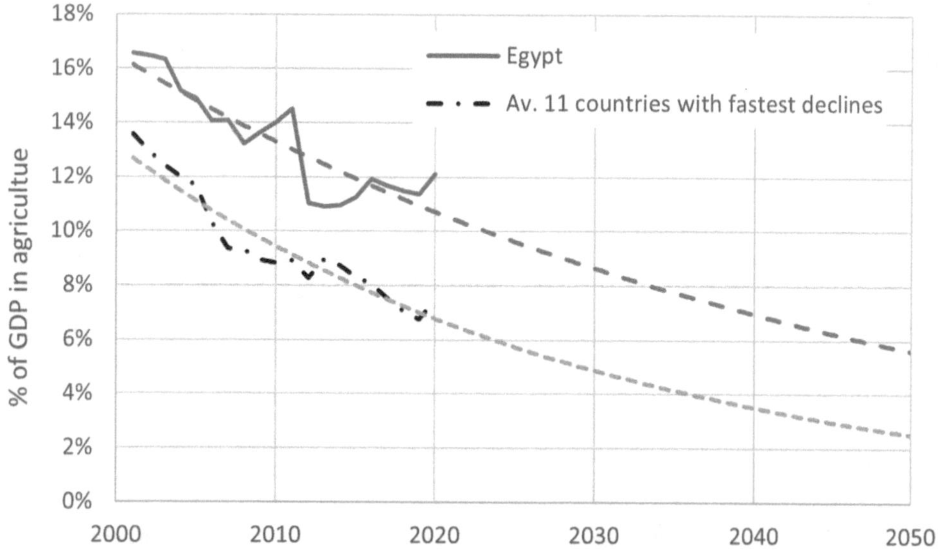

Figure 10. Decline in % of GDP in agriculture.

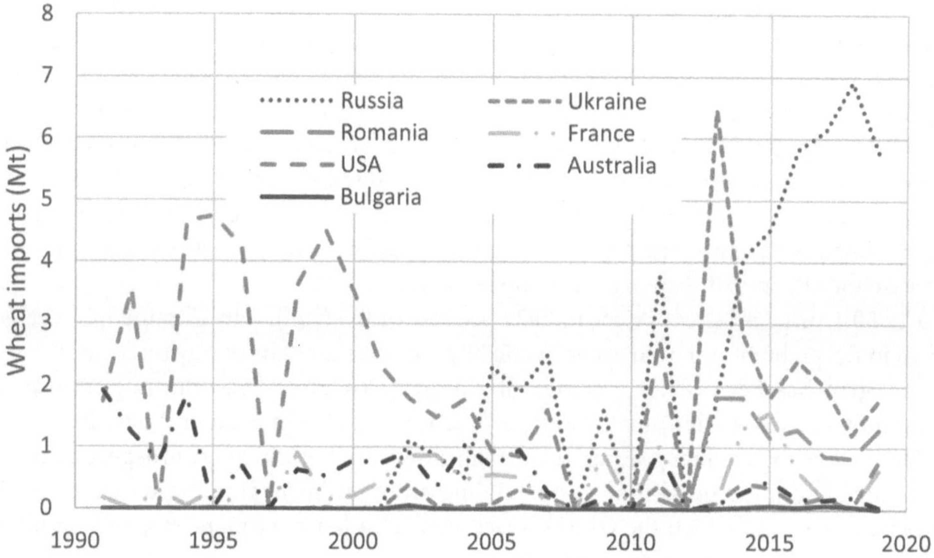

Figure 11. Wheat imports by origin.
Source: FAO (2021).

greatly in their projections for the end of the century (Roth et al., 2018). However, all agree that, for both periods, a lower proportion of the annual flow will occur in the five-month period June–November.

Higher temperatures will offset much of any gain in water availability, and climate change may well reduce not only Egypt's domestic food production but also the world's ability to produce the wheat that Egypt imports. Egypt has always skilfully sourced imports at good prices, varying its suppliers in much the same way as did the Roman Empire (Boccaletti, 2021, pp. 71–72) to avoid disruption from conflict or weather. Over the last 20 years, imports from the United States, Australia and France have declined while those from Russia, Ukraine and Romania have increased (Figure 11), but the present conflict in Ukraine is a warning against dependency on just a few suppliers.

Climate change could also divert maritime traffic away from the Suez Canal as shorter Arctic routes become more attractive.[11] This is not an immediate threat but could become so since the canal contributes almost 15% of Egypt's gross domestic product (GDP). There are many other ways that climate change could affect food security in Egypt, few of them predictable, but a robust and diverse economy is likely to be able to manage the impacts far better than one with a focus on agriculture.

Conclusions

The water deficit on the Nile in Egypt is rising just as the supply is becoming increasingly uncertain due to climatic, political and economic change in the upstream countries. In these circumstances, minor adjustments to agricultural water use in Egypt will not be enough and major reform becomes paramount. Nearly 30 years ago, Tony Allan suggested that, for Egypt, home-grown food security is unobtainable and distorts its hydropolitical behaviour (Howell & Allan, 1994, p. 9). Since then, Egypt has achieved food security, not by expanding its agriculture into the desert but by applying the WINE approach: expanding and transforming its economy so that it can import the food it needs; and engaging with the upper riparians. While its economy grows, food security is likely to increase rather than decrease. Domestic production can be expected to increase – but not enough to match population growth. The greatest risks come from a burgeoning population and hydrocentric thinking that prioritizes water security and can only increase the chances of disputes with upper riparians.

The 50% surge in wheat prices in 2022 because of the conflict in Ukraine has created alarm in Egypt, but it has been accompanied by a similar surge in energy prices and, since 2019, Egypt has been a net exporter of energy. This underlines the importance of maintaining diversity, not only in sources of wheat but also in economic activity.

The Nile is experiencing a period of high flows. There is a good prospect of them continuing and it is a good time to be building more resilience into the system through regional projects, of which the GERD is clearly one, rather than increasing vulnerability by expanding desert irrigation. *Carpe diem.*

Notes

1. This was led by the author for the first two years.
2. For the main report of the NWRP and relevant technical reports, see the WOCAT network (https://wocatpedia.net/wiki/Portal:Library_-_Publications_and_Good_Practices).
3. Flows arriving at Aswan, corrected for abstraction and losses in Sudan. The KGW extension is based on estimates prepared by Kevin Wheeler (personal communication).
4. This was signed to meet a benchmark for release of funds by USAID under the Matching Irrigation Supply and Demand programme, but does not have the force of law.
5. Updated using construction cost indices (Engineering News Record, 2021).

6. Measurements from Howell et al. (1988, p. 51).
7. If the British bouncing bomb attacks had succeeded in breaching dams in the Ruhr in the Second World War, they would have caused a 200-year flood. Nevertheless, water levels had to be drawn down because of cracking, reducing hydropower output.
8. At present Sudan is abstracting only about 16.9 bcm of its share of 17.5 bcm.
9. Countries on the World Population Prospects database for which there are 25 years of records of TFRs starting around 3.0, the current rate in Egypt (UN, 2020).
10. Algeria, Libya, Morocco, Tunisia, Turkey and Western Sahara.
11. For a discussion, see https://www.bbvaopenmind.com/en/science/environment/arctic-shipping-routes-new-suez-canal/.

Disclosure statement

No potential conflict of interest was reported by the author.

ORCID

Stephen Brichieri-Colombi http://orcid.org/0000-0002-6695-9084

References

Abdel-Aziz, T. M. (2004, October 18–21). Safe visual navigation on the Nile River. *9th International Symposium on River Sedimentation*, China, Yichann.

African Development Bank (AfDB). (2021). https://www.afdb.org/en/news-and-events/sudans-record-wheat-harvest-proof-bank-agricultural-transformation-program-will-boost-breadbasket-goal-38388

Allan, A. J. (2011). *Virtual water: Tackling the threat to our planet's most precious resource*. I.B. Taurus.

Al-Youm, A.-M. (2015, January). 22 reasons for rejecting project to connect the Nile and the Congo rivers. *Egypt Independent*.

Azim, R. A., & Viala, E. (2011). *Matching irrigation supply and demand in Egypt*. Colorado State University. mountainscholar.org/bitstream/handle/10217/207277/125-206-uscid-Boise_Azim.pdf?sequence=1&isAllowed=y

Boccaletti, G. (2021). *Water: A biography*. Pantheon.

Brichieri-Colombi, J. S. A. (1966). Equitable use and the sharing of the Nile. *Proceedings of the IVth Nile 2002 Conference*, Kampala.

Brichieri-Colombi, J. S. A. (2009). *The world water crisis: The failures of resource management*. I.B. Tauris.

El-Gafy, I., El-Ganzori, A. M., & Mohamed, A. I. (2014). Decision support system to maximize economic value of irrigation water at the Egyptian governorates meanwhile reducing the national food gap. *Water Science*, *27*(54), 1–18. https://doi.org/10.1016/j.wsj.2013.12.001

Engineering News Record. (2021). *Construction cost indices*. https://digital.bnpmedia.com/publication/?i=654892&article_id=3633447&view=articleBrowser&ver=html5)

ENTRO. (2014). *Investment opportunities analysis – Country consultation report*. Eastern Nile Technical Regional Office. https://hdl.handle.net/20.500.12351/527

FAO. (1977). *Crop water requirements (Irrigation and Drainage Paper 24)*.

FAO. (1997). *Crop evapotranspiration* (*Irrigation and Drainage Paper 56*).

FAO. (2020). *Food security indicators*.

FAO. (2021). *Trade matrices, Rome*.

Gad, A.-A., Ali, R. R. (2009). Water rationalization in Egypt from the perspective of the virtual water concept, and E. Moujabber L. Mandi, G. T. Liuzzi, I. Martin, A. Rabi, & R. Rodriguez

(Eds.), *Technological perspectives for rational use of water resources in the Mediterranean region* (pp. 301–310). CIHEAM Options Méditerranéennes: Série A. Séminaires Méditerranéens 88.

Howell, P. P., & Allan, J. A. (Eds). (1994). *The Nile: Sharing a scarce resource*. Cambridge University Press.

Howell, P., Lock, M., & Cobb, S. (Eds). (1988). *The Jonglei Canal: Impact and opportunity*. Cambridge University Press.

MWRI. (2005). *Water for the future: National water resources plan for Egypt, 2017*. Mineral Water Resources and Irrigation.

NWRP. (2005) . *National Water Resources Plan for Egypt - 2017*. MWRI.

NWRP-CP. (2011). *TR57 matching water supply and demand*. MWRI.

Ouda, S., & Zohry, A. E.-H. (2015). *Crop rotation: An approach to save irrigation water under water scarcity in Egypt*. https://www.semanticscholar.org/paper/Crop-Rotation%3A-An-Approach-to-Save-Irrigation-Water-Ouda-Zohry/193b2c710edf261310f9201f7cf3a972f6634433

PIDA. (2021). Ethiopia–Sudan transmission interconnector (Ethiopia section). *Programme for infrastructure development in Africa*. https://www.au-pida.org/view-project/312/

Roth, V., Lemann, T., Zeleke, G., Subhatu, A. T., Nigussie, T. K., & Hurni, H. (2018). Effects of climate change on water resources in the upper Blue Nile Basin of Ethiopia. *Heliyon*, *4*(9), e00771. https://doi.org/10.1016/j.heliyon.2018.e00771

Sabry, M. (2021, April 9). Egypt–Sudan hold military exercise amid crisis with Ethiopia. *Al Monitor*.

Schimmelfennig, F. (2003). *The EU, NATO and the integration of Europe: Rules and rhetoric*. Cambridge University Press.

Suez Canal Authority. (2015). *Rules of navigation*. Government of Egypt.

United Nations. (2020). *World population prospects. 2019 revision*. UN Dept Economic and Social Affairs, Population Division. population.un.org/wpp/Publications/Files/WPP2019_Highlights.pdf

Valadi, S. (2020, August 18). VICMED: A feasible bid for smooth-sailing Nile basin ties? *Egypt Today*.

Wheeler, K. G., Jeuland, M., Hall, J. W., Zagona, E., & Whittington, D. (2020). Understanding and managing new risks on the Nile with the Grand Ethiopian Renaissance Dam. *Nature Communications*, *11*(1), 5222. https://doi.org/10.1038/s41467-020-19089-x

Whittington, D. (2022). Contested baselines and transboundary water resources management, with illustrations from the Nile. *Water International*, *47*(6), 934–951. https://doi.org/10.1080/02508060.2022.2123611

World Bank. (2021). *World development indicators*.

Contested baselines and transboundary water resources management, with illustrations from the Nile

Dale Whittington

ABSTRACT

The effect of a policy intervention is estimated as the difference between a state of the world without the intervention – the dynamic baseline – and with the intervention. Negotiations on managing transboundary water resources confront three kinds of problems in specifying the dynamic baseline: unexamined baselines, uncertain baselines and contested baselines. This paper focuses on contested baselines. Controversy may arise from different ethical or political assessments of the appropriate choice of the state of the world without the policy intervention. The Grand Ethiopian Renaissance Dam on the Nile highlights the importance of understanding contested baselines involving transboundary water resources.

Introduction

The effect of a proposed or planned policy intervention is estimated using some performance indicator that is associated with the difference between (1) a state of the world without the policy intervention and (2) a state of the world with the policy intervention. Just as the traveller in Robert Frost's poem *The Road Not Taken*, must compare and then choose between which road to take, a decision-maker must compare and then choose between two or more states of the world (Frost, 1915).

Because the effects of most policy interventions play out over time, this comparison requires a forecast of the state of the world without the policy intervention (the *dynamic baseline*). This means the policy analyst must look down the road not taken as far as one can see – or until the effects of the policy intervention are no longer significant. The forecast of the dynamic baseline typically begins with a description of current (status quo) conditions which, itself, can be a major contribution to policy analysis. Certainly, Tony Allan spent much of his career carefully describing to both technical and non-technical audiences (Allan, 2019a, 2019b):

- The concepts of *food water* and *non-food water*, *green water* and *blue water*; and *watersheds* and *problemsheds*.

- How these concepts can help us understand the baseline quantities of water used for food and non-food production, and where this water comes from.
- The importance of understanding baseline patterns of the global food trade for addressing the problem of water scarcity.[1]

This paper identifies three kinds of problems that policy analysts confront in specifying the dynamic baseline and their importance for negotiations on managing transboundary water resources. The first – *unexamined baselines* – results from the failure of stakeholders to understand the need for a dynamic baseline.[2] The second is a failure of stakeholders to agree on how to account for uncertainties in the forecast of the dynamic baseline – *uncertain baselines*. The third is a failure of stakeholders to agree on the dynamic baseline because of disagreements over whether an extension of status quo conditions provides a reasonable or fair basis from which to forecast the dynamic baseline – *contested baselines*.

Here, the focus is on the third issue: contested baselines. If dynamic baselines are contested by stakeholders, controversy over the outcome of an ex-ante policy analysis may be due to different ethical or political assessments of the appropriate choice of the state of the world without the policy intervention, not to differences of opinion about the magnitude of the effect of the policy intervention or parameter values used for the estimation of the treatment effect.[3] I use the current controversy over the Grand Ethiopian Renaissance Dam (GERD) on the River Nile to illustrate this concept of contested baselines and to provide a concrete example of their potential importance in understanding disagreements about policy interventions on transboundary water resources. I conclude with three messages about dynamic baselines for policy analysts working on transboundary water resources.

Background

Policy analysts have long understood the need for a dynamic baseline for both ex-ante and ex-post assessment of interventions. Economists conducting ex-ante cost–benefit analyses have included the task of predicting baseline conditions in their guidance documents and have recognized that these forecasts will be uncertain (Robinson et al., 2019; United States Department Health and Human Services, 2016; United States Environmental Protection Agency, 2010). However, many cost–benefit textbooks and guidance documents pay hardly any attention to the issue of forecasting the dynamic baseline and, to the best of my knowledge, none has clearly identified the problem of contested baselines.

Scholars of environmental impact assessment and environmental risk assessment have also recognized the importance of baselines in understanding the effects of policy interventions (Aagaard 2011; Alagona et al., 2012; Barandiarran, 2019; Bento et al., 2016; Braverman, 2020; Craig, 2014; Hilborn & Walters, 1981; Kinchy, 2019; Knowlton & Jackson, 2008; Pauly, 1995, 2019; Ureta, 2018; Ureta et al., 2020). They make the important distinction between *a baseline* that is a component of an environment impact assessment, and *baselining*: the bureaucratic, political and social practices used to produce or create an environmental baseline (Ureta, 2018). Although these scholars work in numerous disciplines (ecology, biology, law, sociology, geography), their interest in environmental baselines has focused primarily on two types of evaluation. The first is ex-ante comparison of (1) baseline environmental conditions without a policy intervention with (2) a new state of the world with the proposed policy intervention. In this comparison, the static baseline is a snapshot in time before the intervention; Barandiarran (2019) describes this baselining process as documenting 'what is there'.

The second type of evaluation is ex-post calculation of an environmental loss. It involves a comparison of current degraded environmental conditions that have resulted from, for example, an existing polluting activity (status quo condition) with a baseline condition where the polluting activity does not exist. Environmental impact assessment practitioners have argued that status quo or current conditions outside the project area may not be good baselines for the assessment of restoration efforts and have looked back in time to find appropriate historic benchmarks for natural habitats. Often, they have tried to define *natural* or *undisturbed* conditions as both a reference baseline and a desired state-of-the-world from which to measure the detrimental impact of a polluting activity or development (Bonebrake et al., 2010; McClenachan et al., 2006; Ureta et al., 2020).

But looking back into the past can be problematic. Pauly (1995) introduced the notion of the *shifting baseline* in the context of fisheries, which he described as a process by which each new generation of fishery scientists *forgets* what past fishery stocks were like before they started their career and, thus, recalibrates what is natural or considered normal. He argued that through this process we have forgotten how abundant marine fisheries were in the past. Braverman (2020) gives an example of a shifting baseline from the study of the bleaching of coral reefs due to global warming; scientists realized that what they understood to be a healthy reef today would have been a depleted reef in the recent past. Pyle (2003) described the problem of the shifting baseline as the 'extinction of experience'.

Although both types of evaluation necessitate an estimate of an environmental baseline, neither comparison requires the analyst to forecast a dynamic baseline without a policy intervention, which is the focus of this paper. The problem of the shifting baseline looks to the past to establish a baseline from which to assess damages, not to the future to determine ex-ante the benefits and costs of a proposed or planned intervention.

Three challenges in specifying the dynamic baseline

Unexamined baselines

The identification revolution in economics has focused economists' attention on something that physical and life scientists have long understood: the importance of accurately identifying the proper counterfactual (Angrist & Pischke, 2019b; Banerjee & Duflo, 2009). In a randomized controlled trial (RCT), the control is allowed to evolve without the intervention for the duration of the experiment and, thus, provide the dynamic baseline against which the change in one or more performance indicators or outcome metrics is measured. As public health scientists and, later, economists took RCTs out of the laboratory and into the field, the proper specification of the dynamic baseline became more challenging both politically and ethically, especially when the target of the intervention was people. Some stakeholders often question the need for a counterfactual in such situations: a central concern has been the withholding of the intervention from control subjects when the presumption is that the treatment will be effective but the magnitude of the treatment effect is unknown. One solution sometimes deployed has been to give the control group something else of similar value to the intervention that is being withheld (Quinn et al., 2019).

For many types of policy problems it is not possible to conduct an RCT to test the effectiveness of an intervention. The construction of a dam is an obvious example. Even if there were two identical rivers, one could not practically or ethically build a dam on one river and withhold a dam on the other (control) river to test the effectiveness of an investment in a dam. In such situations, the policy analyst needs an analytical model (perhaps coupled with results from a natural experiment) to estimate ex-ante how a system would evolve with and without the proposed policy intervention.[4] However, stakeholders often fail to recognize the importance of forecasting the dynamic baseline. They may be satisfied with a simple comparison of two states of the world: before and after the construction of a dam. Nevertheless, a dynamic counterfactual remains an essential component of a sound evaluation strategy (Aldy, 2014; United States Department Health and Human Services, 2016).

Figure 1 illustrates this case of *unexamined baselines*. Suppose there is an unregulated river with a history of past floods and droughts causing loss of lives. Before time t there is no dam on the river. If a dam were built and completed at time t, the loss of lives from floods and droughts would be reduced in the future.[5] The magnitude of this reduction in lives lost in time $t + 1$ is shown in Figure 1. Using a before-and-after comparison, this change in the number of deaths due to the construction of the dam is measured by the size of the shift in the past trend line and the new situation after construction (shown as the before versus after treatment effect with a magnitude $[x_3 - x_1]$ in Figure 1).

Such an *interrupted time-series* analysis of before and after conditions may be compelling but it ignores the question of what would have happened if the dam had not been built. Suppose in the dynamic baseline that the number of lives lost due to floods and droughts would have grown over time due to population growth and urban expansion. In this case, the dam would be more valuable than the before versus after analysis indicates. This is shown in Figure 1 (case A) as a forecast of the dynamic baseline in which deaths increase over time.[6] Thus, the benefits of the dam would be underestimated in time $t + 1$ by a before versus after comparison that fails to examine the dynamic baseline in the policy analysis (unexamined baselines).

In both cost–benefit analysis and cost-effectiveness analysis, the discounting of future streams of changes in performance indicators can obscure the dynamic baseline. Cost-effectiveness analyses are especially susceptible to this problem of unexamined baselines because the outcome measure of a cost-effectiveness analysis is a ratio that reports a change in the performance indicator (PI) per dollar spent, that is, ΔPI/$.[7] Because the numerator of the cost-effectiveness ratio is a change in the performance indicator, a dynamic baseline is required to calculate this change. In each future period of the analysis there is a difference between the dynamic baseline and the forecast of the state of the world with the policy intervention. This time stream of differences is typically discounted back to obtain a present value of the changes in the performance indicator. This present value is then divided by the cost of the intervention to obtain the cost-

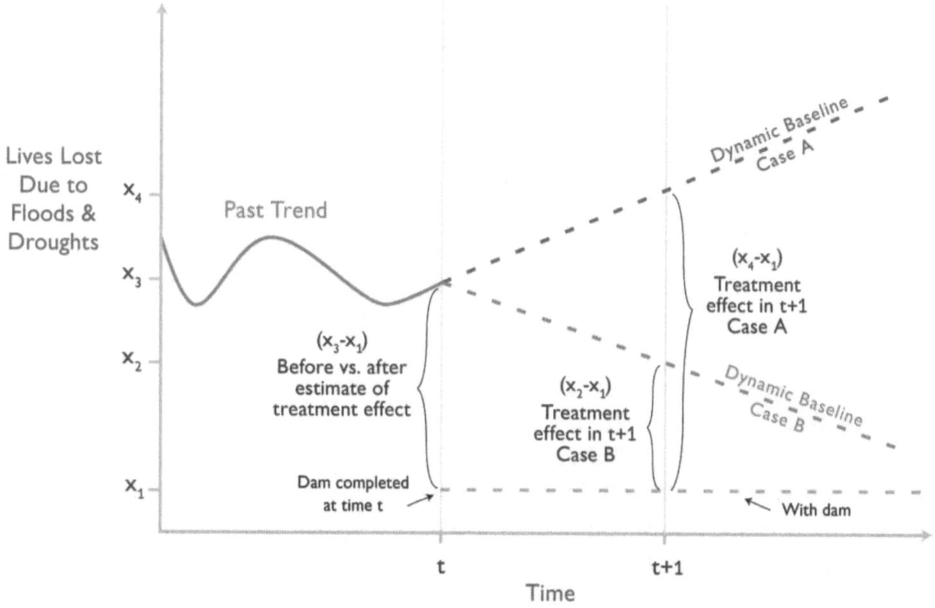

Figure 1. Forecasting the dynamic baseline.

effectiveness ratio. Because of this two-step transformation of the differences between the dynamic baseline and the state of the world with the policy intervention (i.e., discounting the differences and then division by the costs), the dynamic baseline that is used in the cost-effectiveness calculation may be less apparent or hidden in plain sight.

Uncertain baselines

In formal decision analysis, the dynamic baseline is a possible future. There are typically many uncertain states of the world and decision-makers compare the outcomes in the state of the world with the policy intervention with outcomes in these possible futures. However, forecasts are always uncertain (Schlee & Smith, 2019). The dynamic baseline will be affected by many factors that will occur in the absence of the policy intervention, for example, population and economic growth, demographic changes, climate change, technological innovation, and rural-to-urban migration. Analysts may have honest differences of opinion about the importance and magnitude of such drivers of change in the absence of the policy intervention under consideration. For example, a forecast of the number of lives that may be lost due to floods and droughts in the river basin may show a decrease from status quo conditions due to a more optimistic assessment of technological innovations in early warning systems and an estimate of higher rural-to-urban migration that leads more people to move out of harm's way. This is shown in Figure 1 (case B) as a forecast of the dynamic baseline in which deaths decrease over time. In this case the change in the number of lives lost due to the dam at $t + 1$ is $[x_2 - x_1]$, which is less than the change in the before-and-after comparison $[x_3 - x_1]$.

Which forecast an analyst chooses to use for the comparison with the state of the world with the policy intervention (case A or case B in Figure 1) will affect the estimate of change in the performance indicator. As shown in Figure 1, in time $t + 1$ it could be either $[x_4 - x_1]$ or $[x_2 - x_1]$, depending on which forecast of the dynamic baseline is used for the ex-ante policy analysis.

For an ex-ante evaluation of a dam investment, the effect of the dam usually will be estimated as a comparison of alternative runs of a river basin simulation model: (1) a simulation without the dam (the dynamic baseline), and (2) a simulation with the dam (Basheer, 2021; Blackmore & Whittington, 2008). If different forecasts of the dynamic baseline have different hydrological characteristics due, for example, to different assessments of the effect of climate change on a river's hydrology, then the magnitude of the treatment effect (i.e., the ability of the dam to control hydrological fluctuations) will change depending on the dynamic baseline selected. In other words, the dam could be more or less valuable depending on one's forecasts of how climate change affects future flows of the river (Jeuland & Whittington, 2014).[8]

One approach to dealing with uncertainty in forecasts of the dynamic baseline is to perform sensitivity analyses with different baselines – similar to the one-way and Monte Carlo sensitivity analyses now commonly used in cost–benefit analysis for uncertainty in parameter values (Boardman et al., 2018; Jeuland & Pattanayak, 2012). The use of probabilistic hydrological methods in river basin simulation models is another approach for dealing with uncertainty in the dynamic baseline (Farmer & Vogel, 2016; Fiering & Jackson, 1991).

The reports of the Intergovernmental Panel on Climate Change (IPCC) provide scenario analyses of carbon dioxide emissions and temperature change in the absence of mitigation efforts. The IPCC reports then show the effect of different policy interventions (mitigation efforts) on performance indicators such as greenhouse gas emissions or temperature change relative to these multiple dynamic baselines. These IPCC scenario analyses provide an example of how to display and communicate uncertainty in forecasts of the dynamic baseline (Intergovernmental Panel on Climate Change, 2021).[9] However, the use of any of the different types of sensitivity analyses to deal with uncertainty in forecasts of the dynamic baseline is still relatively rare in policy analyses.

Contested baselines

Another challenge in the specification the dynamic baseline arises not due to uncertainty in the forecast but to disagreement over the legitimacy or fairness of status quo conditions. Such disagreement can arise from one stakeholder group's feelings of past injustices and their resulting perception of unfair or unequitable resource allocation in the status quo conditions. Kinchy (2019) uses the term 'contentious baselines' to describe baselines that are controversial and notes that some groups in a controversy may benefit by fostering 'strategic ignorance' about the appropriate baseline to use in an analysis (Lind et al., 2019). Ureta (2018) notes that the choice of the appropriate baseline may involve power struggles among parties.

Here, I use the term *contested baselines* to refer to dynamic baselines on which the parties or stakeholders cannot agree over the future evolution of conditions.[10] Ideally, a policy analyst evaluating a possible investment in a dam would have an analytical model and set of assumptions to use to forecast the dynamic baseline that are mutually agreed upon by the stakeholders. However, in a rapidly evolving, complex political environment such an analytical model may not be available. Even if such a model exists, there may be no agreement on its acceptability or the assumptions to be used to forecast the dynamic baseline.[11] In such a situation, some stakeholders may have information or intuition about other policy interventions that could occur in the dynamic baseline. Whether these alternative policy interventions are endogenous in this policy process – that is, under one or more stakeholders' control – may be ambiguous to the some or all stakeholders. They may decide to include them in the dynamic baseline and deal later with the political feasibility of influencing a policy intervention assumed to be embedded in the dynamic baseline. Including an alternative policy intervention in the dynamic baseline could be a politically strategic decision, that is, a stakeholder may know how this choice will influence the comparison of the state of the world with the intervention and the dynamic baseline with embedded alternative policy actions. But a stakeholder may also perceive this possibly endogenous dynamic baseline as the most likely counterfactual.

In Figure 1, the case A dynamic baseline shows a future state-of-the-world in which past trends evolve and conditions worsen, that is, the lives lost due to flooding and drought increase. If one stakeholder group perceives this deterioration in future conditions to be due in part to a continuation of bad behaviour by another party, then the case A dynamic baseline may be rejected as an appropriate counterfactual because it implicitly assumes that this bad behaviour is allowed to continue in the future. One way to understand this problem of contested baselines is that one party believes that a prior policy change is needed before the policy intervention under consideration can be properly evaluated.[12] In effect, this party is arguing that a different policy mix should be evaluated: a change in the status quo conditions plus the intervention. However, in a rapidly moving political environment, some stakeholders may prefer a dynamic baseline with embedded policy decisions, possibly as a practical or heuristic device. They may not perceive or have time to discriminate between an exogenous dynamic baseline without alternative policy decisions and an array of different combinations of policy options. Thus, the dynamic baseline with possibly endogenous alternative policy options embedded in it can become the main counterfactual. Such an endogenous dynamic baseline may be contested by other stakeholders.

Returning to the example of a proposed dam, the benefits and costs of a new dam will depend on the existing pattern of water uses in the basin. For example, some irrigated areas may be short of water and the new dam may provide an increase in the quantity and reliability of water to them. Due to climate change, these status quo conditions may get worse without the dam but, from the perspective of these irrigators, the dynamic baseline logically should start with these initial conditions.

However, another group of a stakeholders may perceive the existing pattern of water use to be unfair and inequitable. From this perspective, the dynamic baseline should not assume that these inequities will continue. This second party may argue that the dynamic baseline should assume a different, more just resource allocation, and that the policy intervention should be assessed relative to this new pattern of water use. For example, many tribes in the Southwestern United States feel that their water resources have been usurped by non-native American irrigators and urban water users. Tribes believe that this status quo condition is unfair and should be rectified. Future interventions such as climate change mitigation and adaptation measures should be evaluated once this inequitable distribution has been addressed.

Tribes in the United States view themselves as sovereign entities within the national borders of the United States, and a reallocation of water resources from non-native irrigators to a tribe is similar in some respects to a transboundary water resources conflict. If one looks at this reallocation through the lens of international water law and applies the principles of *no significant harm* and *equitable use*, the measurement of significant harm presents a difficulty. This is because the dynamic baseline is contested. There is no agreement on the reference condition from which significant harm should be measured.

From the perspective of the non-native irrigator, the reallocation of water to a tribe will be experienced as a welfare loss which should be measured by the minimum compensation that the irrigator would accept to make them as well off as if the reallocation had not occurred (Table 1).

Because people value welfare losses much more than gains, from the irrigator's perspective the *appreciable harm* resulting from the water reallocation could be very large. From the perspective of the non-native irrigator, the benefits of the water reallocation should be measured by the tribe's willingness to pay (WTP) for the water. However, from the tribe's perspective, the irrigator's welfare loss is a reduction in an unjust or even illegal gain. Any estimate of significant harm that the non-native irrigator experiences should be valued much less than the minimum compensation that the irrigator would accept to make them as well off as if the reallocation had not occurred. Similarly, from the tribe's perspective, the welfare gain from reallocating water to the tribe should be valued as a reduction in a loss, not as a gain from the current status quo conditions. This welfare gain should be measured as the minimum compensation the tribe would accept not to have the reallocation occur. In this example, the willingness-to-accept measure would be much higher than the amount the tribe would be willing to pay for the water.

Thus, the two parties have different baselines from which to measure significant harm. The non-native irrigator views the dynamic baseline as a continuation of the current status quo conditions in which they hold the water rights. The tribe views the dynamic baseline as a situation in which past injustices have been rectified. From a welfare economics perspective, these different dynamic baselines imply that the parties would use different welfare measures to assess significant harm to the non-native irrigator and the welfare gain to the tribe resulting from the water reallocation.[13]

Table 1. Measuring *no significant harm* in the presence of contested baselines: a water reallocation from a non-native American irrigator to a native American Tribe.

How should welfare change be measured from the perspective of ...	Welfare loss to non-native irrigator	Welfare gain to Tribal members
Non-native irrigator	WTA compensation to incur loss of water	WTP for increased water supply
Tribe	WTP to avoid a loss of water	WTA compensation to forego the increase in water supply

WTA, willingness to accept compensation; WTP, willing to pay. For many types of interventions, WTA measures of welfare changes are approximately three times WTP measures.

Nile illustrations: contested baselines

Background

In March 2011, Ethiopia announced that it would build a large hydroelectric dam (subsequently named the Grand Ethiopian Renaissance Dam – GERD) on the Blue Nile near the Sudanese–Ethiopian border (Figure 2). In 2008, the report of the Scoping Study Team to the Eastern Nile Council of Ministers had recommended that investments in storage should begin on the Blue Nile in Ethiopia (Blackmore & Whittington, 2008), but Egyptian policymakers disagreed with this conclusion and argued that investments on the White Nile were more attractive. Even among planners who favoured infrastructure development on the Blue Nile, there was no consensus on the size and location of such investments. Thus, Ethiopia's unilateral decision to construct the GERD caught Egypt and Sudan, as well as the international community, by surprise.

Ethiopia began construction on the GERD immediately after its announcement without the independent engineering, economic, environmental, and social impact studies that would have been required if international financing had been involved. In November 2011, Ethiopia agreed to form an International Panel of Experts (2013) with two representatives from each of the Eastern Nile riparians (Ethiopia, Egypt and Sudan) and four international experts to investigate the existing plans for the GERD and the impacts of the dam. The International Panel of Experts (2013) issued its report in May 2013, but many questions remained unresolved, including policies for both filling and operating the GERD.

In 2014, the three countries agreed to establish a tripartite National Committee composed of representatives from the three Blue Nile riparians to address questions raised by the IPoE about the impacts of the GERD. The plan was for the Tripartite National Committee to hire and jointly supervise an international consulting firm to conduct these studies. The Tripartite National Committee issued a request for proposals and eventually the contract was awarded to two French firms: BRL Ingenierie (BRLi) and Artelia.

Contested baselines: Nile example 1: GERD impact studies

The studies by BRLi and Artelia were never completed because Ethiopia and Egypt could not agree on the baseline from which to measure the hydrological, environmental and economics impacts of the GERD.[14] The construction of the GERD and its filling and operation will change the hydrology of the Nile downstream, essentially eliminating the flood in Sudan on the Blue and Main Nile, improving both hydropower generation and irrigation in Sudan, and eventually generating approximately 15,000 gigawatt-hours (GWh) of electricity in Ethiopia annually (Basheer et al., 2021; Wheeler et al., 2020).[15] However, if a multi-year drought were to occur during the filling period, Egypt could suffer irrigation deficits and, thus, economic losses depending on the filling and operating rules for the GERD. But from what baseline conditions should such losses to Egypt be measured?

For Egypt the answer was obvious: the changes resulting from the construction and operation of the GERD should be measured relative to a continuation of the status quo conditions before the dam was built. The 1959 Nile Waters Agreement between Egypt and Sudan defined the legal status quo and allocated an assumed mean annual flow of

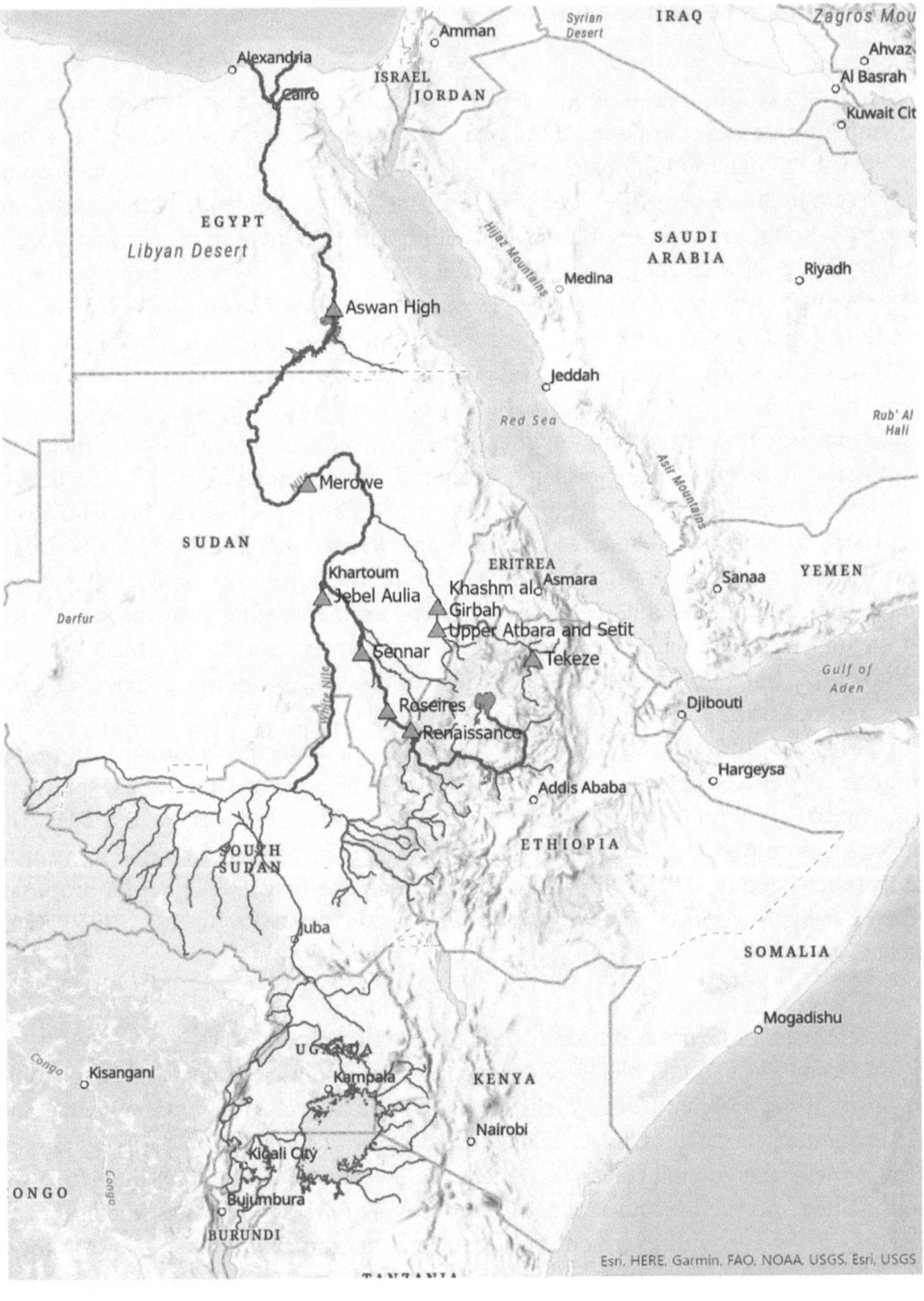

Figure 2. Map of Nile basin showing location of the Grand Ethiopian Renaissance Dam (GERD), which is shown as 'Renaissance'.

84 billion cubic metres (bcm) measured at Aswan between Egypt (55.5 bcm) and Sudan (18.5 bcm), with 10 bcm assumed to be lost to evaporation from the Aswan High Dam Reservoir (Whittington et al., 2014). Ethiopia was not a party to the 1959 Nile Waters Agreement and always objected to being left out of this bilateral allocation of Nile waters.

In the Tripartite National Committee, Egypt argued that the existing Egyptian and Sudanese uses of Nile waters were simply a fact, and that the purpose of the BRLi and Artelia study was to measure the economic, environmental, and social changes that would result compared to a continuation of status quo conditions. From Egypt's perspective, the baseline without the GERD was uncontroversial and there was no alternative dynamic baseline available to the French consultants from which to measure the impacts of the GERD. Egypt was especially concerned about its ability to continue to release 55.5 bcm annually from the Aswan High Dam and argued that any irrigation deficits that would result during a prolonged multiyear drought should be measured from this target release that was codified in the 1959 Nile Waters Agreement.

Since Ethiopia had never agreed to Egypt's and Sudan's allocations in the Nile Water Agreement, it did not agree that the French consultants should calculate any Egyptian irrigation deficits as a shortfall from 55.5 bcm. From Ethiopia's perspective, if Egypt were to release, say, only 50 bcm during a drought, this was still the majority share of the Nile flow. From Ethiopia's perspective, the difference of 5.5 bcm from what Egypt wanted should not be perceived as a deficit because the 55.5 bcm baseline was itself illegitimate. Ethiopia feared that accepting a continuation of current water uses as the dynamic baseline would confer legitimacy to the 1959 Nile Waters Agreement and preclude future water use in Ethiopia upstream of the GERD.

Ethiopia could have proposed a different dynamic baseline and then measured the impacts of the GERD relative to this hypothetical or alternative water allocation. However, to the best of my knowledge, it never did so. Egypt and Ethiopia never reached a resolution of this problem of the contested baseline. Without an agreed upon dynamic baseline, the French consultants could never start the impact assessment work specified in their terms of references and the consultancy was abandoned.

Contested baselines: Nile example 2: a proposal for cooperative filling and management of the GERD

Despite the inability of the Tripartite National Committee to oversee the GERD impact studies, construction of the dam continued. Attention increasingly focused on its filling and operation (Wheeler et al., 2016). At the 2018 Cairo Water Week conference, Mr Karol Heijnert from the Dutch consulting firm Deltares gave a presentation in a session entitled 'Transboundary water governance and benefit sharing' in which he shared some insights from Deltares' work for Egypt on the impacts of the GERD (Heijnert, 2018).[16] Heijnert strongly advised against filling the GERD in a fixed period of years regardless of the inflows to the GERD, which he called a 'non-cooperative strategy' and said that this was not a wise approach because it could result in severe water shortages in Egypt, that is, cause 'significant harm'. Instead, he argued for 'cooperative' or 'adaptive' filling strategies and listed three possible components of a cooperative filling strategy: (1) filling the GERD with a fixed percentage of the annual inflow to the GERD, (2) maintaining a minimum outflow from the GERD and (3) maintaining a minimum level in the Aswan High Dam Reservoir. He described the third component – maintaining a minimum level in the Aswan High Dam Reservoir – as the 'most cooperative' because 'it is really important for Egypt to have sufficient water in Lake Nasser to irrigate the country. If you apply the right (cooperative) filling strategy, average hydrological conditions would not cause problems in Egypt', that is, Egypt will be able to continue to release its 55.5 bcm allocation under the 1959 Nile Waters Agreement.

From the perspective of the conceptual framework depicted in Figure 1, Heijnert assumed a dynamic baseline that was largely a continuation of status quo conditions. The intervention to be assessed relative to this dynamic baseline was the construction of the GERD. If Ethiopia followed a non-cooperative GERD-filling strategy such as filling in a fixed number of years, this intervention could result in severe water shortages in Egypt if filling occurred during a multi-year drought. However, Heijnert argued that if Ethiopia adopted a cooperative filling strategy and released water from the GERD to maintain a minimum level of the Aswan High Dam Reservoir, this would enable Egypt to largely return to the conditions in the dynamic baseline in which water shortages were rare. He acknowledged that this cooperative strategy would result in hydropower losses to Ethiopia but argued that these would be much less than the economic benefits to Egypt of avoiding water deficits measured from a target release of 55.5 bcm (conceptualized as a WTP measure of welfare loss, not as the minimum compensation Egypt would accept to incur the loss).

In 2019 Egypt made a proposal – Egyptian Proposal for Technical Aspects of the Agreement on the Filling and Operation of the Grand Ethiopian Renaissance Dam – for operating and filling the GERD that included several of these elements of a cooperative filling strategy. It specified a minimum annual release from the GERD of 40 bcm (except during drought conditions) and that the level of the Aswan High Dam Reservoir be maintained at 165 masl during the filling of the GERD. The Egyptian proposal also introduced the concepts of 'natural flow' and 'average natural flow' at the GERD site and specified that Egypt and Ethiopia should share equally the difference between the natural flow and the average natural flow to refill the two reservoirs after a prolonged drought. The Egyptian proposal required that Ethiopia would use the GERD's 'Strategic Reserve' storage between 595 and 617 masl (approximately 21 bcm) to supplement flows to Egypt during drought conditions, thus reducing Egyptian water deficits as Heijnert had advocated.

From Egypt's perspective, this was an equitable, balanced proposal – but it was based on the assumption that a filling and operation strategy for the GERD should be evaluated using a dynamic baseline that status quo conditions would continue into the future. Ethiopian acceptance of the proposed definitions of natural flow and average natural flow based on historical hydrology, and a commitment to share equally the difference to refill the two reservoirs after a prolonged drought, would have effectively precluded Ethiopia's future use of water upstream of the GERD.

Ethiopia rejected this 2019 Egyptian proposal but the fundamental problem was not the proposal per se but Egypt's assumed dynamic baseline. Ethiopia did not agree that the objective of a cooperative filling and operation strategy for the GERD should be to minimize Egyptian deficits from target releases of 55.5 bcm. This disagreement over the dynamic baseline was the same issue that halted the work of BRLi and Artelia.

Since 2019, the controversy over the GERD has escalated. At the time of writing (February 2022) the construction of the GERD is still unfinished and Ethiopia had only managed to capture about 8 bcm in the GERD Reservoir, 11% of the ultimate target of 74 bcm (Wheeler, 2022). In July 2021, representatives of Egypt, Ethiopia and Sudan debated the risks and rewards of the filling and subsequent operation of the GERD before the UN

Security Council. Egypt wanted assurances that the GERD's filling and operating policies would not infringe on its historic water use ('no significant harm'): Ethiopia argued that an agreement on the filling and operating policies of the GERD should not restrict its rights to develop and use Nile waters originating on its sovereign territory ('equitable use'). These positions were essentially statements about the countries' contested baselines – about implicit visions of future irrigation development in the Nile basin, not only in Ethiopia upstream of the GERD but also in Sudan and Egypt.

Conclusions

This discussion of contested baselines highlights three main lessons for policy analysis of transboundary water resources management and negotiation. First, disputes over transboundary water resources often focus on the trade-off between two principles in international water law: *no significant harm* and *equitable use*. However, the measurement of no significant harm requires agreement of a dynamic baseline – which may not exist. Thus, parties may disagree not only about the allocation principle to share transboundary water resources but also the dynamic baseline to use to make the assessment of no significant harm. This dilemma is likely to be fairly common in the management of transboundary water resources.

Second, stakeholders and users of policy analyses need to develop the habit of searching carefully for both dynamic baseline assumptions and the *baselining process*, even if analysts or stakeholders themselves have not made these clear. Any ex-ante policy analysis that seeks to determine the consequences of a policy intervention should specify a dynamic baseline from which change is measured. However, either for strategic reasons or neglect, policy analysts often draw stakeholders' attention away from the specification of the dynamic baseline.

The two illustrations of the contested baseline controversy from the Nile are unusual because the controversy over the filling and operation of the GERD has revealed the problem of contested baselines so clearly. Typically, the issue is more obscure. Sometimes, the dynamic baseline is hidden in plain sight – but never mentioned. Sometimes, as in Robert Frost's poem, the road not taken is hidden in the undergrowth of the analysis, as often happens in cost–benefit and cost-effectiveness calculations.

Third, in situations where baselines are contested, stakeholders need to discern the source of the disagreement. Dynamic baselines may be chosen by parties to a conflict for strategic reasons, that is, to influence the outcome a policy analysis. However, the choice of a dynamic baseline may not be strictly instrumental but can also reflect deeply held feelings of fairness and justice. One useful approach would be to undertake the policy assessment with the different dynamic baselines proposed by the opposing parties. Doing the analysis with each party's proposed baseline may help clarify their differences and open new possibilities for compromise.

Notes

1. For this work he was awarded the Stockholm Water Prize in 2008.
2. If the policy alternative only has very short-term effects, then a dynamic baseline may not be required; status quo conditions may be sufficiently accurate as a comparator. However, the

management of transboundary water resources typically requires a long planning horizon and a dynamic baseline is necessary.

3. For example, in a cost–benefit analysis the discount rate is often a parameter that affects the estimate of the net benefits of shifting from the dynamic baseline to a new state of the world with the policy intervention.
4. For ex-post evaluations, various non-experimental methods are available for the construction of the dynamic baseline, such as synthetic controls, even if there is only one (or a few) treated units (Abadie, 2021).
5. For simplicity, Figure 1 shows the lives lost after the construction of the dam to be constant. This need not be the case. Technological innovation, population and economic growth, and climate change could cause lives lost to rise or fall after time *t*.
6. This might occur because the dam encouraged settlement below the dam and increased the number of people at risk from the release of excess flood waters or dam failure.
7. Sometimes a cost-effectiveness measure is reported as the inverse of this ratio, that is, the dollars required for a unit change in the performance indicator, $/ΔPI. In this case, the dynamic baseline is hidden in the denominator of the ratio.
8. New approaches may be required for thinking about decision-making under deep uncertainty (Lempert et al., 2013).
9. Careful consideration of uncertainty in the forecast of the dynamic baseline also focuses attention on what is known or being assumed about the magnitude of the treatment effect. Is the magnitude of the treatment effect estimated as (1) a percentage change in the value of the performance indicator in the dynamic baseline, or (2) an absolute magnitude change over a specified baseline? If the treatment effect is estimated as a percentage change in the dynamic baseline, then the size of the change will be dependent on the specific forecast of the dynamic baseline. If it is measured as an absolute magnitude, then it may be possible to apply this estimate to different forecasts of the dynamic baseline. Which approach to use is an empirical question and context specific. For most RCTs, the treatment effect is typically reported as a percentage point change in the performance indicator between the treatment and controls, not as percentage change in the dynamic baseline. Whether this percentage point change can be applied to another baseline is a question of the external validity of the experimental results.
10. I thus make a distinction between *contested baselines* and Kinchy's notion of *contentious baselines*. Contested baselines refers to disputes over future dynamic baselines; Kinchy's notion of contentious baselines refers to disputes over past or status quo conditions.
11. Conceptually, endogenous policy decisions could be incorporated into the dynamic baseline in such an analytical model, but policy analysts typically avoid this approach because it is so challenging. However, stakeholders may not be able to avoid confronting this complexity.
12. Conceptually this problem of *contested baselines* is similar to the use of *hypothetical baselines* in stated preference studies in which respondents are provided with a description of a baseline that is intentionally not the actual state of environmental quality, health or other baseline condition. The stated preference researcher then poses a valuation question or choice task that is contingent not on the existing status quo, but rather on the state of the world described in this new hypothetical baseline (Whittington & Adamowicz, 2011; Whittington et al., 2017). Just as with contested baselines, there may be disagreements among respondents about the appropriate specification of the hypothetical baseline in a stated preference survey due to perceived inequalities in the status quo conditions.
13. Knetsch (2020) refers to both the willingness-to-accept measure to value a loss and the willingness-to-accept measure to value a gain that is viewed as a reduction in a loss as 'the domain of losses' (see also Nguyen et al., 2021).
14. There was also disagreement between Egypt and Sudan over the appropriate baseline. Egypt argued that the baseline should be based on current water use: Sudan felt that the dynamic baseline should be based on the water allocations in the 1959 Nile Waters Agreement because it was not yet using its full water allocation under the treaty.

15. Sudan has two main concerns about the construction and operation of the GERD. First, the elimination of the Nile flood would eliminate the remaining recession agriculture along the Nile in Sudan. Second, Sudan needs timely information on the operation policy of the GERD in order to better operate its downstream dams, especially Roseires.
16. Deltares does water resources modelling throughout the Nile basin, indeed throughout the world, and has had a long relationship working with Egypt on Nile water management issues.

Acknowledgements

The author thanks Mohammed Basheer, Thinus Basson, David Fuente, Giorgio Guariso, Jim W. Hall, Kingsley Haynes, Chris Heady, David Hulme, Marc Jeuland, Martin Keulertz, Jack Knetsch, Lisa Robinson, Aaron Salzberg, V. Kerry Smith, Eelco van Beek, John Waterbury and Kevin Wheeler for helpful comments on drafts of this paper.

Disclosure statement

No potential conflict of interest was reported by the author.

Funding

This work was supported by the UK Research and Innovation Economic and Social Research Council [ES/P011373/1] as part of the Global Challenges Research Fund through the "Future Design and Assessment of water-energy-food-environment Mega Systems" (FutureDAMS) research project.

ORCID

Dale Whittington http://orcid.org/0000-0002-6075-8812

References

Aagaard, T. S. (2011). Environmental harms, use conflicts, and neutral baselines in environmental law. *Duke Law Journal*, *60*(7), 1505–1564. https://scholarship.law.duke.edu/dlj/vol60/iss7/1

Abadie, A. (2021). Using synthetic controls: Feasibility, data requirements, and methodological aspects. *Journal of Economic Literature*, *59*(2), 391–425. https://doi.org/10.1257/jel.20191450

Alagona, P. S., Sandlos, J., & Wiersma, Y. F. (2012). Past imperfect: Using historical ecology and baseline data for conservation and restoration projects in North America. *Environmental Philosophy*, *9*(1), 49–70. https://doi.org/10.5840/envirophil2012914

Aldy, J. E. (2014). *Learning from experience: An assessment of the retrospective reviews of agency rules and the evidence of improving the design and implementation of regulatory policy*. Report prepared for the Administrative Conference of the United States. https://www.acus.gov/sites/default/files/documents/Aldy%2520Retro%2520Review%2520Draft%252011-17-2014.pdf

Allan, J. A. (2019a). Invited opinion interview with Professor Tony Allan: "Water scarcity, food production, and virtual water" (Part 1). *Water Economics and Policy*, 6(1), 1-14. doi: 10.1142/S2382624X19710036.

Allan, J. A. (2019b). Policy note: Invited opinion interview with professor Tony Allan: 'Water scarcity, food production, and virtual water' (Part 2). *Water Economics and Policy*, 6(2), Part 2) https://doi.org/10.1142/S2382624X19710048

Angrist, J. D., & Pischke, J. S. (2010). The credibility revolution in empirical economics: How better research design is taking the con out of econometrics. *Journal of Economic Perspectives, 24*(2), 3–30. https://doi.org/10.1257/jep.24.2.3

Banerjee, A. V., & Duflo, E. (2009). The experimental approach to development economics. *Annual Review of Economics, 1*(1), 151–178. https://doi.org/10.1146/annurev.economics.050708.143235

Barandiarran, J. (2019). Documenting rubble to shift baselines: Environmental assessments and damaged glaciers in Chile. *Environment and Planning E: Nature and Space, 3*(1), 58–75. doi:10.1177/2514848619873317.

Basheer, M. (2021). Cooperative operation of the Grand Renaissance Dam reduces Nile riverine floods. *River Research and Applications, 37*(6), 805–814. https://doi.org/10.1002/rra.3799

Basheer, M., Nechifor, V., Calzadilla, A., Siddig, K., Etichia, M., Whittington, D., Hulme, D., & Harou, J. J. (2021). Flexible collaboration on the Nile increases benefits and enhances economic resilience. *Nature Communications, 12*(1), 5622. https://doi.org/10.1038/s41467-021-25877-w|www.nature.com/naturecommunications

Bento, A. M., Kanbur, R., & Leard, B. (2016). On the importance of baseline setting in markets for carbon offsets. *Climatic Change, 137*(3–4), 625–637. https://doi.org/10.1007/s10584-016-1685-2

Blackmore, D., & Whittington, D. (2008). *Opportunities for cooperative water resources development on the Eastern Nile: Risks and rewards.* Final Report of the Scoping Study Team to the Eastern Nile Council of Ministers. World Bank.

Boardman, A., Greenberg, D., Vining, A., & Weimer, D. (2018). *Cost-benefit analysis: Concepts and practices* (5th ed.). Cambridge University Press.

Bonebrake, T. C., Christensen, J., Boggs, C. L., & Ehrlich, P. (2010). Population decline assessment, historical baselines, and conservation. *Conservation Letters, 3*(6), 371–378. https://doi.org/10.1111/j.1755-263X.2010.00139.x

Braverman, R. (2020). Shifting baselines in coral conservation. *Environment and Planning E: Nature and Space, 3*(1), 20–39. doi: 10.1177/2514848619882560.

Craig, R. K. (2014). Perceiving change and knowing nature: Shifting baselines and nature's resiliency. In K. Hirokawa (Ed.), *Environmental law and contrasting ideas of nature: A constructivist approach* (pp. 87–111). Cambridge University Press.

Farmer, W. H., & Vogel, R. M. (2016). On the deterministic and stochastic use of hydrological models. *Water Resources Research, 52*(7), 5619–5633. https://doi.org/10.1002/2016WR019129

Fiering, M., & Jackson, B. B. (1991). Synthetic streamflows. Monograph. *American geophysical union water resources monograph series* (Vol. 1). Wiley.

Frost, R. (1915, August). The road not taken. *The Atlantic Monthly*, 223.

Heijnert, K. (2018). Paper PS3 presented at *Cairo International Water Week 2018.* In Session on transboundary water governance and benefit sharing. Ministry of Water Resources.

Hilborn, R., & Walters, C. J. (1981). Pitfalls of environmental baseline and process studies. *Environmental Impact Assessment Review, 2*(3), 265–278. https://doi.org/10.1016/S0195-9255(81)80062-5

Intergovernmental Panel on Climate Change. (2021). Climate change 2021 The physical science basis. *Report of Working Group I Contribution to the Sixth Assessment Report of the IPCC.* October. 978-92-9169-158-6. websitewww.ipcc.ch

International Panel of Experts. (2013). *Final report on the Grand Ethiopian Renaissance Dam project.* May 31. Addis Ababa. Available at the International Rivers site. https://archive.internationalrivers.org

Jeuland, M., & Pattanayak, S. (2012, February). Benefits and costs of improved cookstoves: Assessing the implications of variability in health, forest, and climate impacts. *PLOS, 7*(2). https://doi.org/10.1371/journal.pone.0030338

Jeuland, M., & Whittington, D. (2014). Water resources planning under climate change: Assessing the robustness of real options for the Blue Nile. *Water Resources Research, 50*(3), 2086–2107. https://doi.org/10.1002/2013WR013705

Kinchy, A. (2019). Contentious baselining: The politics of 'pre-drilling' environmental measures in shale gas territory. *Environment and Planning E: Nature and Space*, *3*(1), 76–94. doi: 10.1177/2514848619877585.

Knetsch, J. L. (2020). Behavioural economics, cost–benefit analysis, and the WTP versus WTA choice. *International Review of Environmental and Resource Economics*, *14*(2–3), 153–196. https://doi.org/10.1561/101.00000119

Knowlton, N., & Jackson, J. (2008). Shifting baselines, local impacts, and global change on coral reefs. *PLOS Biology*, *6*(2), e54. https://doi.org/10.1371/journal.pbio.0060054

Lempert, R. J., Popper, S. W., Groves, D., Kalra, N., Fischbach, J.R., Bankes, S.C., Bryant, B.P., Collins, M.T., Keller, K., Hackbarth, A.,& Dixon, L. (2013). *Making good decisions without predictions. Robust decision making under deep uncertainty*. Rand Corporation.

Lind, J. T., Nyborg, K., & Pauls, A. (2019). Save the planet or close your eyes? Testing strategic ignorance in a charity context. *Ecological Economics*, *161*(9), 19. https://doi.org/10.1016/j.ecolecon.2019.02.010

McClenachan, L., Jackson, J. B. C., & Newman, M. J. H. (2006). Conservation implications of historic sea turtle nesting loss. *Frontiers in Ecology and the Environment*, *4*(6), 290–296. https://doi.org/10.1890/1540-929520064290:CIOHST2.0.CO;2

Nguyen, K. T., Knetsch, J. L., & Mahasuweerachai, P. (2021). WTP or WTA: A means of determining the appropriate welfare measure of positive and negative changes when preferences are reference dependent. *Environmental and Resource Economics*, *78*(4), 615–633. https://doi.org/10.1007/s106640-021-00546-0

Pauly, D. (1995). Anecdotes and the shifting baseline syndrome of fisheries. *Trends in Ecology & Evolution*, *10*(10), 430. https://doi.org/10.1016/S0169-5347(00)89171-5

Pauly, D. (2019). *Vanishing fish: Shifting baselines and the future of global fisheries*. Greystone Books.

Pyle, R. M. (2003). Nature matrix: Reconnecting people and nature. *Oryx*, *37*(2), 206–214. https://doi.org/10.1017/S0030605303000383

Quinn, A. K., Williams, K., Thompson, L., Rosa, G., Díaz-Artiga, A., Thangavel, G., Balakrishnan, K., Miranda, J. J., Rosenthal, J. P., Clasen, T. F., & Harvey, S. A. (2019). Compensating control participants when the intervention is of significant value: Experience in Guatemala, India, Peru and Rwanda. *BMJ Global Health*, *4*(4), e001567. https://doi.org/10.1136/bmjgh-2019-001567

Robinson, L. A., Hammitt, J. K., Cecchini, M. Chalkidou, K., Claxton, K., Cropper, M., Eozenou, P. H.V., de Ferranti, D., Deolalikar, A.B., Guanais, D.T., Jamison, S., Kwon, J. A., Lauer, L., O'Keeffe, D., Walker, D., Whittington, T., Wilkinson, D., Wilson, and B. Wong. (2019). Reference case guidelines for benefit-cost analysis in global health and development. Report. Harvard T.H. Chan School of Public Health. https://cdn2.sph.harvard.edu/wp-content/uploads/sites/94/2019/05/BCA-Guidelines-May-2019.pdf

Schlee, E. E., & Smith, V. K. (2019). The welfare cost of uncertainty in policy outcomes. *Journal of Environmental Economics and Management*, *98*, 102264. https://doi.org/10.1016/j.jeem.2019.102264

United States Department Health and Human Services. (2016). *Guidelines for regulatory impact analyses*. Office of the Assistant Secretary for Planning and Evaluation.

United States Environmental Protection Agency. (2010). *Guidelines for preparing economic analyses*. National Center for Environmental Economics, Office of Policy. Updated May 2014

Ureta, S. (2018). Baselining pollution: Producing 'natural soil' for an environmental risk assessment exercise in Chile. *Journal of Environmental Policy & Planning*, *20*(3), 342–355. https://doi.org/10.1080/1523908X.2017.1410430

Ureta, S., Lekan, T., & von Hardenberg, W. (2020). Baselining nature: An introduction. *Environment and Planning E: Nature and Space*, *3*(1), 3–19. https://doi.org/10.1177/2514848619898092

Wheeler, K. (2022). *The GERD dam filling impact and mitigation. UCLA Lecture Series – The Nile River Basin in crisis: Water sharing and transboundary conflict or cooperation*. Interview. January 26. http://waterbalance.org/public-presentations/

Wheeler, K. G., Basheer, M., Mekonnen, Z. T., Eltoum, S. O., Mersha, A., Abdo, G. M., Zagona, E. A., Hall, J. W., & Dadson, S. J. (2016). Cooperative filling strategies for the Grand Ethiopia Renaissance Dam. *Water International*, *41*(4), 611–634. https://doi.org/10.1080/02508060.2016.1177698

Wheeler, K. G., Hall, J. W., Abdo, G. M., Dadson, S. J., Kasprzyk, J. R., Smith, R., & Zagona, E. A. (2018). Exploring cooperative transboundary river management strategies for the Eastern Nile Basin. *Water Resources Research*, *54*(11), 9224–9254. https://doi.org/10.1029/2017WR022149

Wheeler, K., Jeuland, M., Hall, J. W., Zagona, E., & Whittington, D. (2020, October 16). Understanding and managing new risks on the Nile with the Grand Ethiopian Renaissance Dam. *Nature Communications*, *11*(1). https://doi.org/10/1038/s41467-020-19089-x

Whittington, D., & Adamowicz, W. (2011). The use of hypothetical baselines in stated preference surveys. In Y. Alem (Ed.), *Environment for development discussion paper series, EfD* (pp. 11). University of Gothenburg, Environmental Economics Unit.

Whittington, D., Adamowicz, W., & Lloyd-Smith, P. (2017). Asking willingness-to-accept questions in stated preference surveys: A review and research agenda. *Annual Review of Resource Economics*, *9*, 317–336. https://doi.org/10.1146/annurev-resource-121416-125602

Whittington, D., Waterbury, J., & Jeuland, M. (2014). The Grand Renaissance Dam and prospects for cooperation on the Eastern Nile. *Water Policy*, *16*(4), 595–608. https://doi.org/10.2166/wp.2014.011b

Water and complex problemsheds in Karamoja, Uganda

Alan Nicol, Liza Debevec and Samuel Okene Ayaru

ABSTRACT
A complex political economy revolves around shared land and water use between Kenyan Turkana and Ugandan Karamojong pastoralists. In response to growing pressure on resources, donors and the Ugandan government are investing in new surface water sources. However, power and political economy issues embedded within societal relationships are rarely factored into water infrastructure development. Drawing on Tony Allan's teaching, we examine studies of two dams recently constructed in Karamoja and argue that a wider view encompassing power and politics within the Karamoja–Turkana Complex would help ensure more sustainable and effective future water supply development. Allan's idea that catchments are part of much wider social, political economic and integrated livelihood systems, or problemsheds, is a key concept. Here we argue that adopting this concept in a complex of pastoral systems can improve future water resources planning and intervention in Karamoja, Uganda and similar contexts.

Introduction: resilience building and competing narratives

> After peace, our next priority is water. Am glad we have done some dams, but they are not enough. I have come to check on what we can do to solve this problem completely. [...] It is not true that there is no water in Karamoja, it comes and goes and you follow it to Teso. This must stop. The water must stay here. We are going to build about 20 dams. (President Museveni to regional leaders in Moroto district, Karamoja, June 2019)[1]

So stated the President of Uganda ahead of elections in 2020. In drylands across East Africa, augmenting water storage is regarded as an important buffer against drought and water insecurity, helping to bridge dry periods for both humans and livestock and enabling dry-season-irrigated agriculture – what is often seen as part of wider efforts at sedentarization of pastoralists. It is also, by extension, politically attractive, especially for governments facing complex state–society relations in often remote pastoral areas. This leads to a supply-led water security agenda that can generate extensive infrastructure construction but without appropriate longer term management arrangements, with significant consequences for the future climate resilience in pastoral areas – and future conflict over resource access.

Augmenting water supplies for livestock has been pursued across the Karamoja subregion (Figure 1) under a succession of donor-funded programmes, starting with the Northern Uganda Social Action Fund (NUSAF) (2003–20),[2] and, more recently, the Karamoja Livelihoods Programme (KALIP) (2010–15) framed under Farming and Peace, Food and Security. The wider focus of these programmes on resilience-strengthening in the face of climate shocks represents a desire to shift from emergency food relief to investments that support sustainable livelihoods for the most vulnerable communities and help in establishing alternative livelihood resources, including small-scale irrigation. Access to water for livestock is regarded as a key element in this policy approach.

In tandem with these approaches, wider issues of water management within catchments have been part of central government water policy objectives. Under the Enhancing Resilience in Karamoja Programme (ERKP)[3] in 2015, this broader water catchment perspective became an additional focus of donor interventions, encompassing water supply as a part of a whole systems approach. This was a significant step towards

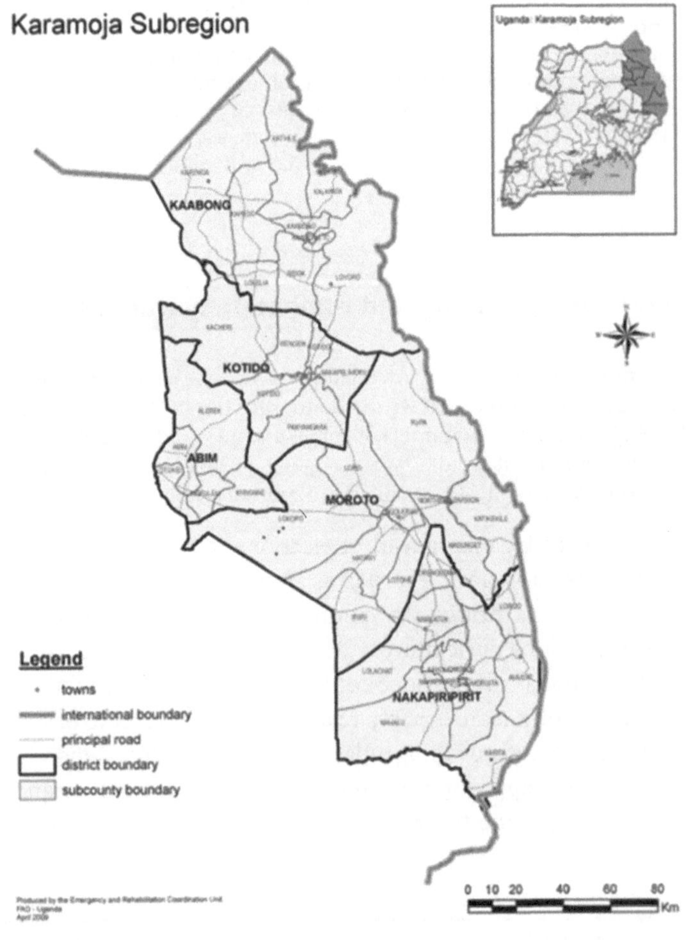

Figure 1. Districts of Karamoja subregion.

integrated planning, including food, land and water systems integration within environments under growing human- and climate-induced stress (e.g., the growing demand for biomass energy outside the region causing rapid deforestation).

At the same time, the catchment boundaries remained the planning unit, reducing management to hydrological limits and leading to the reductionism so apparent in much integrated water resources management. This is where Tony Allan's idea that catchments are part of much wider social, political economic and integrated livelihood systems, or *problemsheds* (Allan, 1999) is a key concept, demonstrably so in the Karamoja subregion. Here, we argue that adopting this concept in a complex of pastoral systems can improve future water resources planning and intervention in Karamoja and similar contexts.

Managing water: power, contest and control

Issues of livestock survival lie at the heart of the Karamoja–Turkana Complex (KTC) - the problemshed at the centre of water management challenges in the region. Movement of livestock between sources of water and grazing is critical to pastoralist livelihoods, especially in light of growing human and animal numbers. Specific seasonal migration corridors provide access to key grazing and water resources as the dry season progresses. People and livestock follow the flow of water along the two major river systems and down into the wetlands of Teso, neighbouring Karamoja to the south-west. Capturing and using some of this flow in a series of dams is, as President Museveni noted, a key objective for government. And the more these sources are made available, the more important the region becomes for the neighbouring Kenyan Turkana, who live in a far drier environment.

In recent years, livestock have been affected by disease[4] and poor veterinary support, as well as historic cattle-rustling. In 2021, cattle-rustling and insecurity revisited Karamoja after a period of relative calm. But at a more structural level, after years of livelihood insecurity within pastoral systems, many young Karamojong began seeking alternative livelihoods that would provide immediate cash income, ranging from wage labour to charcoal production and artisanal goldmining. The latter, in particular, contributes to a growing cash economy in this gold-rich region and has emerged alongside the rapid growth in the availability of mobile phones, enabling wealth storage in the digital realm through mobile banking, in contrast to wealth represented by livestock (the more traditional form of 'mobile' banking).[5]

In the face of changing rainfall patterns in the region, building resilience to water scarcity and flood control has been central to the classic *hydraulic mission* of water infrastructure development. Alongside larger dams holding inter-annual storage, this includes many, smaller valley tanks to capture water sufficient to last three to six months. In this paper, we examine the political economy of the problemshed surrounding two of the three largest dams constructed since 2010, namely Arachet and Longoromit, respectively located in Napak and Kaabong districts. Together, they illustrate how the political-economy of the KTC problemshed is critical to understanding how complex demand for water and, therefore, future management challenges surrounding new water storage infrastructure require a broader view encompassing challenges and pressures that extend beyond the basin boundary.

Introducing new water structures without effective management concentrates people and livestock around water points, potentially generating competition and conflict, and degrading soils and tree cover. In Karamoja, as in other drylands in East Africa, these interventions are often instituted through top-down programmes with little community involvement. The structures are then handed over to the same communities to bear responsibility for managing infrastructural integrity in the long-term and effective regulation and control of the multiple user communities which, in this case, includes Turkana pastoralists from neighbouring Kenya (Mehta et al., 2014; Mollinga, 2008; Nicol & Mtisi, 2003).

This wider scale of analysis, embedded within which are various notions of power, preoccupied Tony Allan's London Water Issues Group which drew distinctions between *soft* and *hard* power, and different power subcategories (Zeitoun et al., 2011, 2022, in this issue). Building on this analysis, we identify within the KTC three major forms that power has taken: first, direct *coercive power*, where physical instruments of state and non-state power exercise control over and access to resources, including water under the *hydraulic mission*; second, *ideational power* where controlling development narratives shape specific forms of resource deployment, positioning of interventions and management arrangements – including how notions of resilience-building are constructed under donor and government interventions and water infrastructural is made a centrepiece of this approach; and third, *economic power*, private or public, which shapes how livelihoods decisions are taken – for example, the growing markets for artisanal gold and charcoal that extend well beyond Karamoja to the wider national level, yet play key roles in shaping livelihoods and ecosystems within Karamoja. To these we add a fourth dimension – *cultural power* – through which, on the one hand, Karamojong society is sometimes characterized by non-Karamojong (and even elites from within the Karamajong) as backward, whilst, on the other, Karamojong exert their strength as a culture through identification with a wider Karamoja Cluster[6] across the region, drawing on affinities in culture and language between peoples from Kenya, South Sudan and Ethiopia.[7] The clash of cultural power with development narratives and shifting to more sedentary livelihoods is a strong current in many donor interventions.

Part of the complexity and schism within the KTC problemshed lies in British colonial demarcation of the border between Kenya and Uganda, which used the edge of the Nile basin and the north-east escarpment of Karamoja as an approximate boundary – hence its distinctly non-straight line. This action bisected peoples who were already sharing a landscape and associated water systems, albeit not always peacefully. Today an estimated 150,000 Turkana herders (about 10% of the total Turkana population in Kenya) move each year across this artificial political divide from drier herding areas to reach richer grazing – in both quantity and quality – and especially at the current time of drought in northern Turkana. This movement is central to the wider problemshed concept (Mollinga et al., 2007) as it involves a complex of factors involving political, economic, small-arms-related violence, and government–government at both local and national levels. In 2017, according to an official in Kaabong[8] (one of the wettest parts of the region), the district hosted about 50,000 cattle and 30,000 people from both Kenya and South Sudan.

Water governance in such a context is unbounded by basin limits and means that management decisions (on siting, access and regulation) impact wider political complexities between these polities and social groups. Political pressures and disputes arising extend across a substantial range of user communities; pressures that may exist transiently (as in the case of seasonal migrations to water points and grazing and, also, due to the changes in availability across wetter and drier seasons under an uncertain climate), and in very different spatial contexts. Understanding this broad, social, political and cultural context of how resources are used and developed becomes key to determining how, where and why investments in water supply structures and management systems can and should be made.

Field studies

Here we build on earlier work in the Karamoja region conducted during the period 2015–19, including specific analysis of the catchment-based planning process in the Karamoja region on behalf of the Department for International Development (DFID) and Deutsche Gesellschaft für Internationale Zusammenarbeit (GIZ) undertaken from November 2016 to June 2017 (Debevec et al., 2017; Nicol et al., 2021). For the two dam studies, we held key informant interviews at the national level (Ministry of Water, Directorate of Water for Production, Directorate of Water Resource Management – DWRM), through water-zone level (Kyoga Water Management Zone), catchment level (with representatives sitting on newly established catchment management committees), down to district level (with the elected and appointed officials in the districts where the dams are located) and subcounty levels. Interviews were held with officials at the dam locations as well as in district offices. To cover issues of gender dynamics and water resources development, we held gender-disaggregated focus groups within the communities in the vicinity of the dams – including age–sex-disaggregated discussions – and with key community members involved in small-scale irrigation schemes adjoining both dams. We also interviewed representatives of various non-governmental organizations (NGOs) and donors working in the subregion.[9]

We carried out four field trips of six to 10 days' duration (October 2017, March–April, August and November 2018). In addition to the field study, we reviewed the relevant literature with a specific focus on social science and development studies. We extended the political economy framework developed by Nash et al. (2006) to include structural features under 'historical factors and processes of change' (A in Figure 2). These factors shape the current institutions and structures of power (B) which, in turn, affect and are affected by actors and agents at different levels (C), including the way they exercise power, construct values and project ideologies. While we tried to include a range of stakeholders in the study sample, time and resources were limited; random sampling or household-level survey work was not possible as *kraals* (pastoral settlements) sprawl over large areas, so we relied on focus group discussions and key informant interviews. The selection of focus group participants may have been biased by individuals and communities being suggested by the district and subcounty-level officials and coordinated by our local research assistant, so we triangulated and corroborated results from and between focus groups and with our key informant interviews.

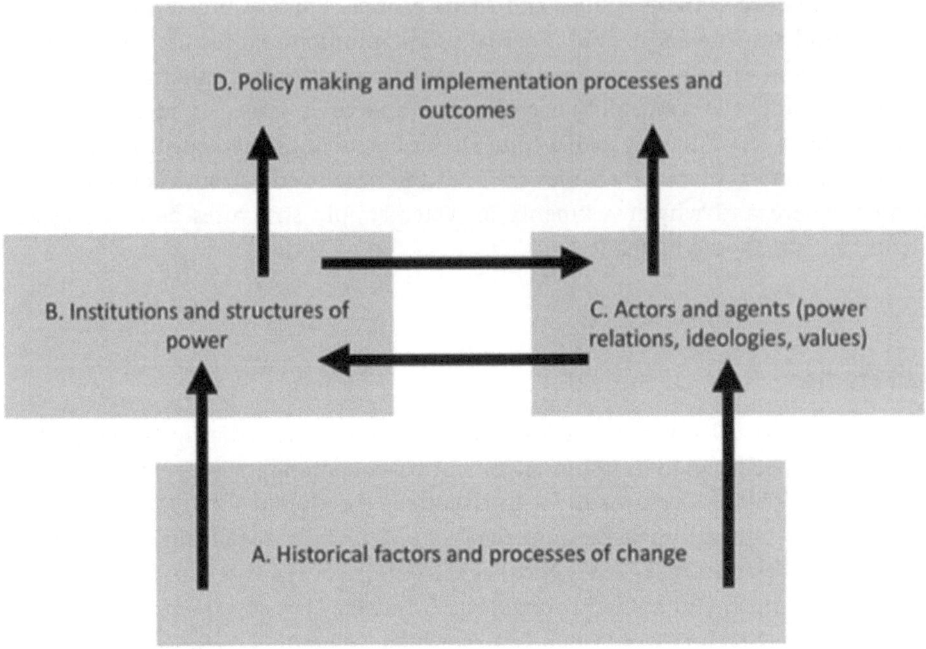

Figure 2. Political economy framework.

A new structuring of power and authority

The Ministry of Water and Environment (MWRE) (2014) has been rolling out integrated water resources management (IWRM)-style catchment-based planning, including the issuance of permits, for over a decade (Nicol & Odinga, 2016, p. 630) under which Uganda's eight river basins have been grouped into four water management zones (WMZs). The Kyoga basin zone (KWMZ) includes the study area and both the Lokok and Lokere catchments (Figures 3 and 4).

This new institutional arrangement envisages a role for districts located within catchments (Nicol & Odinga, 2016, p. 636). By late 2016, ten district water resources management organizations (DWRMO) had been established.

In Karamoja, the ERKP programme identified strengthening of integrated catchment-based planning and management as key to the development of water resources, including support to the development of catchment management plans for the Lokok and Lokere river basins, the two major catchments in the KWMZ (Figure 4).

Small water, big narratives

The three dams, Arachet, Longorimit and Kobebe, which have been constructed in the last 10 years, all lie within these two basins and are built to be multipurpose, covering livestock watering, domestic supply, and a source of fish and irrigation water. In global terms they are small, but in the context of Karamoja are huge in their scope and influence on the surrounding social and economic systems – and therefore on the wider KTC problemshed.

Figure 3. Water management zones.

Arachet dam (Figure 5) is located in Napak district within the Lokere catchment. It is sited near a much smaller dam known as *Natapar ka Emomwai* (literally, dam of the sorghum) which silted up and disappeared decades ago. The new structure was built in 2009–10 by Pearl Engineering/BEK Consulting for the Ministry of Water and Environment at a cost of Ush 6 billion (US$1.6 million at current rates). It is designed to serve both adjacent communities with water for irrigation (see Figure 6) and to provide water for migrant herds.

Napak district also has 18 much smaller valley tanks, but these empty in the dry season. Arachet is one of very few perennial water sources in Karamoja and services the Bokora, Jie and Matheniko Karamojong and, also, crucially, visiting Turkana pastoralists who may come directly or move to Arachet from Kobebe dam to the north – the third dam mentioned above – especially during March and July. Prior to construction of Arachet dam, herders would move their livestock in search of water to neighbouring districts – 'chasing the water' as they describe it – sometimes as far as Teso, Katakwe and Amuria where the wetlands of the Lokok and Lokere catchments provided dry-season grazing.

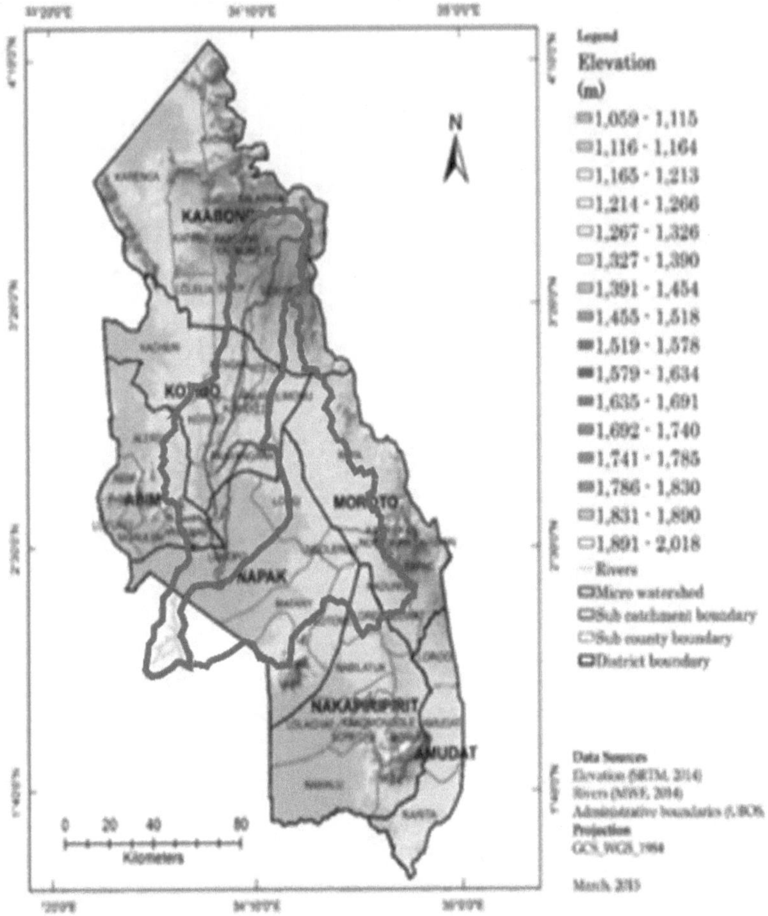

Figure 4. Lokok and Lokere catchments.
Source: International Union for the Conservation of Nature.

Arachet dam was constructed with a storage capacity of some 5 million m^3. A water point next to the dam, provided at the time of construction, serves as the domestic supply for Nakichumet parish which previously had none. This attracted people from as far as Matany subcounty and Lotome county, bordering Nakapiripirit.

Management is carried out by a water user committee (WUC) of nine, with a caretaker paid for by the district before this responsibility shifted to the subcounty. In discussions with committee members, the largest reported challenge is taking care of the heavily used cattle troughs (Figure 7). The committee asks people to contribute when watering their cattle and sought to charge Ush 5000 a month for large livestock users, mostly Karamojong herders,[10] and a smaller fee for domestic users of the existing standpipes for household use. During the 2016 national elections, according to local informants, this was contested by the LC5 (district chairperson) who urged people not to pay user fees for livestock to try to gain local votes. Another issue arising from the dam's construction is the landowner claiming compensation for land lost. According to

Figure 5. Arachet dam, October 2017.

Figure 6. Arachet dam irrigation scheme, March 2018.

local informants, much of land where the dam is sited belonged to a former British colonial agent family.

As a new water structure, the dam comes under Uganda's national water policy framework. The competing claims on the water from the reservoir reveal a lack of clarity regarding rules for issuance of permits for water abstraction. In a small but significant way, this case demonstrates how water remains highly contested even at this very local level, and how strong narratives on water and development can come to dominate decision-making, sometimes disregarding the essentially politically contested nature of water resources management in drylands such as these.

In 2017, as part of the Soroti–Moroto road construction project, an important national project that opens up remote Karamoja to heavy road traffic (essential, amongst other things, for future mining in the region), the international contractor involved requested water for construction. As per regulations, the request was submitted to the permit-issuing office at the DWRMO and was received by officers at the KWMZ in Mbale for appraisal. The KWMZ approved a permit for up to seven trucks carrying 25,000 litres

Figure 7. Cattle troughs, Arachet dam, October 2017.

per day, but a subsequent field visit by the permit officer revealed that the contractors were drawing more than the requested amount, that they had done so before receiving the permit, and had started abstracting water at night. Moreover, it appeared that they had done a deal at a local level directly with the district water office and other district officials and had bypassed the DWRMO and KWMZ.

According to community members, abstraction was in the region of 20 trucks a day, each capable of removing some 20 m^3. As abstraction proceeded, the local population noticed declining water levels – including the exposure of the new fish-farming cages. After complaining to local councillors, they took the complaint directly to district officials, and then Nakichumet subcounty community members blocked the road to prevent further abstraction of water. The KWMZ informed district officials that the permit had been officially rejected as the amount of water requested was not available. The district intervened to stop the contractors and instructed them to find alternative solutions. During October 2017, a series of water ponds and tanks dug by the contractors appeared along the road as an alternative to water abstraction.

During the next field trip, in March 2018, the contractor had begun pumping water again with the knowledge of district officials, under an memorandum of understanding (MoU) agreed with the district in December. The pumping took place through valves leading from the dam and no longer directly from the reservoir. No protests took place this time. One of the explanations given by the dam caretaker was that there was enough water; another was that since the water was coming from a valve not linked to the cattle trough, this was 'not the same water' and therefore did not appear to be in competition. In conversations with local officials, it became apparent that the money was at this time collected at the district and not the DWRMO level. A second MoU with the district, signed in July 2018, stipulated that the contractor could draw water when there was no rain, and the contractor was to pay the district Ush 3 million per month. The dam caretaker was required to note each time the contractor loaded the water bowser. His notes for March 2018 showed a total of 223 trucks were loaded, equivalent to over 4000 m^3 if each truck was the standard 20 m^3. Although the local population was not consulted about these arrangements and received no compensation, the dam caretaker confirmed having received money personally from the contractor informally. He was also responsible for turning on the tap that connected the dam to the cattle troughs and for preventing misuse of the reservoir, including bathing by local people.

Jurisdiction over the dam had therefore become a point of conflict between the subcounty and the district, including who should benefit financially from payments for abstraction. In many ways, these management challenges represent a microcosm of wider water governance challenges across drylands in East Africa where opacity in ownership rights combines with complex land–water interactions, including emerging relationships between government, private sector contractors and local communities. Some of the opaqueness surrounding the issuance of permits became apparent in discussion with the DWRMO office in Entebbe. We were told that, given the special status of Karamoja following disarmament, it was possible that 'normal permit rules did not apply'. Officials stated that in the case of Arachet dam, which was built to support the subregion and district development,[11] permit fees did not need to be paid (implying some kind of special status). However, we uncovered no official documentation of such status, which implied flexibility in the application of water permitting rules leaving open the possibility of misuse and rule-breaking. Ostensibly, all permit revenues go to the Treasury, and are then reallocated, including to the DWRMO. The permit issuing office at the KWMZ receives no direct funds and this, as their staff explained, reduced their motivation to follow up breaches in the rules governing permits. Given that there is no official line for local revenue use, there is district-level motivation to make a direct agreement with the contractors.

According to the Water for Production Office,[12] revival of the former irrigation scheme in Nakichumet next to Arachet dam came under a wider initiative by government to establish 40 small-scale irrigation schemes in northern Uganda, which was soon abandoned. A key challenge was the sharing of benefits – but not costs – between scheme users, as well as securing the site and its valuable produce from theft. It was also reported that political interests, appreciating the value of the land, then claimed it back after selling it to the scheme. The number of people associated with the scheme (around 30) was also a challenge, leading to a lack of cohesion and common 'buy-in' across the group coupled with a general lack of irrigation experience. As a result, the scheme came to be seen as a 'rent extractor', rather than investment asset, so money did not flow back into maintaining the irrigation, only away from it to the users.

Longoromit dam in Kaabong district (Figure 8) was constructed in 2012 at a similar cost and also financed by the World Bank. It was intended to provide a permanent water source in the dry season for the district population of some 400,000 people as well as pastoralists from other areas, including Turkana in neighbouring Kenya.[13] An adjoining water pump fed from shallow groundwater provides for domestic use. As part of efforts at livelihood diversification, the production department introduced fish into the reservoir, although local communities actively shun fish consumption.[14] To the west, a small-scale irrigation plot was established. A detachment of troops is stationed nearby to oversee security, given the contestation over access to the reservoir, and made use of the fish stock and water pump.

A management committee was established but its governance, constitution and remit remained unclear at the time of the visit; the committee represented only one community surrounding the dam (Lobonge), with members chosen from among

Figure 8. Longoromit dam.

LC1s and LC2s, according to local informants. Local stakeholders stated that their remit was mainly to safeguard the micro-catchment around the reservoir and to prevent cultivation around the dam (beyond a small, irrigated plot). The committee also managed access to the cattle troughs including preventing direct access to the reservoir for cattle, and called community meetings when there were clashes at watering points. The neighbouring pastoralists from Jie (Kotido district) or Turkana (Kenya) would seek permission from the host community through the management committee and subcounty officials.

The Turkana 'threat', as some informants described it in Kaabong district, included their use of water provided by Longoromit dam. Community members mentioned that their presence had restricted artisanal goldmining, an increasingly important source of cash income, in particular for youth. During 2016, a year of lower-than-average rainfall, the reservoir was used for the whole year by Turkana, with locals resorting to the Kaabong River (the traditional water source for livestock) under an agreement reached between the two groups. The agreement followed a tradition of dialogue on water and pasture usage between Turkana and Karamajong communities, in spite of years of cattle-rustling. Nevertheless, in 2017, there were deaths in Kaabong over disputed grazing areas near Longoromit, according to local leaders. The same local leaders consulted during the research claimed that they had suggested a more distant location and the proximity of Longoromit to Kaabong town was a factor possibly contributing to the incident.

Part of the wider political economy in relation to access to resources in Karamoja concerns dialogue and reaching of agreements over grazing and access to water, including the establishment of two peace corridors at Lomia–Moroto and Lokink–Kweata. Discussions with both Karamojong communities and district officials, as well as with Turkana representatives, revealed the lived reality of one system shared between two communities. A further example of this wider institutionalization of cooperation is the Loyolo Resource Sharing Agreement signed between Dodoth of Kaabong and Turkana, enabling Turkana to access pasture and water in Uganda. The agreement included a bush market and land allocation to the Turkana for cultivation. Under USAID, PEACE III is also supporting the peace process between the Dassanech of Ethiopia to access water and pasture in the Todonyang area of Turkana–Kenya. The Dassenach can now access these resources, which greatly helps their coping with drought and dry seasons. In other words, this is not a simple resource-scarcity/conflict relationship. Arguably, across what we call the KTC, there are many complex relationships that each has an impact on the management of resources in specific geographies and at particular times.

At the extremities of this system, changes in access to pasture and water may heighten competition and exacerbate tensions. Elders in Kaabong[15] describe tensions surrounding Turkana overgrazing and felling of trees for firewood and shelter in their district. Jie people from the drier Kotido district also come to Kaabong to cut trees for firewood. Additional external pressures include the impact of the Gibe III dam in Ethiopia, which has changed the flow of the Omo River into Lake Turkana affecting key wetland areas for cattle-grazing; particularly affecting the Nyangatom from South Omo who are related to the Karamojong and part of the Karamoja 'Cluster'. The Turkana are further affected by oil exploration in southern Turkana and associated land enclosures in important grazing areas. Exploitation by extractive industries is likely to occur as the LAPSETT corridor from Lamu to South Sudan is constructed[16] with this longer term infrastructure development opening up Karamoja's rich mineral resources to greater exploitation.

Changing pressures on livelihoods in Karamoja are likely to bear down on water management as complex political economies shape and change the interest groups with a stake in future water management and governance. As catchment planning is institutionalized in the region under recent donor interventions, understanding these political economies should be a key part of refining the water management process through understanding the broader political-economic pressures that determine patterns of demand within the KTC problemshed. Local officials expressed knowledge of meetings in Gulu and Kitgum under the WMZ related to catchment planning in the Lokok catchment but, short of requests received to establish WUAs, these key political actors maintained only a tenuous sense of engagement in wider catchment planning processes.

The implications of forms of power within a complex problemshed

The problems associated with these two small dams reflect wider challenges and processes surrounding water-supply investment, power and politics in the East Africa region. They show that, beyond the technical challenge of construction, attention needs to be given to the long-term governance of surface water resources taking account of political pressures. Narratives on resilience are often reduced to making water more available – a supply-driven 'hydraulic mission' – but, in fact, supply without effective governance can exacerbate resource competition and reduce resilience.

At the outset, we described forms of power within this problemshed and how these can help unpack relationships between users, managers and legal entities. Figure 9 helps interpret these in relation to the evidence presented.

Dam construction often acquires a political logic of its own. It is a symbol of progress that donors and local leaders can parade – especially at election time – with the promise of future benefits. This is amplified by national media, suggesting that local objections to siting dams nearer to towns and district offices can be overridden in the name of development. This ideational power helps to overcome local resistance and can present this resistance as 'barriers to change'. As a local key informant observed of the Longoromit siting, 'sometimes politicians don't want things located technically, but for political reasons'.[17] This ideational power is further reflected in the coupling of dam construction with irrigation development – a challenge to established livelihood norms in the region based on pastoralism and rainfed agriculture.

In the cases of both the Arachet and Longoromit dams, post-completion management structures appeared inadequate, with confusion over responsibility for overseeing and maintaining structures and operations – including of benefit streams. As soon as water abstraction and permit issues arise – and the monetary value of water becomes apparent – struggles for control over these value streams may surface, even between formal government institutions demonstrating where de facto economic power lies. If such dams are to contribute to real resilience and sustainable livelihoods, their design, siting and management needs to be improved – and the economic power they represent needs to be managed and disbursed under effective governance structures. Otherwise, and especially in times of scarcity, competition for this power may lead to conflict. Whilst their purpose is simple – to provide dry-season stock water – their siting, management and inclusion in

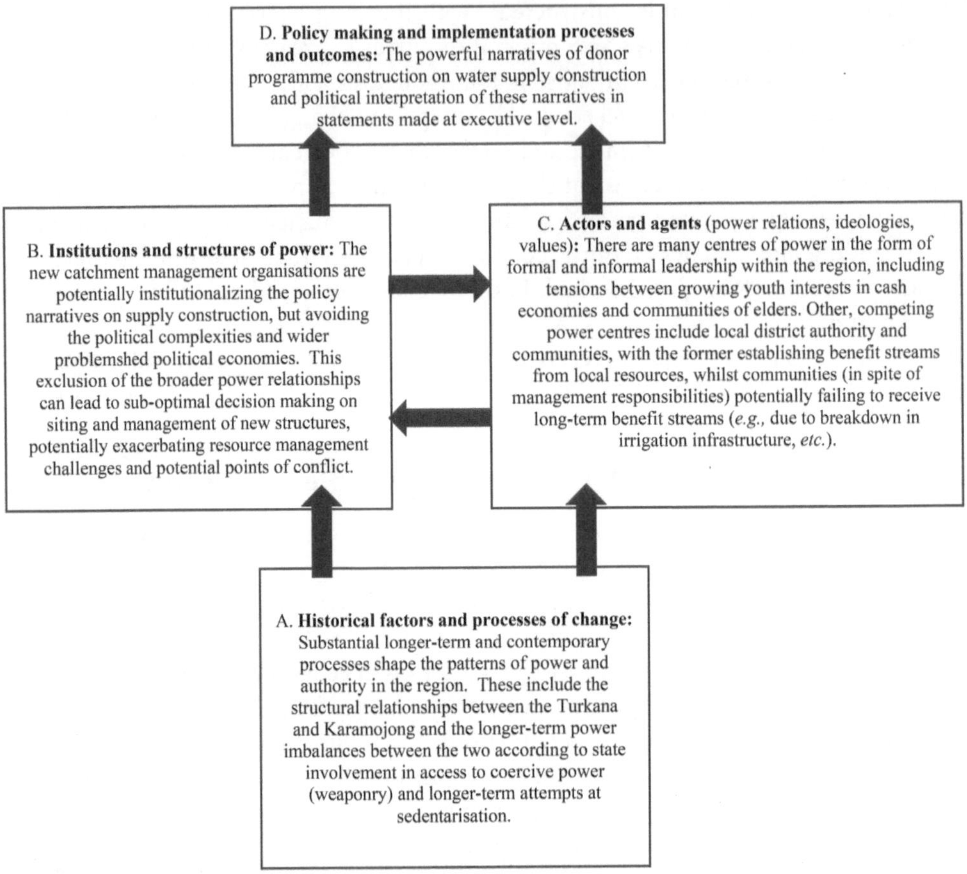

Figure 9. Factors in the political economy of the Karamoja-Turkana Complex.

broader narratives for water production is complex. It presents new governance challenges, including multiple layers of decision-making. Indeed, informal coercive power in the form of small-arms violence around these structures can arise.

This suggests the need for a broader set of criteria in designing and implementing dams such as these in complex problemsheds, and criteria that include ways of understanding competition between narrow, sectional interests and wider development objectives. As the race to exploit the region's resources accelerates, and given the government's stated objective of multiplying these structures across Karamoja, these interests and powers may well find themselves in conflict. More generally, the decline of pastoralism is a fundamental shift in the political economy of Karamoja, hitherto home to 20% of Uganda's total livestock (Office of the Prime Minister (OPM), 2014). And as land use changes, conflicts over land are increasing as communality gives way to private land grabs. Those who have 'gone to school', as Karamojong elders term it, understand the value of land (including in relation to farming, minerals and urbanization) and join a new elite seeking greater individual ownership at the expense of established communal management. In response, a process of customary land certification has been initiated by central government, but this will be a long process with no guarantee of success.

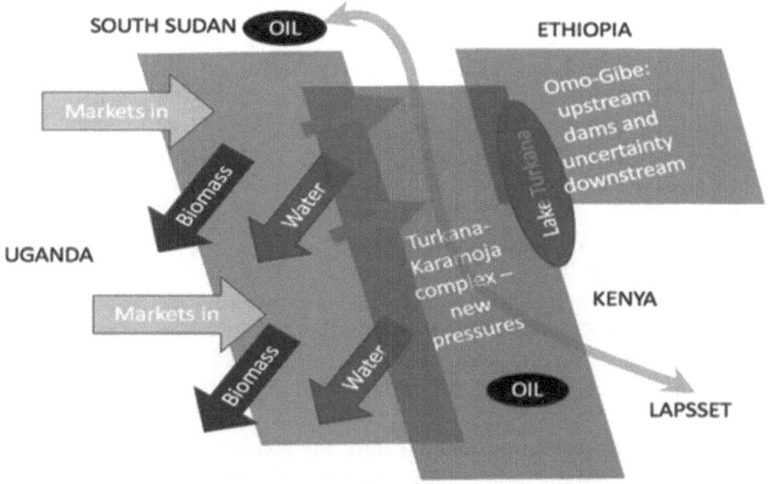

Figure 10. Karamoja–Turkana Complex (KTC).

The major assumption of water resources development in Karamoja remains that water can be isolated from surrounding complex polities and social structure, and be treated as a single, contained system. Our analysis suggests, instead, that future planning and investment should be examined within the context of the KTC. Figure 10 depicts how some of these processes and pressures apply to water development in the region.

Economic power drives markets for firewood and charcoal from Karamoja leading to the rapid depletion of woodland, in turn affecting land degradation and flood management in the Lokok and Lokere catchments.

The demand for water will continue, and possibly grow if livestock numbers recover, yet the narrative on making water available without understanding the wider problemshed complexities continues to affect siting and management. Catchment planning must factor in future power and political-economy issues specific to water management to strengthen the next generation of planning and investment processes in the region and, in so doing, respond to the reality of wider political-economic landscapes beyond the basin shaping future water resources development within this remote patch of the Nile basin.

Notes

1. See https://kampalapost.com/news/government-build-20-valley-dams-karamoja/.
2. NUSAF was largely funded by the World Bank; KALIP was funded by the European Union (EU).
3. Funded by the Department for International Development (DFID).
4. Muhanguzi et al. (2017) report that African animal trypanosomiasis and tick-borne diseases in particular have a serious impact on communities in the region. This aggravates poverty and malnutrition and undermines wider development efforts.
5. The extension of mobile telephone services has revolutionized cash income-earning. Herders use small panels to charge their mobile devices directly and, in one community visited near Kaabong town, nearly all men had mobile phones which they used for banking

and security. Few, if any, rural women have access to phones and the network remains unreliable.

6. See https://resilience.igad.int/clusters/igad-cluster-1-karamoja-cluster/.
7. Many researchers use the term 'Karamojong' to refer in a political sense to the wider citizens of Karamoja (e.g., Caravani, 2018; Scott-Villiers & Karamoja Action Research Team, 2013). However, this masks great heterogeneity, including people originating from other parts of Uganda. This is important for claims to land and access to resources, particularly as there is a trend towards more individual land ownership. Karamojong inhabiting southern districts consist of three clans: Matheniko in Moroto district, Bokora in Napak district and Pian in Nakapiritpirit district (Czuba, 2011). The name Karamojong is often applied to the Jie and Dodoth further to the north due to similarities in language and culture. Jie live in Kotido, while Dodoth live in Kaabong district. Within these main groups there are nine separate tribal subgroups, all of which speak Ngakaramojong (Adoch & Ssemakula, 2011).
8. Personal Communication with the chief administrative officer, Kaabong district, October 2017.
9. Oral consent was given by informants before the recording of interviews. Focus group discussions were held in the Karamojong language with immediate translation into English.
10. Other specific challenges faced include upstream catchment management around the dam, such as loss of trees in 2016 and grass-burning in the dry season. In that year, grass-burning damaged the plastic pipes that carried water from the reservoir to the small, irrigated area.
11. For an example of the water construction narrative on Karamoja related to Arachet, see this contemporary blogpost (https://msserwanga.blogspot.com/2012/).
12. Interview with Mr Patrick Okotel, Mbale.
13. Unfortunately, this dam was almost dry in 2022 (https://www.africa-press.net/uganda/all-news/shs12b-longoromit-dam-in-kaabong-district-dries-up).
14. Communities spoken to in Kaabong referred to it as akin to a 'consuming snake'.
15. 14 October 2017 focus group discussion with elders, Kaabong district.
16. Interviews with NGO officials from Turkana county, Kenya, September 2018.
17. District official, Kaabong town, September 2018.

Disclosure statement

No potential conflict of interest was reported by the authors.

ORCID

Alan Nicol http://orcid.org/0000-0002-7237-8825
Liza Debevec http://orcid.org/0000-0001-6757-7435

References

Adoch, C., & Ssemakula, E. G. (2011). *Killing the goose that lays golden egg. An analysis of budget allocations and revenue from the environment and natural resource sector in Karamoja Region* (ACODE Policy Research Series No. 47). Advocates Coalition for Development and Environment.

Allan, J. A. (1999). *Water in international systems: A risk society analysis of regional problemsheds and global hydrology* (SOAS Water Issues Group Occasional Paper 22). School of Oriental and African Studies, University of London. https://www.soas.ac.uk_water_paper_file38365pdf

Caravani, M. (2018). De-pastoralisation in Uganda's Northeast: From livelihoods diversification to social differentiation. *The Journal of Peasant Studies*, *46*(7), 1–24. https://doi.org/10.1080/03066150.2018.1517118

Czuba, K. (2011) *Governing the Karamojong: Tradition, modernity and power in contemporary Karamoja* (Project Evaluation Report). BRAC.

Debevec, L., Nicol A, A., & Nigussie, L. (2017). *Integrated water resources management in the Karamoja sub-region of Uganda – An evidence review and diagnostic study of current investments and best practice* (Unpublished Report for Deutsche Gesellschaft für Internationale Zusammenarbeit).

Mehta, L., Alba, R., Bolding, A., Denby, K., Derman, B., Hove, T., Manzungu, E., Movik, S., Prabhakaran, P., & van Koppen, B. (2014). The politics of IWRM in Southern Africa. *International Journal of Water Resources Development*, *30*(3), 1–15. Retrieved January 1, 2021, from https://doi.org/10.1080/07900627.2014.91620

Ministry of Water and Environment (MWRE). (2014). *Uganda catchment management planning guidelines*.

Mollinga, P. P. (2008). Water, politics and development: Framing political sociology of water resources management. *Water Alternatives*, *1*(1), 7-23. https://www.water-alternatives.org/index.php/volume1/v1issue1/15-a-1-1-2/file

Mollinga, P. P., Meinzen-Dick, R. S., & Merrey, D. J. (2007). Politics, plurality and problemsheds: A strategic approach for reform of agricultural water resources management. *Development Policy Review*, *25*(6), 699–719. https://doi.org/10.1111/j.1467-7679.2007.00393.x

Muhanguzi, D., Mugenyi, A., Bigirwa, G., Kamusiime, M., Kitibwa, A., Akurut, G. G., Ochwo, S., Amanyire, W., Okech, S. G., Hattendorf, J., & Tweyongyere, R. (2017). African animal trypanosomiasis as a constraint to livestock health and production in Karamoja region: A detailed qualitative and quantitative assessment. *BMC Veterinary Research*, *13*(1), 355. https://doi.org/10.1186/s12917-017-1285-z

Nash, R., Hudson, A., & Luttrell, C. (2006). *Mapping political context. A toolkit for civil society organisations*. ODI.

Nicol, A., Debevec, L., & Okene, S. (2021). *Chasing the water: The political economy of water management and catchment development in the Karamoja–Turkana Complex (KTC), Uganda* (IWMI Working Paper 198). International Water Management Institute. https://doi.org/10.5337/2021.214

Nicol, A., & Mtisi, S. (2003). *The politics of water: A Southern African example* (Sustainable Livelihoods in Southern Africa Research Paper 20). Institute of Development Studies.

Nicol, A., & Odinga, W. (2016). IWRM in Uganda – Progress after decades of implementation. *Water Alternatives*, *9*(3), 627–643. https://www.water-alternatives.org/index.php/alldoc/articles/vol9/v9issue3/326-a9-3-13/file

Office of the Prime Minister (OPM). (2014). *The second Karamoja integrated development plan (KIDP2) 2015–2020*. Office of the Prime Minister, Ministry for Karamoja Affairs.

Scott-Villiers, P., & Karamoja Action Research Team. (2013). *Eki and Etem in Karamoja. A study of decision-making in a post-conflict society*. Institute for Development Studies.

Zeitoun, M., Cascão, A., Daoudy, M., Greco, F., Mirumachi, N., & Warner, J. (2022). Power plus: Tony Allan's contributions to understanding transboundary water arrangements. *Water International*, *47*(6), 1001–1015. https://doi.org/10.1080/02508060.2022.2125234

Zeitoun, M., Mirumachi, N., & Warner, J. (2011). Transboundary water interaction II: The influence of 'soft' power. *International Environmental Agreements: Politics, Law and Economics*, *11*(2), 159–178. https://doi.org/10.1007/s10784-010-9134-6

Ozymandias in the desert: irrigation in Saudi Arabia

Elie Elhadj

> *Ozymandias* is the poem by Percy Bysshe Shelley (1792–1822) about the King of Kings, the mighty Pharaoh Ramesses II, whose immense statue – and all his other works – was toppled by the sands of time.

In 1998, after a 30-year career in banking in New York, London and Riyadh, I became a student again – seeking answers to questions about the cultures, politics, economies and prospects for reform in Arab countries. I joined the University of London's SOAS where it was possible to access focused regional expertise and scholarship. I vividly recall my first meeting with Professor Tony Allan; it was in the SOAS Staff Common Room. Four hours later, we were still discussing how equally water-poor but ideologically different political economies in the Middle East, especially Saudi Arabia and Syria, pursue unsustainable water policies. I found myself in the presence of a gifted strategic thinker.

Our discussion became the subject of my 'recreational PhD' as he used to joke about my status among his students. With gratitude and a sense of infinite debt, I shall always remember the weekly meetings and many hours of engaging discussion with Tony Allan over the past two decades. His clarity of vision, perception and capacity to see the big picture, integrity, dedication to teaching, sincerity and loyalty were inspirational. He was always the perfect gentleman.

Setting the stage

Tony Allan taught that food is an encapsulation of water – virtual water (Allan, 2019). While money measures the value of goods, agricultural production may be expressed in terms of the water needed to plant, grow, harvest, package and ship the produce for consumption. For example, 140 litres of water are embedded in a cup of coffee, 1000 litres in a litre of milk, 1300 litres in 1 kg of wheat, 2700 litres in a T-shirt, 3900 litres in 1 kg of chicken, 15,500 litres in 1 kg of beef. Coffee, milk, wheat, T-shirt, chicken and beef are all virtual water. Annually, an individual requires 1 m^3 of water for drinking, 100 m^3 for domestic chores and 1000 m^3 for food.

The global hydrological cycle of evaporation and precipitation supports rainfed vegetation, runoff to rivers and recharge of aquifers. Human intervention, through dams, diversion structures and wells, supplements natural precipitation; 92% of society's water consumption goes to produce food. A country's water is comprised by so-called *blue water*, abstracted from streams and aquifers and managed in farm irrigation, and so-called *green water*, which is rainfall stored in the soil (Allan, 2019). In Saudi Arabia, green water is negligible and, without rivers or lakes, all blue water is extracted from aquifers, which, importantly, are hardly being recharged.

Under these conditions, non-renewable groundwater should be used sparingly and only for the most important societal needs. It should be prioritized for drinking, household needs and industry (jobs) of future generations. Tony Allan maintained that food imports provide global solutions to local water scarcity challenges; the virtual water in imported food supplements domestic water resources and water-poor countries can most rationally import food beyond what their renewable water resources can produce. To earn the dollars to import water-hungry foodstuffs, a water-poor country needs to export manufactured widgets and gadgets that require little water to produce, or else be endowed with exportable commodities such as crude oil.

This article scrutinizes Saudi Arabia's experiment since the early 1980s to achieve food self-sufficiency, especially in wheat. It updates chapter 3 of my doctoral dissertation, supervised by Professor Allan, which addresses Saudi Arabia's quest for food independence (Elhadj, 2004a). I shall outline the enormity of the cost in terms of water, followed by a short history of Saudi Arabia's drive to make the desert bloom. I will guesstimate the financial cost, describe the environmental damage of such folly, examine Riyadh's imports and exports of virtual water, relate how the decision to irrigate the desert was made and contrast it with a criterion for the efficient allocation of scarce economic resources. I will also consider whom might have benefited and explain why Saudi Arabia's quest for food independence is a delusion.

Depletion of Saudi aquifers

Over the millennia, aridity constrained both agricultural production and population in the Arabian Peninsula (except around oases and in parts of Yemen). In the 1960s and 1970s, the Saudi Ministry of Agriculture and Water (MAW) surveyed groundwater resources in the Saudi Desert; the estimated water volumes were quite low. In the early 1980s, two further MAW studies discovered more water than had been imagined and, combined with the steep rise in oil prices in 1973,[1] the desert kingdom embarked upon a grand irrigation adventure to achieve food independence. The centrepiece was wheat self-sufficiency. Riyadh announced its readiness to purchase all the wheat Saudis could grow – at prices that would give growers a handsome return on capital.

Major trading families, with little or no experience in farming, acquired tracts of land in the Central Province, the heart of the desert and home to the ruling House of Saud. From the East and the West, they employed farm managers, engineers, mechanics and labourers, and imported pumps, tractors, harvesters, trucks, seeds, fertilizers, pesticides, and giant centre-pivot irrigation systems famously seen from the air on the approach to Riyadh and the towns of the Qassim Region.

During the first two decades of the food self-sufficiency strategy, 1980–99, the desert kingdom extracted 300 bcm of water, equivalent to six years of water flow of the River Nile into Egypt, irrigated 21 million ha and grew 153 million tons of produce (Central Bank Yearly Statistics, 2020, tabs 19, 20).[2] During this period, the average water consumption was over 14,000 m^3/ha (Central Bank Yearly Statistics, 2020, tabs 19, 20), and the average annual agricultural water consumption was 15 bcm.

During the next 19 years, 2000–18, irrigation extended over 19.2 million ha, producing 203 million tons of crops.[3] Agricultural water consumption data during this period are not available, but if we assume uniform agricultural water consumption during the four decades 1980–2018, it would amount to 700 bcm,[4] an average annual water consumption of 18 bcm.[5] According to MAW, two-thirds of the water was thought to be *partially renewable*. However, according to the Ministry of Planning, the volume of non-renewable water was much greater; the Planning Ministry uses the term *non-renewable* to signify what MAW describes as *partially renewable* (Elhadj, 2004a).

Water may be found at depths ranging from 100 to 500 m, sometimes as deep as 2500 m. There are 27 aquifers in the country, of which six: Saq, Wajid, Minjur, Dhruma, Wasia and Umm ErRadhuma, are considered to be major. Due to the unavailability of means and money, they remained largely undisturbed until the early 1980s – but continued abstraction at the current level will, sooner or later, deplete the non-renewable groundwater. No one knows with accuracy how much water remains in the aquifers. No one knows whether new sources of water might be found. In one important agricultural area, Wadi ad-Dawasir in the Central Province of the Riyadh region, NASA's Earth Observatory concluded that the water table has dropped by as much as 6 m (20 feet) per year since the 1980s and that the aquifer could be depleted within a few decades (Voiland, 2015). From a peak irrigated area of 1.6 million ha in 1993, some 592,000 ha, or 37% of the arable, has already turned to dust.

A short history

Between 1980 and 1992, the Saudi irrigated area increased from 610,000 to 1,600,000 ha and wheat production increased 29-fold to 4.1 million tons (Central Bank Yearly Statistics, 2020, tab. 1, pp. 2; tab. 2, p. 11), more than twice what was needed for self-sufficiency of a country of 17 million inhabitants, as well as beyond the storage capacity of the government's Grain Silos and Flour Mills Organization (2.4 million tons). Pretending to be a land of milk and honey, Saudi Arabia became the world's sixth-largest wheat exporter (Richards & Waterbury, 1998). But not for long. It merely proved that pouring money and water into the sand can make the desert bloom until either the money or the water runs out. In the event, both ran out.

The cost of the 1991 Gulf War and persistent budget deficits following a damaging drop in oil prices after 1986 forced the government to delay payments of US$72 billion to contractors and farmers. They were eventually paid, without interest, in medium-term government bonds which the recipients discounted at Saudi banks for cash. Not only were local businesses affected: Boeing, for example, stopped delivery of the first 777 aircraft to government-owned Saudi Airlines until the payments were made. Throughout most of the 1990s, I was a first-hand witness to these events as chief executive officer of the Arab National Bank in Riyadh.

The liquidity crunch signalled the beginning of the end of the wheat project. Quickly, government purchases dropped by 75%, from US$1.9 billion in 1991 to US$480 million in 1996; wheat production decreased by 70%, from 4 million tons in 1991 to 1.2 million tons in 1996; and 476,000 ha of private sector investment in farmland was abandoned between 1993 and 2000 (Elhadj, 2004a, p. 11). Concerned over dwindling water reserves, Riyadh announced, in January 2008, a running down of the wheat programme over the next eight years. In 2016, all of Saudi wheat needs (3.6 million tons) were imported (Hallman, 2018).

And yet, agricultural water consumption increased. The abandonment of 476,000 ha growing wheat should have saved around 6.7 bcm of water annually,[6] but instead the water saved from growing less wheat was used to irrigate lucerne to feed livestock for domestic consumption as well as exports to neighbouring Gulf Cooperation Council (GCC) states. While the average water usage between 1990 and 1994 was about 12,000 m^3/ha, it increased to an average of more than 15,000 m^3/ha between 1995 and 1999 because lucerne consumes two to three times as much water as wheat (Jenson, 1981). By 2016, the government decreed that lucerne was to be phased out over three years and Saudi al-Mara'i Company, the largest food company in the Middle East, acquired agricultural lands in Argentina, Arizona and California to grow stockfeed, which is shipped back to Saudi Arabia to feed its herd of 170,000 cattle (Al-Mara'i, 2018).

The financial cost

Saudi wheat was expensive. Between 1984 (when the data became available) and 2000 some US$20 billion was spent on buying 40 million tons of domestic wheat, an average of US$500/ton. During the same period, the international market price averaged US$120/ ton; had the wheat been imported, the government could have saved US$15 billion. Between 1984 and 2000, government and private sector investments were estimated at US $84 billion, excluding three cost elements for which data are unavailable: (1) government subsidies to the fuel and electricity used to operate farm machinery; (2) concessionary borrowing rates of interest from the Saudi Agricultural Bank on around US$10 billion; and (3) cost of administering the programme.

The opportunity cost

Irrigation schemes are not the best use of fossil groundwater in the desert – especially if the agricultural production is uneconomic compared with any imported alternatives: better to reserve it for future domestic and industrial use.

The Riyadh and Qassim regions are in the heart of the Arabian Desert. They encompass around 40% of Saudi agricultural land, or 400,000 ha. Despite the fact that these two regions have good quality groundwater, desalinated water is piped from the Gulf Coast, 450 km away, climbing 800 m over the Najd Plateau to reach their 10 million residents (Central Bank Yearly Statistics, 2020, tab. 38). Had the local aquifers in Riyadh and Qassim supplied water to local peoples' homes and industry (1.25 bcm in 2018) instead of irrigation (5.6 bcm in 2018), only 22% of that water would have been needed.[7]

It is hard to justify running down good-quality groundwater to irrigate farms in the searing heat of the desert while local homes and factories are supplied with desalinated water from far-away processing plants. Nationwide, had the local aquifers supplied water to local peoples' homes and industry, Saudi society would have saved not only the investment and expense of operation and maintenance of desalination plants, but also the investment and expense of operation and maintenance of 8400 km of pipelines to carry this desalinated water (US–Saudi Business Council, 2021). The cost of desalinating and delivering the water is estimated at around US$2/$m^3$ (Elhadj, 2004b).

Saudi Arabia has 33 desalination plants, eight on the Gulf in the Eastern Province and 25 on the Red Sea. It is the world's leading producer of desalinated water; in 2019, it produced 1.9 bcm (Central Bank Yearly Statistics, 2020, p. 31). Desalination is hugely damaging to the environment: Saudi plants use 1.5 million barrels of oil per day, around 15% of Saudi oil production (Ferzly, 2018); they pollute the air and contribute to global heating; they discharge brine that kills marine organisms. It is astonishing that this assault on the Earth's environment continues unabated in the age of global environmental awareness.

The direct cost of Riyadh's food self-sufficiency adventure and ongoing programme of fodder production[8] comprises: (1) the extra cost paid by Saudi consumers for domestically produced food as a result of Riyadh's trade barriers against food imports in order to protect local food producers (US Department of Commerce, 2019); (2) the costs to society of the subsidies supporting the programme; (3) the difference between the cost of supplying local groundwater and the cost of supplying piped desalinated water, including the unquantified environmental damage caused by desalinating 1.9 bcm per annum in terms of consumption of fossil fuels and hyper-salinization of local coastal areas; and (4) at some point in the future the need to desalinate sea water to replace depleted groundwater supplies.

Saudi virtual water trade

Imports of virtual water

Food is virtual water. Importing 1 ton of wheat instead of growing it at home could generally save about 1300 m^3 of water, more in the desert. Importing 1 ton of meat could save about 15,500 m^3 of water, a lot more in the desert (Allan, 2001a). Tony Allan considered water scarcity a local problem and trade in commodities the global solution. He characterized virtual water trade as economically invisible and politically silent, convenient for politicians who prefer to keep food imports secret. Where domestic food production falls short of a country's requirements, surely it is better to fill the gap with imports than to invest in financially non-viable and economically and environmentally damaging local farming schemes.

Saudi Arabia's food independence strategy failed to achieve its own goals; the contribution of agriculture and fishing to gross domestic product (GDP) dropped from 4.9% in 2000 to 2.2% in 2018. Fortunately for Saudi Arabia, it has the dollar revenues from oil exports to pay for food imports. In 2018, it paid US$4.8 billion for live animals and animal products, US$8.8 billion for vegetable products and US$7.8 billion for animal or vegetable fats and oils, prepared foodstuffs, beverages and tobacco, for a total of US $21.4 billion (Saudi Arabian General Authority for Statistics, 2019).

Export of virtual water

Following Tony Allan's reasoning, food exports are like shipping water away. While Saudi Arabia was spending over US$21 billion on food imports in 2018, it was also exporting foodstuffs to neighbouring states. In 2018, exports of live animals and animal products were US$1.6 billion. Assuming an average price of US$3,000/ton and 4500 m^3 water to grow 1 ton of this category of exports, then the embedded water would be around 2.4 bcm. Also, in 2018, Saudi Arabia exported vegetable products, animal or vegetable fats and oils, prepared foodstuffs, beverages and tobacco valued at US$-2.1 billion. Assuming an average price of US$1500/ton, and an average of 2000 m^3 of water to grow 1 ton of products in this basket, then their embedded water would be 2.8 bcm.

The total value of Saudi food exports in 2018 was US$3.7 billion, or 17% of the value of food imports. In terms of water, 5.2 bcm were exported, representing almost 30% of the irrigation water consumed that year. Saudi exports of virtual water are not new; between 1997 and 2001, exports to neighbouring GCC city-states of meat, fruits and vegetables averaged 2.5 bcm annually, exclusive of lucerne exports which stopped in 2000.

How was the decision taken to make the desert bloom?

The policy goals of desert farming were economically and environmentally unsustainable. The dramatic rise and equally dramatic fall in cultivated land, wheat production, exports of forage and foodstuffs, and the accompanying supply of desalinated water to cities and villages with relatively easy access to good quality groundwater were an extreme case of a politically determined ecological policy formulated by misguided politicians enjoying rentier economic circumstances. Their strategy lacked both environmental consideration and safe political processes to combine the natural, human and financial capitals to achieve economic security.

An absolute monarchy, Saudi Arabia's governance is non-representative and non-participatory. Foreign suppliers, in this case agribusiness, desalination and pipeline executives, closely associate themselves with powerful members of the ruling elite. Obscene and illicit commissions generate dubious schemes. Exorbitantly expensive white elephant projects, such as food self-sufficiency in the desert or supplying desalinated water to areas rich in groundwater, were attractively packaged with nationalistic slogans. In the absence of political parties, a free press, environmental groups or other concerned groups such as egalitarian nongovernmental organizations, it is impossible to introduce a balancing economic or environmental perspective into water policy. Consequently, there has been no effective dissent against desert farming. On the contrary, the government and its defenders at home and apologists abroad have praised this folly as a judicious and patriotic strategy.

Once the wrong strategy is in place, it becomes progressively more difficult to reverse. Tactical decisions to make a wrong strategy less damaging, like raising the operational efficiency of irrigation, or shifting emphasis from one crop to another, will not correct the original flawed decision – modifying a wrong strategy is like slowing down on a road, while still going in the wrong direction.

A criterion for efficient allocation of scarce resources

Investment in irrigation and land reclamation should be justified on a purely economic basis, with full costing of water and accounting for the negative and positive externalities of production and consumption. Tony Allan advocated proper valuation of water and environmental protection and devoted chapters 3 and 4 in his book *The Middle East Water Question* (Allan 2001b) to this controversial and sensitive subject. His approach makes water extraction and delivery a central factor in the cost of production. Accounting for water would raise the price consumers pay for red meat, for example. Higher red meat prices would reduce the demand for beef, reduce the planet's herd of some 1.4 billion cattle (Borunda, 2019), reduce dangerous methane emissions and thereby mitigate global heating. Tony Allan himself became vegetarian, but he was also a realist; he accepted that proper accounting for water would be a major departure from attitudes developed over centuries when poverty and age-old customary practices expected water to be free of charge.

Had Riyadh applied a rate-of-return-on-investment criterion in allocating its scarce economic resources, desert irrigation would never have been pursued. But in Riyadh, as in many other capitals, public policy is dominated by political convenience, not science. There is a profound difference between economics and political economy, statistics and political statistics, investment theory and political investments, accounting theory and political accounting. In the last two years we have seen comparable differences between Covid-19 science and Covid-19 politics, facemask science and facemask politics, vaccine science and vaccine politics. In public policy, theory is utopia, science is truth, while politics is Machiavellian. As Marc Reisner asserts in *Cadillac Desert* (1986): 'water flows uphill toward money'. In short, political economy considerations explain bad public policy decisions.

Whom might have benefited from desert agriculture?

Why did Saudi Arabia pursue desert agriculture? Was it for food independence or private enrichment?

Food independence

In reaction to calls in certain US media that Washington should impose a food boycott on Saudi Arabia in retaliation for its oil embargo on the United States and the quadrupling of oil prices during the 1973 Arab–Israeli war, Saudi Arabia promoted the notion that wheat independence was necessary to protect the economy and national security. This allegation does not stand scrutiny:

- Had Riyadh taken the American threat seriously, it would not have waited seven years to react. In an absolute monarchy, decisions can be made quickly. The king controls the three branches of government and the fourth estate; his wish is law.
- Food self-sufficiency does not provide political or economic security. Washington possesses many weapons to retaliate other than food boycotts. Removing US military protection of Riyadh would be catastrophic in a volatile Middle East; and a boycott of Saudi banks and many of the essentials that Saudi Arabia imports, such as military hardware and spare parts, desalination plants, construction machinery, pharmaceuticals, etc., would be as effective.

- Saudi food independence is impossible. High population growth rates with finite water resources will eventually deplete non-renewable aquifers.

Enriching the ruling group

The Saudi ruling elite is composed of four groups. In return for benefits and privilege, they support the five generations of the more than 10,000 direct descendants of King Abdulaziz. These elites are the Islamic *ulama*, military leaders, tribal chiefs and major merchant families. The *ulama* lend the ruling family religious legitimacy by threatening the populace with hell's fire if they ignore Quranic and Hadith injunctions to obey the monarch blindly. The military class defends the regime against domestic dissent. The tribal leaders and major merchant families control the rank-and-file tribesmen and employees of the private sector. In a country of 13 million expatriate workers out of a total population of around 34 million (Central Bank Yearly Statistics, 2020), such a societal construction provides strong protection to an absolute system of governance. Furthermore, there are no political parties, labour unions, civil society organizations, student and women associations, social clubs, let alone a parliament.

Saudi Arabia's agricultural strategy benefited the ruling group. The pioneers of desert agriculture were traders, not farmers. Without their participation, the strategy would not have materialized. The promise of quick and high financial returns enticed some of the country's richest business families to add farming to their investment portfolios. Businessmen from the Riyadh and Qassim regions, home to the al-Sauds, dominated the scene. They brought in the necessary men and materiel and enjoyed high returns on investment until the government was forced to abandon the dream of wheat independence in 2008.

The leading trading families are also the exporters of food. In 2018, one-third of Saudi Arabia's US$3.7 billion in virtual water exports were carried out by al-Mara'i company (Al-Mara'i, 2018). Ownership of the publicly traded stocks of al-Mara'i is concentrated in a few hands: 23.7% is owned by a member of the ruling family, with a small number of leading businessmen owning substantial shareholding in the company. In its 2014 Annual Report, the company disclosed that 31 shareholders owned 87% of the issued shares (Al-Mara'i, 2014); the 2015 Annual Report disclosed that 33 shareholders owned 88% of the issued shares (Al-Mara'i, 2015); the disclosure has been removed from subsequent annual reports.

The history of the food self-sufficiency strategy since the rise in oil prices of 1973, almost 50 years ago, suggests that the main beneficiaries of the food independence strategy have been the powerful men of the ruling group. Three facts attest to this: the strategy had to wait seven years to emerge; it produced, in the early 1990s, more than twice the volume of wheat needed for self-sufficiency; and the government continues to allow Saudi food companies to export one-third of the country's annual agricultural water to neighbouring states. Food independence was merely a politically convenient slogan to rally the populace behind an imagined national security threat.

The illusory quest

It is impossible to attain food independence in the desert kingdom. Renewable water resources are insufficient to feed the burgeoning population. Given that an individual needs 1000 m^3 of water annually for his/her food needs, a population of 35 million (in 2020) needs roughly 35 bcm of water. At present, Saudi Arabia abstracts around 50% of the water needed to feed itself and imports the other 50%. As the size of the population grows, so must food imports.

If you live in a desert, water in the desert is more valuable than water elsewhere. As non-renewable groundwater is depleted, desert agriculture will be abandoned. Its legacy will be the same as that of Ozymandias. Future Saudi generations will tell the story of rulers who squandered gargantuan volumes of non-renewable water – simply because they could.

Notes

1. Between 1968 and 1973, the average price for crude oil was US$3.64 per barrel. Between 1974 and 1980, the price rose to US$19.53 per barrel (InflationData.com, 2021).
2. Central Bank Yearly Statistics (2020): Table 19: Cultivated area size; and Table 20: Agricultural production (https://www.sama.gov.sa/en-US/EconomicReports/Pages/YearlyStatistics.aspx).
3. 300 bcm/20 years = 15 bcm/year.
4. Since 153 million tons of produce between 1980 and 1999 consumed 300 bcm water, then 203 million tons of produce between 2000 and 2018 could have consumed 398 bcm (300 bcm/153 million tons = 1960 m^3/ton × 203 million tons = 398 bcm) and the total water extraction for the 39 years is 698 bcm (about 700 bcm).
5. 698 bcm/39 years = 17.9 bcm, about 18 bcm.
6. Assuming that 1 ha of wheat uses a similar volume of water as the rest of the cultivated area (476,000 ha × 14,000 m^3/ha = 6.7 bcm). In 1992, cereals occupied 72% of the cultivated area and wheat production was 88% of all-cereal production (Elhadj, 2004a, tab. 1, p. 2; tab. 2, p. 11).
7. In 2018, the Saudi cultivated area was 1.004 million ha and the Riyadh and Qassim regions comprised 40%, that is, around 400,000 ha. With an average annual water use of 14,000 m^3/ha, the two regions' use of irrigation water would have been around 5.6 bcm. In 2018, Riyadh's use of drinking and household water was 1.08 bcm and Qassim's was 167,000 m^3, for a total of about 1.25 bcm (Saudi Arabian General Authority for Statistics, 2019).
8. In 2018, total agricultural production was 14,905 million tons – fodder 9134 million tons, fruits 2311 million tons, vegetables 1748 million tons and grain 1440 million tons (Central Bank Yearly Statistics, 2020, tab. 20).

Acknowledgements

I am grateful to my friends Chris Perry and David Dent for reviewing the paper. I benefited from their expert comments, but I assume full responsibility for the contents.

Disclosure statement

No potential conflict of interest was reported by the author.

References

Allan, J. A. (2001a). Virtual water – Economically invisible and politically silent – A way to solve strategic water problem. *International Water & Irrigation*, *21*(4), 29–31.

Allan, J. A. (2001b). *The middle east water question. Hydropolitics and the global economy.* IB Taurus.

Allan, J. A. (2019). Policy note: Invited opinion interview: 'Water scarcity, food production and virtual water'. Professor Allan was interviewed in 2019 by Professor Dale Whittington and Dr Duncan Thomas for University of Manchester's Coursera. https://www.worldscientific.com/doi/abs/10.1142/S2382624X19710036

Al-Mara'i. 2014. *Annual Report*, 38. https://www.almarai.com/en/corporate/investors/annual-report-financial-statement/#annual-reports

Al-Mara'i. 2015. *Annual Report*, 42.

Al-Mara'i. 2018. *Annual Report*, 2, 5, 78, 253.

Borunda, A. (2019, January 23). *Methane explained.* National Geographic. https://www.nationalgeographic.com/environment/article/methane

Central Bank Yearly Statistics. (2020). *Gross domestic product by kind of economic activity, Table 1; Cultivated area, Table 19; Agricultural production, Table 20.* https://www.sama.gov.sa/en-US/EconomicReports/Pages/YearlyStatistics.aspx

Elhadj, E. (2004a). *Camels don't fly, deserts don't bloom: An assessment of Saudi Arabia's experiment in desert agriculture (Occasional Paper 48, Section 5 – Economics).* Centre for Water and Development, School of Oriental and African Studies, University of London. https://www.soas.ac.uk/water/publications/papers/#3Financingwaterservices

Elhadj, E. (2004b). *Household water and sanitation services in Saudi Arabia: An analysis of economic, political and ecological issues (Occasional Paper 56, p. 14).* School of African and Oriental Studies, University of London.https://www.soas.ac.uk/water/publications/papers/#3Financingwaterservices

Ferzly, N. (2018June11). Jubail plant and desalination, Saudi Arabia. *Environmental Justice Atlas.* https://ejatlas.org/conflict/water-desalination-in-saudi-arabia

Hallman, A. (2018). *Grain and feed annual.* GAIN Report SA 1802. US Embassy/USDA Foreign Agricultural Service

InflationData.com. (2021). *Historical crude oil prices.* https://inflationdata.com/articles/inflation-adjusted-prices/historical-crude-oil-prices-table

Jenson, M. E. (1981). *Design and operation of farm irrigation systems.* American Society of Agricultural Engineers, St Joseph MI, Table 6.6.

Reizner, M. (1986). *Cadillac Desert.* Viking. p.12 in Penguin edition, 1993.

Richards, A., & Waterbury, J. (1998). *A political economy of the Middle East.* Westview Press.

Saudi Arabian General Authority for Statistics. (2019). *Agricultural production survey.* Tables 1, 5, 7 and 8. https://www.stats.gov.sa/en/1060-0

US–Saudi Business Council. (2021). *Water in Saudi Arabia: Desalination, wastewater, and privatization.* https://ussaudi.org/water-in-saudi-arabia-desalination-wastewater-and-privatization

US Department of Commerce. (2019). *Saudi Arabia – Country commercial guide, import tariffs on food/agricultural products.* https://www.trade.gov/knowledge-product/saudi-arabia-trade-barriers

Voiland, A. (2015). *Desert crops thrive as the aquifer shrinks.* NASA Earth Observatory. https://earthobservatory.nasa.gov/images/145975/desert-crops-thrive-as-the-aquifer-shrinks

OPEN ACCESS

Locating the channel and other tales from the river bank: constants and change in river boundary delimitation

Richard Schofield

ABSTRACT
Nowhere is the contrast between the top-down process of drawing a boundary and the ground-up reality of living with it so obvious as in delimitations along rivers – live features not just physically but also, potentially, navigable arteries. Illustrations are drawn from the chequered history of two famous colonial river boundaries along the Shatt al-Arab and Jordan rivers; and in the invocation of history in legal settlements of the status of Kasikili/Sedudu Island in the Caprivi Strip (1995–99) and the Abyei region on the Bahr al Arab between Sudan and South Sudan (2008–09).

Introduction

In legal terms, inter-state territorial delimitations running along rivers are classified as international land boundaries, but this prominent subcategory – whose origins always go back deep into history – poses a unique set of complex management challenges. Land boundary delimitations are essentially statist abstractions, legal lines of coordinates that are only given width and presence on the ground by the functions, facilities and (increasingly) the fortifications states choose to place there. Rivers are live physical features whose fluidity contrasts with sovereign connotations of territorial permanence and whose dynamism often results in changes to channel course and form (Prescott & Triggs, 2008, p. 215).

If the above amounts to an admonition that rivers are poorly suited for adoption as international boundaries, then there are also more fundamentally broad considerations at play here that relate more broadly to the history of human experience – as Bouchez alluded in a comparative review of river boundary delimitation following France and Germany's Rhine agreement of April 1960:

> Rivers, [. . .] are natural links for the people who live adjacent to them. From history it is evident that both sides of a river valley were often areas of identical cultural and economic development. Thus, in relation to cultural and economic development, river valleys must be considered as a unit. (Bouchez, 1963, p. 789)

This position brings into sharp relief the oft-observed remote and marginal character of borderlands surrounding land boundaries that lay beyond a state's economic and political core regions, frequently observable in a colonial context. For navigable rivers adopted as international boundaries the position was often quite different, especially economically where the banks could be a hive of activity, if not always hugely significant politically beyond their local context. The potential for contradictions and clashes between the top-down nomination of rivers as territorial limits and the ground-up socio-economic organization of riverain borderscapes was and is always there. While this article primarily seeks to further contribute to an existing body of literature that details the evolution of river boundary delimitation from a historical, technical and legalistic standpoint (Bouchez, 1963; Prescott & Triggs, 2008; Donaldson, 2009, 2014), it acknowledges that the current critical border studies literature has potentially much to offer in unveiling the ground-up human geographies of international river boundaries, with its phenomenological focus on the 'lived experiences' of borderlanders (Bialasiewicz and Minca, 2010).

If the illogic of nominating rivers as international boundaries is apparently clear, conversely there has been a clear historical logic to adopting them; as territorial outcomes of conflict; for the strategic benefits this confers; to reflect a fad for 'natural' boundaries and lastly; for their superficially easy recognition on maps. When one also considers the pronouncedly historical character of most international river boundaries, these considerations would have been instrumental in their adoption. Bear in mind, too, that the classical delimitation expert, Stephen Jones, did not see their adoption as likely to stop following the Second World War: 'rivers will probably be adopted as boundaries even though the geographer or engineer inveigh against them' (Jones, 1945, p. 108). And who knows, with state fragmentation and conflict providing ever more fertile grounds for secession in 2022, their nomination may continue to figure.

This article traces in varying detail levels four case studies - Shatt al-Arab (Iran/Iraq); Jordan (Israel/Jordan); Chobe (Botswana/Namibia); and Bahr al-Arab (South Sudan/Sudan) - to show how definitional and operational challenges were broached historically, how the thought process about river boundaries developed and how history was used retrospectively in the dispute resolution process.[1]

- An 11 March 2019 announcement suggested that Iran and Iraq had agreed *inter alia* to respect the *thalweg* boundary delimitation introduced by bilateral treaty in June 1975. The arrangements struck that year arguably constitute the most sophisticated and comprehensive arrangements concluded in international law for dealing with a river boundary, with every conceivable safeguard contained for managing future dispute over delimitation status and alignment. Arguably, the specifications went too far, however – over-egging the pudding in trying to cater for the possibility of future physical change. Our focus here is to chart the *evolving definition of the thalweg* (Figure 1).
- The 26 October 1994 Israel–Jordan Peace Treaty reaffirmed a vaguely worded 1922 colonial territorial limit (Palestine/Transjordan) as the international boundary north and south of the West Bank (Figure 2). For the most part the peace agreements retained the basis for dealing with physical change in the boundary river decided upon by the British Colonial Office in 1927 but, pragmatically, left all options open for dealing with future instances of sudden changes of river course (avulsion). Our chief concern here is to consider

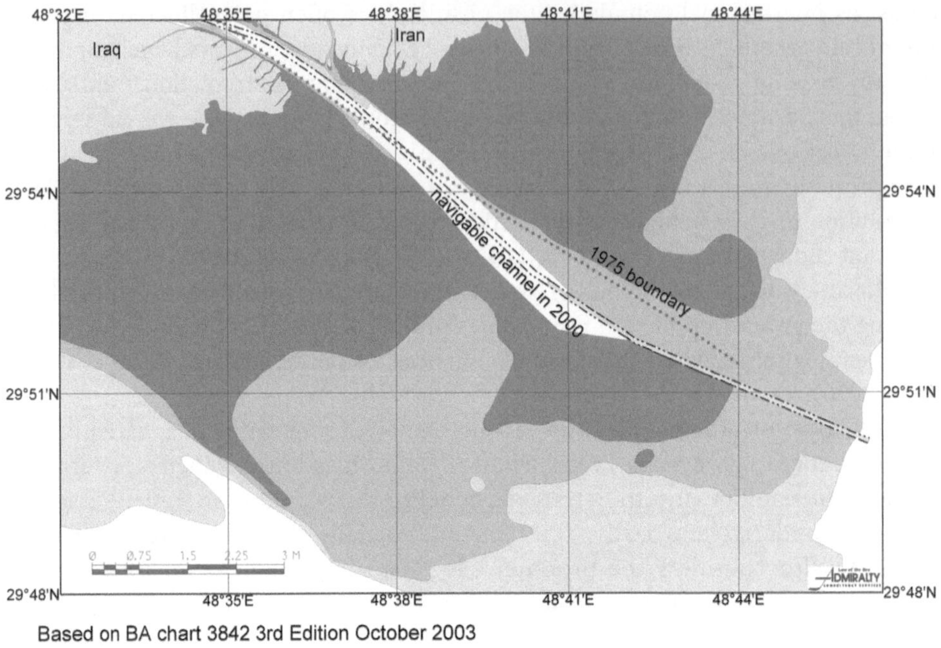

Figure 1. Movement of the main navigable channel of the Shatt al-Arab since 1975.
Source: Kind acknowledgements are extended to John Brown, formerly Head of the Law of the Sea Division of the UK Hydrographic Office, Taunton, for producing this diagram and allowing its publication here.

Figure 2. Looking eastwards across the Jordan River from northern Israel.
Source: Photo by the author.

attitudes to *physical change in boundary rivers* and their effect on the alignment of delimitation – so the decisions taken during 1927 are clearly central.

- A 13 December 1999 judgment of the International Court of Justice (1999) held that the river boundary between Botswana and Namibia should run in the channel north of Kasikili/Sedudu island in the Chobe River. Yet, in something of a curve-ball, the court decreed at the same time that there should be equal rights of access north and south of the feature. There had been few indications that the sovereignty of the insular feature was a major issue on Namibian independence in 1990, but the dispute was forwarded for judicial treatment by special agreement in February 1995. Historical concerns were essentially retrospective therefore in the process of dispute resolution – specifically concerning issues of territorial allocation in an 1890 spheres of influence agreement between Britain and Germany. Our focus here is on *identifying the major or principal channel* in a boundary river.
- A 22 July 2009 Permanent Court of Arbitration award established the territorial extent of Abyei province as an effective starting point in defining the international land boundary between Sudan and its new neighbour of Southern Sudan. The location and identity of rivers and the recognition of riverain terrain constituted a key component of this hurried case and award. The chief historical concerns here were again retrospective, centring around the transfer of one tribal group from one province of the Anglo-Egyptian Condominium of Sudan to another in 1905 – with the presumption that there must have been some inkling of territorial definition for that transaction to take place. Our concern here is to highlight *prevailing historical uncertainty over the identity of rivers*, amidst questions about the reliability and admissibility of colonial evidence.

Drawing international boundaries along rivers

The rules of the game for defining river boundaries were broadly developed with Article 30 of the 28 June 1919 Treaty of Versailles with a median line stipulated for non-navigable rivers and division along the principal channel recommended for navigable rivers (Prescott & Triggs, 2008, p. 216). Yet this functional rationale had come all too late for the many and large variety of river borderscapes established by this time. Early international river boundary treaties often granted sovereignty over the whole water course to one riparian with the bank specified as the inter-state divide, obviously inequitable and generally the result of one power having the upper hand politically or militarily over its neighbour, a ratification of the territorial outcome of conflict. An example here was a 1773 Prussia–Poland treaty that set the boundary along the Polish bank of the Netze. Twelve such treaty arrangements survived into the post-Second World War period (Sevette, 1952).

It was especially during the European overseas colonial expansions that peaked in the last quarter of the 19th century that rivers were adopted as imperial boundaries. With rivers (and mountain ranges) not only representing opportunities for defence but also supposedly easy to identify on the crude cartography of the day, this reflected a coterminous fad for 'natural' boundaries, territorial limits that were coincident with prominent surficial features of the Earth. They would be almost reified by some prominent imperial boundary-makers, such as Government of India Surveyor-General Thomas Holdich, who posited – looking back on his own handiwork in the Asian subcontinent – that there could be no better boundary than that consisting of riverain and mountainous sections combined (Holdich, 1916, p. 46; Schofield, 2015, p. 137) While looking at things top-down, there is much to support Jones's observation that '[t]he extensive use of river and geometrical boundaries testifie[d] to the prevalence of absentee boundary-making' (Jones, 1945, p. 39); there have also been some efforts – for instance, in legal cases – to posit that equivalent notions of rivers acting as natural divides and barriers to movement held sway from a more localized, ground-up perspective. This would become a minor issue in the PCA's 2008–09 Abyei case (see above). Here the Bahr al-Arab had possibly formed a loose Ottoman administrative division during the late 19th century, but to what degree was it seen as a boundary by groups on either side in a more localized sense? Did it constitute an effective local barrier to movement and, if so, was this more of a seasonal phenomenon?

The vogue for 'natural' boundaries would soon pass, certainly any intellectual case for their adoption. Yet, while no one would argue seriously for their continued adoption, they have disappeared neither from the legal process of boundary-making nor academic discussion. Support for natural boundaries holds in certain situations, such as in international law's application of inter-temporal law in the adjudication of colonial boundary cases. An example can be found with the judicial resolution of the Honduras-El Salvador 1992 land boundary dispute at the International Court of Justice (2017-22). Here, in attempting to second-guess what the colonial power would have understood by the line of *uti possidetis juris* it was bequeathing back in 1821, the ICJ seemingly tried to get into the imperial boundary-making mindset that would have prevailed at the time. The assumption in this instance that natural features would have been utilized in delimitation led to a presumption in favour of rivers as constituting the land boundary (Prescott & Triggs, 2008, p. 226). So, maybe there is life in that old natural boundaries dog yet. Recently, also, critical border scholars from the social sciences have begun to reimagine why boundaries (or any features that might constitute them) might be seen as 'natural' by those who encounter or have to negotiate them (van Houtum, 2005). The concept of natural boundaries in the round has been revisited imaginatively (Fall, 2010), including within an imperial boundary-making context (Tillotson, 2020).

History is omnipresent in the consideration of any international boundary, resonating most loudly and dramatically when invoked and mobilized for political gain during the course and conduct of disputes and the issue and pursuit of territorial claims (Murphy, 1990; O'Dowd, 2010). Regulation of the territorial framework enshrines the record of treaty settlement to bolster the stability and finality of international boundaries, while the dispute resolution process prioritizes the historical (and often colonial) evidentiary record – leading to charges from postcolonial law that it is essentially neo-colonial in nature (Anghie, 2006).

However, history tends to matter even more for rivers than for other classes of land boundary. After all, more history has been accumulated since they have tended to be around for longer. The varying ways in which boundaries have been drawn along rivers generally reflects a balance of top-down considerations in their nomination and the specific complexities of the regional environment into which they have been introduced. Consequently, river borderscapes tend to be very distinctive in their territorial arrangements and not easily subject to comparative study. If Jones caution that '(e)ach boundary is unique and therefore many generalisations are of doubtful validity' (Jones, 1945, p. vi) holds true in a general sense, it is perhaps even more pertinent for river boundaries. If the logic for sharing the main channel of a navigable river was appreciated (if not always applied) long before the principle's codification at Versailles, there were other ways in which boundaries were drawn along riverain features observable by the early 20th century. In instances where median lines were employed for non-navigable rivers, their adoption tended to be relatively less problematic, requiring less attention and management going forward since the river would generally not be so active an economic zone. Yet there was not any appreciable consistency in the nomination of navigation channels and median lines as river boundaries before 1919 and there would not be any significantly enhanced uniformity going forward.

In addition to riverbanks, navigable channels and median lines being adopted as boundaries, other arrangements were also in evidence – ones that acknowledged the specific natural structure of the river in question. In those rare cases where there were two equally important channels, the boundary might be fixed so that each state possessed its own, and this was how Portugal and Spain treated the River Guidiana in 1895 (Bouchez, 1963, p. 795; Lauterpacht, 1960). To introduce stability to the boundary along the St. Croix River by effectively rendering irrelevant variations in water flow, Canada and the United States arrived at a delimitation comprising arbitrary straight-line segments that were actually demarcated in this instance by pillar (Bouchez, 1963, p. 795). Complex river boundary delimitations arose when different methods of fixing a river boundary were employed for different stretches of the water course (p. 796). No one would ever devise such a scheme in introducing a delimitation; rather, it would usually eventualize in response to the economic and strategic value of the waterway and the local and international use made of it – developing in piecemeal fashion. Sometimes, too, regional circumstances needed to change before the ideal delimitation could be introduced, colonial and postcolonial. The Shatt al Arab was the most obvious example here (see below) (Schofield, 2004). By the early 1960s, legal experts would be championing a co-imperium over waterways as their preferred arrangement for international boundary rivers, with sovereign state limits fixed at the low-water mark on each bank (Lauterpacht, 1960). This was how Germany and France delimited the Rhine in their 8 April 1960 treaty.

Beyond how boundaries were drawn along waterways historically, the way in which their surrounding borderlands were altered as a consequence or deliberately configured in accompanying regional infrastructural schemes warrants mention. River borderscapes were sometimes seemingly more obviously created for the benefits of states or, just as often, colonial powers than the local populations themselves. Their architects might aim for showcase diplomatic settlements that provided the requisite definitional clarity to preclude further territorial disputes and enable the projection of economic or strategic regional interests. The St. Croix (US/Canada) settlement of the late 18th century was essentially designed to introduce enhanced territorial definition to decrease the possibility of future interstate conflict – it was not really for the benefit of the locals (Demeritt, 1997). Britain knew from the start what boundary would be fairest for the Shatt al Arab at the head of the Persian Gulf, but also realized that it was constrained by the art of the possible if it wanted to project its economic interests in southern Mesopotamia from the mid-19th century forward. So it pragmatically arrived at a mixed fudge of vague delimitation and explanatory notes in its attempts to keep the Ottoman Empire and Persia on board – and not wholly successfully (Schofield, 2004). Moving to East Africa a half century later, Britain and Italy came up with a novel scheme by agreement in 1911 to recognize that the river (Juba) separating their colonial territories was highly unstable, frequently cutting new courses including new openings into the Red Sea. Basically, the two sides agreed to regard the whole river valley as a river boundary zone – quite prescient in many ways (Field, TNA, 1927, CO 733/142/1).

The last point to underline in this extended introductory section is that certain conventional wisdoms about inter-state riverine limits, like any international boundary, have been around for a long time. It was also in 1911 that the Institute of International Law issued their Madrid Resolution to decree the following. States with a common river boundary were in 'a position of permanent physical dependence on each other which precludes the idea of the complete autonomy of each State in the section of the natural watercourse under its sovereignty' (Prescott & Triggs, 2008, p. 216). Such a statement was meant to apply to both divided and successive river disputes, would soon find an echo quite soon in the Versailles rules but also, much later, in the 1997 Convention on the Non-Navigational Uses of International Waterways. Old truisms tend to resurface in statements about international boundaries.

Working out how to deal with river boundaries: lessons from history

Shatt al-Arab

It would take a full 128 years of regulating treaty history for boundary delimitation along the estuary to move from eastern (Iranian) bank, the limit Britain had meant to introduce with the 1847 Treaty Of Erzurum, to the *thalweg* line introduced by the package of agreements and treaties agreed between Iran and Iraq during March–July 1975 (Schofield, 2004). Yet the suitability of a mid-channel delimitation had been known from the start, as no less a figure than British Foreign Secretary Lord Palmerston had demurred in March 1847, two months before the Erzurum treaty had been signed:

> [. . .] I have to state to you with reference to the pretension which has been advanced by the Porte [Ottoman Empire] to an absolute right of sovereignty over the Chat-el-Arab, that when the opposite banks of a River belong as they will do in this case of the lower portion of the Chat el Arab, to different Powers, it would be contrary to International usage to give to one of the two Powers the exclusive sovereignty of that portion of the course of such River, and that therefore this proposal of the Turkish Government seems to be inadmissible. (Palmerston, The National Archives (TNA), 1847, FO 78/2716)

The ideals expressed here echoed the legal logic articulated early in the 17th century by Hugo Grotius whereby: 'the jurisdictions on each side reach to the middle of the river that runs betwixt them' (Prescott & Triggs, 2008, p. 215). Yet Britain, after all only a mediating power with Russia on the quadripartite commission that sought to lay down a boundary from the Gulf northwards to Mount Ararat, ultimately had no choice but to expediently accommodate Ottoman claims. It regarded the 1847 treaty as having laid down a boundary along the Persian bank without the document having specified as much (Schofield, 2004). The ensuing four-party project to lay down in stone the line supposedly outlined in the Erzurum deal over the next few years ended in near farce, with ambitions soon relegated to collecting data that would allow for the mapping of the borderlands (Schofield, 2008a). A much more pragmatic Palmerston would soon comment that Britain and Russia could only lay down the boundary they were urging if they possessed full arbitrating powers. These would ultimately be afforded in advance of a further quadripartite treaty of November 1913 that sought to adjust, update and refine the Perso-Ottoman territorial deal of 1847. Another stab at demarcation followed during 1914. Though far from a glowing success, Britain and Russia had at least in their own minds stabilized this classic frontier zone, albeit nearly seven decades later than originally planned. No wonder Ernest Hubbard, Britain's Secretary on the 1914 demarcation commission, described the episode as constituting '[a] phenomenon of procrastination unparalleled even in the chronicles of Oriental diplomacy' (Hubbard, 1916, p. 2).

If expediency had dictated that the mid-channel did not get nominated as the Perso-Ottoman boundary along the Shatt during the mid-19th century, it would be the local observations in the early 20th century of a young Arnold Wilson, then a junior British official at Muhammara (now Khorramshahr) and the future architect of imperial Iraq, that redirected attentions to the question of delimiting the river boundary in the early 19th century (Schofield, 2004). He noted how the local authorities on both sides of the river recognized the mid-stream as the boundary in the vicinity of Muhammara and that not only commerce and shipping apparently respected it as a matter of course but the inhabitants of either bank. The local observations of imperial agents and their suggestions for the application of European spatial norms to areas of the colonial world is a frontier experience that has come under increasing scrutiny in historical border studies in recent years (Simpson, 2021; Tillotson, 2020). There were also notable examples where locals would point out that Britain was not even implementing its own delimitation logic properly when demarcating the territorial limits it had introduced – usually when mistaking or misinterpreting the natural features that supposedly governed them (Jones, 1945, p. 39; Schofield, 2008b, pp. 419–420).

Despite Wilson's observations that Britain would hold to the line that the Persian bank constituted the proper boundary until such time as wider economic circumstances changed more fundamentally – namely the discovery in 1908 of oil in the foothills of the Zagros Mountains. This got their officials thinking about improving Muhammara's anchorage facilities for the future export of oil, specifically replacing the outdated facilities on the River Karun with modern ones on the Shatt. So Wilson's observations would be used as the justification to extend the river boundary to the median line – but note not to the *thalweg* line – of the Shatt in Britain's recommendation for a delimitation, but elsewhere this would remain as running along the low-water mark of the Persian bank in the 1913–14 settlement (Schofield, 2004, 2008b).

Following the specification of the Versailles *thalweg*/median line rules for navigable and non-navigable rivers in 1919, Iran would push fairly consistently for a *thalweg* boundary along the whole length of its boundary with Iraq along the Shatt from the mid- to late 1920s onwards. It was something of a surprise therefore when Iran and Iraq agreed a further treaty in 1937 that extended the river delimitation to the *thalweg*, but only for the short stretch alongside the Iranian oil port of Abadan, home to the world's largest oil refinery in the early 20th century (Schofield, 2004). There were some complex linkages to other bilateral, regional and security questions, but no one has ever really got to the bottom of why Reza Shah agreed to an adjustment that resulted in the Shatt boundary being defined variously along the Iranian bank, median line (Khorramshahr) and *thalweg* (Abadan). So much for the Versailles rules clarifying things for we now had the complex river boundary delimitation mentioned above.

River Jordan

The western boundary of Transjordan, as specified in Article 25 of the Palestine Mandate, was defined in the following manner:

> The territory lying to the east of a line drawn from a point two miles west of the town of Aqaba on the Gulf of that name up the centre of the Wadi Araba, Dead Sea and River Jordan to its junction with the Yarmuk: hence up the centre of that river to the Syrian frontier. (Article 25 of the Palestine Mandate, Geneva, 23 September 1922)

For its sheer vagueness, this rivalled Britain's threadbare definition of the Iraq–Kuwait boundary or the line it would initially set for Iraq and Syria with France in 1920 – where hundreds of miles of boundary were supposedly specified in three or four lines of text (Schofield, 2008b). Admittedly, this was not seen at the time as a burning issue with Palestine and Transjordan coming under the same mandate and, partly for this reason, there was not the same two-stage process of colonial boundary delimitation we had witnessed from Britain elsewhere in the Middle East, where a territorial deal hatched in European corridors of power would be followed invariably by its reconciliation to local surficial realities by commission or the like (Schofield, 2008b). Here, no surveys were undertaken, no lines on maps had been seriously considered and no consideration made of what was mean by the centre of the succession of hydraulic divides that constituted the boundary.

The ambiguities and difficulties resulting from such a lack of exactitude would soon make themselves apparent. Within five years, an instance of avulsion in the Jordan River led Lord Plumer, Britain's High Commissioner in Jerusalem, to ask what precisely was meant by a boundary running 'up the centre of the [. . .], River Jordan [. . .]'. For in February 1927 the Jordan burst its banks and, pretty much overnight, cut a new course for some 500 m in the Beisan subdistrict. Consequently, an area of some 800 *dunums* that had originally belonged to Palestine now found itself east of the river's new course. With the original riverbed dry and debris-ridden, the British authorities now sought clarification as to whether the 'transferred territory' remained Palestinian or Transjordanian (Law, TNA, 1927, CO 733/142/1).

So Plumer commissioned the Colonial Office to conduct a survey of other river boundaries in an effort to establish whether it was the norm in such instances for the old channel course to be maintained as the boundary or for the new shifted course to be adopted. On 27 July 1927, the Colonial Office reported back as follows:

> It will be observed from this memorandum that according to a general rule of International Law, the gradual shifting of the thalweg from one side of the river to the other by reason of imperceptible erosion of accretion of its banks, has the effect of changing the boundary to a corresponding degree. If, however, the river should, by a process known as avulsion, suddenly become diverted from its regular channel, the boundary line remains where it was before the change. (Oliphant, TNA, 1927, CO 733/142/1)

Yet Field's survey of past state practice presented anything but a clear picture. The general rule of international law referred to in the above quotation followed Judge Caleb Cushing's famous 1853 opinion on the effect that a change of course in the Rio Grande had on the coincident international boundary between the United States and Mexico.[2] However, Field reviewed other examples and contexts where there had been change to the course of rivers that formed international boundaries and where this principle had not been followed: Belgium/the Netherlands and the Scheldt, 1842; Germany/(Polish) Russia, 1888 (River Drewens); Germany/Austria, 1898 (River Prezemsa); and Britain/Italy, 1891–1911 (River Juba). These outcomes reflected a pronouncedly individualistic 'horses for courses' approach with little concurrence in inter-state or colonial practice (Field, TNA, 1927, CO 733/142/1).

With no definitive guide as to how the Jordan should be treated, the Colonial Office went along with the view earlier expressed by the Survey of Palestine:

> the Jordan is unique and should be treated as a special case in which it is necessary to adopt the natural boundary of the river itself, wherever it is? The fact that the river describes a course of some 200 miles in a distance of 65 miles is sufficient evidence of its unstable and eccentric nature. (Leys, TNA, 1927, CO 733/142/1)

So an interpretation was made of the vague 1922 Jordan River delimitation that seemed the most practical to manage at a local level. Plumer's decree of 12 September 1927 flew in the face of most conventional legal interpretations. First, a *thalweg* was prescribed for a non-navigable river and, second, it was being proposed that the boundary moved with any shift in the main river channel, even when this was being occasioned by sudden, violent physical change (i.e., avulsion): '[t]his decision will involve periodical minor changes in the boundary as the "thalweg" of the river shifts, but will probably prove to be the most practical solution of the question' (Kirkbride, 12 September 1927).

The Chobe and Bahr al-Arab rivers

The International Court of Justice gained jurisdiction to treat the dispute between Botswana and Namibia over boundary alignment and island sovereignty in the Chobe River by a special agreement of 15 February 1995 (Shaw & Evans, 2000), the kind of reference the court prefers since not only was the substance of dispute clearly articulated but the main legal source of evidence it wanted interrogating also clearly identified:

> The Court is asked to determine, on the basis of the Anglo-German Treaty of 1 July 1890 [an agreement between Great Britain and Germany respecting the spheres of influence of the two countries in Africa] and the rules and principles of international law, the boundary

> between Namibia and Botswana around Kasikili/Sedudu island and the legal status of the island. (International Court of Justice, 13 December 1999)

It would soon become clear that the case was likely to revolve around the identity of the 'major channel' mentioned in the text of the 1890 spheres of influence agreement, one of whose chief effects was to provide Germany with access to the Zambezi River and Victoria Falls through the extended land corridor created for that purpose: the Caprivi Strip.[3]

How might one recognize the major channel when a river divides into two or more channels. Here is what the ruling in the 1966 Chile/Argentina arbitration case had to say:

> Where an instrument (such as a treaty) has laid down that a boundary must follow a river, and that river divides into two or more channels and nothing is specified in that instrument as to which channel the boundary shall follow, the boundary must normally follow the major channel. (United Nations, 2006)

American boundary delimitation expert Stephen Jones had earlier commented that:

> (t)he allocation of islands in rivers has been handled in a number of ways. The Paris treaties provided that the boundary should follow the 'principal branch' about islands and shoals. 'Principal' might mean widest, deepest, most used or carrying the most water. (Jones, 1945, p. 113)

With no detailed specification in its text as to what was meant by the major channel, it was always most likely that the intent of the parties to the 1890 agreement would be unearthed in the records of negotiations that led to the instrument being included.

The 2008–09 Abyei case at the Permanent Court of Arbitration would soon effectively fixate upon the status of the provincial boundary between the Sudanese provinces of Kordofan and Bahr al Ghazal in 1905, just six years into the establishment of the Anglo-Egyptian Condominium Administration. This was because the key historical mechanism for identifying the territorial extent of Abyei revolved around the transfer that year of the Ngok Dinka tribe from Bahr al Ghazal to Kordofan with a presumption that either the boundary between the provinces or the extent of territory inhabited by the tribe (both north and south of the Bahr al-Arab River) must have been known.[4] Abyei had played host to some of the worst violence witnessed in two bloody Sudanese civil wars and the Comprehensive Peace Agreement of 9 January 2005 tried symbolically to make territorial definition of the province a bridge going forward to the peaceful coexistence of two Sudans – one old and established and one new. Once its territorial extent was established, the inhabitants of Abyei could vote in a referendum to join one or the other state – at least, that was the idea (Born & Raviv, 2017). Later in 2005, the findings of the Abyei Boundary Commission (ABC) – a group of recognized experts commissioned purposefully to define Abyei territorially – that the province should extend some distance north into Kordofan were rejected by Sudanese President Bashir, with reports that his immediate reaction was to suggest their report be thrown back into the Sudd from whence they came (confidential source). Hence, the reference of Abyei to the Permanent Court of Arbitration – with the missive being to establish whether the ABC had exceeded its mandate in defining the province so extensively and, should that be found to be the case, to lay down a more appropriate one and a more suitable basis for doing so (Born & Raviv, 2017).

On the face of it, the territorial argument maintained by the Government of Sudan during the case that Abyei province should extend no further north than the Bahr al Arab River seemed to possess some validity. For there were references, albeit generally vague ones, to the Bahr al-Arab as constituting the provincial boundary made by the governors of Kordofan and Bahr al-Ghazal provinces in the first half-decade of the 20th century, while it was even itemized as such in more than one colonial administration report around this time (Schofield & Allan, 2009, p. 14). Bahr al-Ghazal province under Ottoman–Egyptian rule had been characterized spatially as follows: 'This "mudirieh" was vaguely defined, but may be described as enclosing the entire district watered by the southern tributaries of the Bahr al Arab and the Bahr al Ghazal Rivers' (Gleichen, 1905, pp. 2, 151). Even at provincial level, the colonial mindset would probably have worked for the presumption of rivers as boundaries around this time, even if the above quotation hinted at the possibility that the entirety of the riverine area might lie within the province of Bahr al-Ghazal. Whatever, it is fair to say no agreed delimitation was in place along the Bahr al-Arab in 1905 – something the 1911 Handbook on the Sudan specifically confirmed a half-decade on (Government of Sudan Memorial, 2009, p. 376). Nothing had been agreed centrally in Khartoum or back in London.

Ottoman territorial limits on whatever scale were often indeterminate, shifting and impermanent (Schofield, 2018) but what contributed most to the prevailing uncertainty in 1905 concerning the status of any Bahr al-Arab provincial boundary was Britain's confusion as to where the riverine feature lay. In particular, the Ragaba as-Zarga (Ngol) and Bahr al-Arab (Kiir) rivers were frequently confused with the former often mistaken for the latter by all levels of British officialdom in Sudan, not to mention a great number of travellers to this largely unexplored region. The same governors who thought the Bahr al-Arab was probably the provincial boundary demonstrably did not know where it was, even Governor-General Wingate made the same mistake of confusing the two rivers. Logically, therefore, or so the south (SPLM/A) argued in the case, any putative Bahr al-Arab boundary must have been indeterminate and indefinite in 1905 – there had been no central allocation, never mind delimitation while significant leading Condominium personnel did not know where it was (Schofield & Allan, 2009, p. 2).

As to whether the Bahr al-Arab might have constituted a barrier to local movement, historical research into the physical geography unearthed in service of the SPLM/A's case was used to argue that the riverine 'Bahr region' was both discernible, indeed measurable and presented no such impediments:[5]

> the Kiir/Bahr al-Arab is not a physical barrier, and would not present a meaningful obstacle to the movement of the Ngok Dinka and their cattle, save perhaps in very limited parts of the rainy season [...] the whole of the Bahr region (as well as some parts of the goz) would support the Ngok Dinka agro-pastoral culture and way of life. (Schofield & Allan, 2009, p. 2)

Uncertainty was the watchword here though. Britain's geographical knowledge half a decade into the Condominium administration of a region that exhibited huge seasonal variations was unsurprisingly threadbare. No wonder the identity of water course was so confused. Not that much had been explored by this stage, never mind surveyed and very few stretches of channels that were potentially navigable had been cleared. In reality, negotiating the Bahr al-Arab depended on whether the dry or wet season was being experienced.

> the river bed of the Kiir/Bahr el Arab varies in its course through the Abyei Region from continuous flow to areas of discontinuity along its length during the dry season. [. . .] In the high wet season during July/August it will flood, and its banks and areas around them will be under water. During this the river would in some years prevent passage by humans. However, outside of those limited times and particularly in the driest months from November to May, the Kiir/Bahr el Arab is a comparatively gentle one that is not particularly deep. (Schofield & Allan, 2009, pp. 27–28)

Contemporary outcomes: old and new in river boundary dispute settlement

Shatt al-Arab

The complex regional geopolitics that were in large part responsible for the conclusion of the breakthrough package of territorial settlements between Iran and Iraq during 1975 cannot disguise the fact that with agreement on a *thalweg* delimitation for the Shatt, a long-standing Iranian positional demand had been satisfied and Palmerston's original mid-19th-century recommendation realized (Kaikobad, 1988; Schofield, 2004). The 1975 package constituted a sophisticated legal river boundary settlement by any reckoning. Let us concern ourselves here with how the *thalweg* was elaborated and how the contingency of future physical change in the boundary estuary would be regulated and managed. The geopolitics have been covered elsewhere (Schofield, 2021).

The 6 March 1975 Algiers Accord simply announced that Iran and Iraq had 'decided to delimit their river frontiers on the thalweg line' (United Nations, 1975). A fortnight later it was announced that arrangements had been made for the establishment of a tripartite committee (including technicians from Algeria, who had mediated the accord) 'to determine the water border in the Shatt al Arab estuary on the basis of the thalweg principle' (telegram from Tehran, TNA, 1975, FCO 8/2547). Within a couple of months, a *thalweg* delimitation had been established on the basis of surveys undertaken by the said committee in time for the conclusion in Baghdad on 13 June 1975 of the 'Treaty Relating to the State Boundary and Good Neighbourliness between Iraq and Iran' (United Nations, 1975). The boundary treaty was supplemented by three protocols, including a 'Protocol Relating to the Delimitation of the River Frontier between Iraq and Iran'. Its first article declared that 'the State river frontier between Iran and Iraq in the Shatt al-Arab has been delimited along the *thalweg*' (United Nations, 1975). Attached to the Article 1 were four (British Admiralty) charts upon which the delimitation was plotted (Schofield, 2009, vol. 18, maps 27–30).

Article 2 was where things got interesting and perhaps where the pudding got over-egged – most pointedly in anticipating future physical change in the waterway. More so than in any other river boundary treaty signed before this time, the *thalweg* was defined in exacting terms in its first paragraph (though the mixture of its key terms – median line and *thalweg* – could be seen as a little confusing):

> The frontier line in the Shatt al-Arab shall follow the 'thalweg', i.e., the median line of the main navigable channel at the lowest navigable level, starting from the point at which the

land frontier between Iran and Iraq enters the Shatt al-Arab and continuing to the sea. (United Nations, 1975)

Article 2's second paragraph specified that the new *thalweg* line would 'vary with changes brought about by natural causes in the main navigable channel' unless the parties agreed otherwise, with the third establishing that technical teams from both states would have to confirm and attest to such occurrences. This was anticipating gradual change via erosion and accretion as foreseen by the 1919 Versailles rules. Things got more specific and detailed in dealing with the contingency of sudden physical change in the riverbed. Paragraph 4 held that:

> [a]ny change in the bed of the Shatt al-Arab brought about by natural causes which would involve a change in the national character of the two States' respective territory or of landed property, constructions, or technical or other installations shall not change the course of the frontier line, which shall continue to follow the thalweg in accordance with the provisions of paragraph one above. (United Nations, 1975)

Paragraph 5 of the River Protocol's Second Article went on to specify what ought to be done in such a contingency, with its bizarre provision for the possible transfer of the river back to the boundary:

> Unless an agreement is reached between the two Contracting Parties concerning the transfer of the frontier line to the new bed, *the water shall be re-directed at the joint expense of both parties to the bed existing in 1975* – as marked on the four common charts listed in article one [. . .] – should one of the Parties so request within two years after the date on which the occurrence of the change was attested by either of the two Parties. Until such time, both Parties shall retain their previous rights of navigation over the water of the new bed. (United Nations, 1975, emphasis added)

Article 6 of the river boundary protocol committed the two sides to jointly survey the main river channel afresh every 10 years, or even more frequently should one of the parties request as much. War and conflict pretty much ever since has meant that acting on its provisions was never a practicable possibility, at least until recently. Prospectively, the announcement of 11 March 2019 on Hassan Rouhani's first visit to Baghdad as Iranian President that agreement had been reached to renew the 1975 river boundary arrangements is significant (Schofield, 2021, p. 11). For some time before then, though, it had seemed that remaining disputes were less about whether the 1975 regime is applicable – both sides broadly agreed that it was – and more about how details might be implemented in any new settlement.

As for recent physical change in the Shatt's main navigable channel, a willingness to broach any associated problems has accompanied not just better relations between Baghdad and Tehran but also acknowledges a post-conflictual context of regional reconstruction. Nonetheless, Iraq is known to be frustrated that channel movement has all been to the south and west of the coordinates officially nominated for the Shatt's southernmost reaches in 1975, with some law-makers casting doubt on whether this has all been down to nature. It was maritime incidents of summer 2004 and spring 2007, involving the capture of British sailors by Iranian patrols near the estuary mouth, that also drew attention to the absence of a territorial waters boundary between Iran and Iraq south of the terminal point of the 1975 river boundary delimitation. One important consequence of the first incident had been that Iran and the occupying powers (the US and UK) would apparently agree that the current position of the *thalweg* as it existed at any one time along the Shatt would serve as an operative international boundary (Schofield, 2021, pp. 14–15).

River Jordan

In famously signing their Common Agenda in Washington on the White House lawn on 14 September 1993, Israel and Jordan agreed *inter alia* the basis upon which they would negotiate a full boundary agreement over the following year. Paragraph 5 of the Israel–Jordan Common Agenda on Borders and Territorial Matters read as follows:

> Settlement of territorial matters and agreed definitive delimitation and demarcation of the international boundary between Israel and Jordan with reference to the boundary definition under the Mandate, without prejudice to the status of any territories that came under Israeli Military Government control in 1967. Both parties will respect and comply with the above international boundary. (Schofield & Allan, 2010)

Just over a year on, 'The Treaty of Peace between the State of Israel and the Hashemite Kingdom of Jordan' was signed by Israeli Prime Minister Yitzhak Rabin and Jordanian Prime Minister Abdul Salam Majali on 26 October 1994. Article 3 of a long document (with its annexe and appendices) defined and regulated their international boundary. The first four paragraphs (of nine) in the article provided for boundary definition, regulation and future demarcation. Paragraph 1 reiterated that the vague old 1922 colonial delimitation remained the definitional basis of the Israel–Jordan boundary.

> The international boundary between Israel and Jordan is delimited with reference to the boundary definition under the Mandate as is shown in Annex I (a), on the mapping materials attached thereto and co-ordinates specified therein. (United Nations, 1995)

The first annexe to the treaty was entitled 'Israel–Jordan International Boundary Delimitation and Demarcation' and described the basis of delimitation for four specified stretches of the boundary, respectively: the Jordan and Yarmuk rivers; the Dead Sea; Wadi Araba, and the Gulf of Aqaba:

For the Jordan river, Britain's definitional clarification of 1927 was basically retained for, without employing the term specifically, the boundary was defined as running along the *thalweg* of the Jordan river.

> The boundary line shall follow the middle of the main course of the Jordan and Yarmouk rivers. (United Nations, 1995)

More flexibility, however, was granted to Jordan and Israel in dealing with future natural change in the main channel of the river than had been provided all those years ago by the Colonial Office in 1927. While it was confirmed that the boundary would move with gradual shifts in the main channel occasioned by accretion, a newly appointed Joint Boundary Commission would meet to consider the most appropriate course of action in future cases of avulsion, with all options remaining open to them.

Chobe River

On 13 December 1999, the International Court of Justice announced its ruling with respect to the sovereignty of Kasikili/Sedudu island in favour of Botswana by 11 votes to four, thereby also decreeing that the northern channel of the Chobe River was the main one. Yet, even so, both states would have equal navigation rights in each of the river's channels north and south of the island (International Court of Justice, 1999). Three items seem worthy of discussion, given the concerns of this article: the local human context, identification of the main channel, and a judgment that seemingly wanted to address both top-down and ground-up aspects simultaneously.

Traditional, seasonal usage of the insular feature in question – which is only 3.5 km^2 in size and is generally flooded for a third of the year from March onwards – led to Namibia advancing a claim of prescriptive title based upon the Caprivi-based Masubia tribe's use of it for herding cattle. The two disputing parties had agreed a four-stage definition of prescription prior to the case which, in the International Court of Justice's ruling, the Masubia's historical actions did not satisfy (Shaw & Evans, 2000).

> The Court finds that while the Masubia of the Caprivi Strip (territory belonging to Namibia) did indeed use the Island for many years, they did so intermittently, according to the seasons, and for exclusively agricultural purposes, without it being established that they occupied the Island à titre de souverain, that is, that they were exercising functions of State authority there on behalf of the Caprivi authorities. The Court therefore rejects this argument (International Court of Justice, 1999).

As for identification of the Chobe's main channel, there were hints and implications in the textual negotiation record of the 1890 Anglo-German spheres of influence treaty that it would have laid to the north of the island feature but nothing definitive – neither was there any mapping of suitable scale and detail that could indicate the intentions of the rival imperial powers. Likewise, a thorough survey of navigational practice up to Botswanan independence in 1966 and Namibia's a quarter century later did not point to any agreed interpretation of the principal channel. This all only underlined that in the scheme of things, local as well as regional, this had not been a particularly significant river boundary dispute. So, the International Court of Justice effectively took it upon it itself to determine the major channel by running through Jones' old checklist of indicators earlier itemized in this article:

> In order to do so, it [took] into consideration, inter alia, the depth and the width of the channel, the flow (that is, the volume of water carried), the bed profile configuration and the navigability of the channel. After having considered the figures submitted by the Parties, as well as surveys carried out on the ground at different periods, the Court concludes that 'the northern channel of the River Chobe around Kasikili/Sedudu Island must be regarded as its main channel'. (International Court of Justice, 1999)

So with its identification of Kasikili/Sedudu as Botswanan in sovereign terms, what lay behind the International Court of Justice's specification here that there would be 'unimpeded navigation for craft of [Namibian] nationals and flags in the channels of Kasikili/Sedudu island [. . .] a treatment equal to that accorded by Botswana to its own nationals and to vessels flying its own flag' (International Court of Justice, 1999). Was it to soften the blow of a sovereign decision when the historical indicators had not been overwhelming in a (relatively) obscure dispute and which, in many ways, contradicted local patterns of traditional usage? Or did it genuinely reflect more progressive thinking with respects to the local management and implementation of inter-state boundary awards? After all, fairly prominent reference was lent to the 1997 Convention on the Non-Navigational Uses of International Watercourses, which defines a watercourse as a 'system of surface waters and ground waters constituting by virtue of their physical relationship a unitary whole flowing into a common terminus' (United Nations, 2014).

Abyei/Bahr al-Arab

The PCA's award of 22 July 2009 would subsequently be sharply castigated by the lead in the SPLM/A's legal team during the Abyei case: 'Decisions like the Abyei Award provide sobering reminders of the shortcomings of the international legal system and its capacity to resolve interstate disputes' (Born & Raviv, 2017, p. 224). Such comments referenced the tribunal's procedural deliberations more widely but also, more specifically, its rejection of the northern limits for the 'Abyei area' earlier recommended by the Abyei Boundary Commission – with the tribunal's seeming dismissal of their supporting arguments. It will be recalled that there were competing tribal and territorial arguments at work in the case supporting Abyei's areal definition and, specifically, interpretation of the 1905 tribal transfer. On one hand, the 2009 Abyei award concluded that the ABC had not exceeded its mandate in calculating northern limits on the basis of its tribal formula but, on the other, it moved these significantly southwards, truncating the defined territorial extent of the 'Abyei area' by some 40% from 18,559 to 10,460 km^2 (Born & Raviv, 2017, p. 1.

The above quotation might also be extended more widely to refer to what was being attempted with the specified basis of dispute resolution here – establishing what lay in place on the ground over a century earlier in a seasonally flooded area of central Sudan and whether a provincial boundary could be held to be in existence there. We have already seen just how vague and ambiguous imperial boundary delimitations could be but here we were talking about a provincial divide and colonial evidence was always likely to be even more threadbare, flimsy and uncertain (Schofield, 2015, p. 140). Hence, a wider array of evidence needed to be relied upon and considered in the Abyei case than is usual in territorial cases – mapping surveys, travellers' reports, burial sites and oral histories, testimonies and traditions.

For the reasons elaborated above, it was little surprise when Khartoum's claim that the Bahr al-Arab constituted an inter-provincial colonial boundary in 1905 (and therefore the northern territorial limits of Abyei) failed to find favour in the July 2009 Permanent Court of Arbitration award. For starters, no delimitation could be held to have been in place at the time along the river,[6] even had the Condominium government known with more certainty which riverain feature it comprised. If the prevailing lack of geographical knowledge of the Abyei region at the turn of the 20th century on Britain's part worked against any prospect of a river boundary materializing with the Permanent Court of Arbitration July 2009 award, late 19th-century characterizations of the possible extent of the Ottoman province of Bahr al-Ghazal as an essentially riverain area perhaps found partial expression. Any quick perusal of the territorial outcome of the award would confirm that the south ended up with pretty much all of the rivers whose identity and configuration had been at the heart of the territorial arguments put forward by Sudan in the case.

Conclusions

Despite the development of guidelines for drawing boundaries along rivers at Versailles during 1919, there would continue to be wide variability in the manner by which delimitations were selected or agreed upon by states for specific physical contexts and regional constellations – horses for courses, if you like. Pronounced individuality in river boundary regimes in international law reflects, after all, the innate complexity of this class of delimitation: the challenge of reconciling the top-down desiderata that dictated their nomination in the first instance with the live ground-up realities and challenges posed by physical change in channel alignment and the evolving socio-economic development of the adjoining areas.

The Shatt and Jordan cases demonstrate obviously different legal approaches to refining and regulating an international river boundary, which is to be expected since the former is navigable and the latter non-navigable. Although the regional geopolitics that drove their conclusion were wholly decisive, the 1975 package of Iran–Iraq agreements constitute in many ways a model river boundary settlement, with the most appropriate delimitation from a legal and regional perspective defined in exacting terms and the contingency of future physical change dealt with at length. Conversely, Jordan and Israel kept with a threadbare colonial delimitation and management approach in their territorial settlement of 1993–94, even though these arrangements ran contrary to the Versailles rules. Instead of specifying an elaborate regime for sudden, avulsive change – as did Iran and Iraq in their June 1975 river boundary protocol, Israel and Jordan simply agreed to meet up should this eventualize – and decide upon the most appropriate course of action. The approach taken to the Jordan seemingly affords more flexibility and the prospect of sudden changes in the riverbed has always been much greater here than with the main channel of the Shatt al-Arab, where recent change – though significant – has been of a gradual accretional character.

The use made of history in guiding the dispute resolution process in the 1995–99 Kasikili/Sedudu island and 2008–09 Abyei cases at the International Court of Justice and Permanent Court of Arbitration, respectively, demanded the clarification of the major channel of the Chobe River in the first instance and suggested the possibility for a river boundary delimitation along the Bahr al-Arab in the second. Yet the status of the rivers was anything but a live, regional issue for the colonial powers in 1890 and 1905, respectively, the key historical dates decided upon for dispute resolution. Given that evidence was resultantly threadbare for the existence of the Chobe's major channel back then and the Abyei region was essentially unexplored, such prevailing uncertainty had to be acknowledged thoughtfully in resolving these disputes. The PCA's spatial characterization of Abyei as a riverain area tallied to some extent with late 19th-century descriptions and also nodded in some ways, without saying as much, to the arrangements reached by Britain and Italy over the Juba in 1911 – where the greater river valley area was seen as an operative border region. As for the Chobe River, the court decided that through Botswana had argued the merits for sovereign ownership over Kasikili/Sedudu, use of the channels on either side would be free to nationals of both states. This stipulation was couched progressively with reference made to the 1997 UN Non-Navigational Use of Waterways Convention – perhaps the Institute of International Law's 1911 Madrid Resolution could also have been invoked! Maybe this effective instruction to cooperate over the boundary river reflected a slight uneasiness on behalf of the court over its rejection of Namibia's prescriptive arguments and the lack of definitive evidence for what had been its major channel in 1890. River delimitation settlements have always required more local management than other classes of international boundary but here was a legal resolution that took steps to guide it.

The recent legal settlement and management of complex international river boundary questions shows signs of balancing localized human and environmental context with the specific individual histories and geopolitics of the territorial limits in question. Perhaps there is an opportunity for more critical academic engagement with their socio-spatial characteristics, both historically and contemporaneously – one that addresses their localized human dynamics more fully than traditional approaches.

Notes

1. The author and Tony Allan were involved together with the dispute resolution process in two of these cases. In conjunction with Tony's late colleague, Keith McLachlan, we put together a Report on Jordan's Western Boundary during 1993–94 in conjunction with Jordan's Royal Scientific Society (Geopolitics Research Centre, 1994) with Tony leading on water and myself on territorial issues. Fast forward one-and-a-half decades to February 2009, we later co-authored a report entitled The Boundaries and Hydrology of the Abyei Area, Sudan (referred to as the Menas Borders report in subsequent court proceedings) (Schofield & Allan, 2009) to support the Sudan People's Liberation Movement/Army's (SPLM/A) case at the Permanent Court of Arbitration in the Abyei case. We would then defend this under cross-examination a couple of months later at the hearings stage of the case when appearing as expert witnesses for the SPLM/A.
2. This general rule would later be summed up neatly by a 1970 US Supreme Court decision (Arkansas vs. Tennessee, 1970): 'It is settled beyond the possibility of dispute that where running streams are the boundaries between States, the same rule applies as between private proprietors, namely, that when the bed and channel are changed by the natural and gradual processes known as erosion and accretion, the boundary follows the varying course of the stream; while if the stream from any cause, natural or artificial, suddenly leaves its old bed and forms a new one, by the process known as an avulsion, the resulting change of channel works no change of boundary, which remains in the middle of the old channel, although no water may be flowing into it, and irrespective of subsequent changes in the new channel' (US Supreme Court, 25 February 1970).
3. Imperial access to the continent's prominent physical features was seen as a prestige indicator, hence the Kenya/Tanzania boundary was reputedly defined in a manner that shared colonial control of East Africa's two highest mountains, Mount Kenya and Mount Kilimanjaro – a sort of natural boundaries mindset writ large!
4. Hence during the Permanent Court of Arbitration case, the SPLM/A (South) went for a tribal interpretation that reflected the extent of territory inhabited by the Ngok Dinka in 1905, while the Government of Sudan went for a more conventional territorial interpretation that argued for the existence of a known provincial river boundary (Born & Raviv, 2017).
5. 'The Bahr region extends south from the boundary of the 'go'z area in the west of Southern Kordofan at a latitude of approximately 10 degrees 0 minutes N, and south from Lake Keliak in the east of Southern Kordofan, and encompasses the areas around the Ngol/Ragaba ez Zarga and Kiir/Bahr el Arab' (Schofield & Allan, 2009, p. 1).

6. In its definition of delimitation, the Permanent Court of Arbitration relied upon what this author had said in court during the hearings stage in April 2009: 'As explained by [...] Schofield, there are three stages in a boundary's evolution: allocation, delimitation and demarcation. Allocation deals with allocating territory and not the actual boundary, while demarcation simply physically marks out the boundary on the ground. Delimitation, quite differently, is when the line is established and specified. It requires "an executive act" of determining where the actual boundary line should be, and calls for a detailed description of the location of a boundary line' (SPLM/A Oral Pleadings, 22 April 2009, Transcr.121/03-122/02).

Acknowledgement

I would like to thank Matthew Tillotson for his useful comments on an earlier draft and also three individuals who, like Tony Allan, are sadly no longer with us but with whom I have enjoyed so many great conversations about river boundaries over the years: Mike Burrell, Kaiyan Kaikobad and Keith McLachlan.

Disclosure statement

No potential conflict of interest was reported by the author.

Funding

The author reports there is no funding associated with the work featured in this article.

ORCID

Richard Schofield http://orcid.org/0000-0002-9112-5057

References

Anghie, A. (2006). The evolution of international law: Colonial and postcolonial realities. *Third World Quarterly*, *27*(5), 739–753. https://doi.org/10.1080/01436590600780011

Bialasiewicz, L., & Minca, C. (2010). The `border within': Inhabiting the border in Trieste. *Environment and Planning D: Society and Space*, *28*(6), 1084–1105. https://doi.org/10.1068/d2609

Born, G., & Raviv, A. (2017). The Abyei arbitration and the rule of law. *Harvard Journal of International Law*, *58*(1), 177–224.

Bouchez, L. J. (1963 July). The fixing of boundaries in international boundary rivers. *International and Comparative Law Quarterly*, 12(3), 789–817. https://doi.org/10.1093/iclqaj/12.3.789

British Embassy, Tehran to FCO. 18th March 1975 in TNA file: *FCO 8/2457*.

Demeritt, D. (1997). Representing the 'true' St. Croix: Knowledge and power in the partition of the Northeast. *William and Mary Quarterly*, *54*(3), 515–548. https://doi.org/10.2307/2953838

Donaldson, J. (2009). Where rivers and boundaries meet: Building the international river boundaries database. *Water Policy*, *11*(5), 629–644. https://doi.org/10.2166/wp.2009.065

Donaldson, J. (2014). Re-thinking international boundary practices: Moving away from the 'edge. In F. McConnell, N. Megoran, & P. Williams (Eds.), *Geographies of peace* (pp. 89–108). Tauris.

Fall, J. (2010). Artificial states? On the enduring geographical myth of natural borders. *Political Geography*, *29*(3), 140–147. https://doi.org/10.1016/j.polgeo.2010.02.007

Field, J. W. (1927). Foreign Office memorandum on the effect of deviation of rivers on boundaries. 7th October 1927 in TNA file: *CO 733/142/1*. Unpublished.

Geopolitics Research Centre, SOAS. (1994). *Report on Jordan's western boundary*.

Gleichen, E. (ed.). (1905). *The Anglo-Egyptian Sudan: A compendium prepared by officers of the Sudan government* (Vol. 2). War Office.

Government of Sudan. (2009). *Memorial*. Government of Sudan v. SPLM/A (Abyei), Permanent Court of Arbitration.

Holdich, T. (1916, June). Geographical problems in boundary making. *The Geographical Journal, 47*(6), 421. https://doi.org/10.2307/1779240

Hubbard, G. E. (1916). *From the Gulf to Ararat, Blackwood, Edinburgh (2016 centenary edition published as from the Gulf to Ararat: Imperial boundary-making in the late Ottoman Empire*. IB Tauris. with a preface by Richard Schofield.

International Court of Justice. (1999). ICJ finds that Kasikili/Sedudu Island forms part of the territory of Botswana. Press release, ICJ/594. Retrieved December 13, 1999 from https://www.un.org/press/en/1999/19991213.icj594.doc.html

International Court of Justice (2017–22). *Land, Island and maritime frontier dispute (El Salvador/Honduras: Nicaragua intervening)*. https://www.icj-cij.org/en/case/75/judgments

Jones, S. P. (1945). *Boundary-making: A handbook for statesmen, treaty editors and boundary commissioners*. Carnegie Endowment for International Peace.

Kaikobad, K. (1988). *The Shatt-al-Arab boundary question: A legal reappraisal*. Clarendon Press.

Kirkbride, A. (1927 September 12). High Commissioner for Palestine and Transjordan to the Secretary of State for the Colonies. Letter, TNA file: *FO 816/23*.

Lauterpacht, E. (1960). River boundaries: Legal aspects of the Shatt al-Arab frontier. *International and Comparative Law Quarterly, 9*(2), 208–236. https://doi.org/10.1093/iclqaj/9.2.208

Law to the Chief Secretary, Government Offices, Jerusalem. 5th April 1927 in TNA file: *CO 733/142/1*.

Leys to the Chief Secretary, Government Offices, Jerusalem. 21st April 1927 in TNA file: *CO 733/142/1*.

Menas Borders. (2010). *Jordanian boundaries and border geopolitics*.

Murphy, B. (1990). Historical justifications for territorial claims. *Annals of the Association of American Geographers, 80*(4), 531–548. https://doi.org/10.1111/j.1467-8306.1990.tb00316.x

O'Dowd, L. (2010). From a borderless world to a 'World of Borders': Bringing history back in. *Environment and Planning D: Society and Space, 28*(6), 1031–1050. https://doi.org/10.1068/d2009

Oliphant to the Under Secretary of State, Colonial Office. 12 July 1927 in TNA file: *CO 733/142/1*.

Palmerston to Bloomfield. 3rd March 1847 in TNA file: *FO 78/2716*.

Prescott, J. R. V., & Triggs, G. D. (2008). *International frontiers and boundaries: Law, politics and geography*. Martinus Nijhoff.

Schofield, R. (2004). Position, function and symbol: The Shatt al-Arab dispute in perspective. In L. G. Potter & G. G. Sick (Eds.), *Iran and Iraq and the legacies of war* (pp. 29–70). Palgrave Press.

Schofield, R. (2008a). Narrowing the frontier: Mid-nineteenth century efforts to delimit and map the Perso-Ottoman boundary. In R. Farmanfarmaian (Ed.), *War and Peace in Qajar Persia* (pp. 149–173). Routledge.

Schofield, R. (2008b). Laying it down in stone: Delimiting and demarcating Iraq's boundaries by mixed international commission. *Journal of Historical Geography, 34*(3), 397–421. https://doi.org/10.1016/j.jhg.2007.11.002

Schofield, R. (ed.). (2009). *Arabian boundaries: New documents, 1966–1975*. Cambridge Archive Editions (Cambridge University Press).

Schofield, R. (2015). Back to the barrier function: Where next for international boundary and territorial disputes in political geography? *Geography, 100*(3 (Autumn 2015)), 133–143. https://doi.org/10.1080/00167487.2015.12093968

Schofield, R. (2018). British bordering practices in the Middle East before Sykes–Picot: Giving an edge to zones. *Égypte/Monde arabe, 18*(2), 149–161. https://doi.org/10.4000/ema.4279

Schofield, R. (2021). Back to 1975 and all that: Linking the technicalities and underlying dynamics of Iraq's border disputes at the head of the Persian Gulf. In G. Joffe & R. Schofield (Eds.), *Geographic realities in the Middle East and North Africa: State, oil and agriculture* (pp. 11–24). Routledge.

Schofield, R., & Allan, J. A. (2009). The boundaries and hydrology of the Abyei Region, Sudan. Expert report submitted to the PCA, signed on 11th February 2009 by Schofield and Allan. Menas Borders. Submitted with the SPLM/A Counter-Memorial.

Sevette, P. (1952 January). *Legal aspects of the hydro-electric development of rivers and lakes of common interests, committee on electric power of the United Nations*. Economic Commission for Europe. (Doc E/ECE/136).

Shaw, M., & Evans, M. (2000). Case concerning Kasikili/Sedudu Island (Botswana/Namibia). *International and Comparative Law Quarterly, 49*(4), 964–978. https://doi.org/10.1017/S0020589300064782

Simpson, T. (2021). *The frontier in British India: Space, science, and power in the nineteenth century*. Cambridge University Press.

Tillotson, M. (2020). Nature, space and distances in an imperial boundary network: The delimitation of the Nyasa–Tanganyika plateau boundary. *Political Geography, 76*(1), 102081. https://doi.org/10.1016/j.polgeo.2019.102081

United Nations. (1975). *Treaty Series, 1017*, 14903.

United Nations. (1995). *The question of Palestine: 'Letter dated 9 January 1995 from the permanent representatives of Israel, Jordan, the Russian federation and the United States of America to the United Nations addressed to the Secretary-General'*. https://www.un.org/unispal/document/auto-insert-179122/

United Nations. (2006). *Reports of international arbitral awards: Argentina–Chile frontier case*. xvi, 109–192. Retrieved December 9, 1966, from https://legal.un.org/riaa/cases/vol_XVI/109-182.pdf

United Nations. (2014). *Convention on the Law of the Non-navigational uses of International Watercourses 1997 [Adopted by the general assembly of the United Nations on 21 May 1997.* [Entered into force on 17 August 2014. See General Assembly resolution 51/229, annex, Official Records of the General Assembly, Fifty-first Session, Supplement No. 49 (A/51/49)], https://legal.un.org/ilc/texts/instruments/english/conventions/8_3_1997.pdf

van Houtum, H. (2005). The geopolitics of borders and boundaries. *Geopolitics, 10*(4), 672–679. https://doi.org/10.1080/14650040500318522

Power plus: Tony Allan's contributions to understanding transboundary water arrangements

Mark Zeitoun, Ana Elisa Cascão, Marwa Daoudy, Francesca Greco, Naho Mirumachi and Jeroen Warner

ABSTRACT

We trace the development of a theory and analytical frames within international political economy that originated from Tony Allan's mainstreaming of *power* as a determining factor in the control of transboundary flows. These include the Framework of Hydro-Hegemony, coexisting conflict and cooperation and Transboundary Water Interaction Analysis, counter-hegemony and virtual water rivers. These contributions are exemplified briefly through cases from around the globe and in particular the Nile. To ensure that this aspect of Tony's legacy lives on, we propose a research agenda that includes the analysis of power interactions at substate levels between central governments and non-state actors.

Inception of transboundary water analysis

Tony Allan's contribution to the analysis of institutional arrangements over transboundary water was made almost in passing. Three-quarters of the way through his *The Middle East Water Question* (2001) he enumerated the factors that influenced international relations and control over water in the Nile, Jordan and Tigris–Euphrates. With his consummate ability to notice significant things that others had overlooked, he noted that which state controlled the headwaters was not what mattered, it was *relations of power*.

Concerned that, until that time, analysis of transboundary waters had been neither theorized nor grounded, he encouraged his 2000s' generation of students to read Antonio Gramsci (Gramsci, 1935/2003) and Steven Lukes (Lukes, 1974, 2005), and to learn from each other. The London Water-Issues Research Group that he had established at the School of Oriental and African Studies (SOAS) and King's College London brought together political scientists and hydrologists, international relations theorists, engineers, lawyers, and ecologists, including the present authors (see Nicol, 2022, in this issue; Warner et al., 2017). We then turned our collective efforts to address a more specific version of Tony's question: if power matters in transboundary waters, to what extent, and how?

Curiosity and energy drove dozens of stimulating seminars, working groups, research projects and workshops, as well as several ideas and analytical frameworks that we draw upon in this review of power and hegemony theory: the Analytical Framework of Hydro-Hegemony, Transboundary Water Interaction Analysis and virtual hydro-hegemony. Each is exemplified from a wide geographical range of transboundary water arrangements before the utility of the theory is then demonstrated in greater detail in the case of the Nile.

Political–economic first steps

With no map to guide us, we knew little more than that there was something about water's biophysical properties and the ideas which people hold about it that had to be figured out. While Gramsci and Lukes seemed to have most of the answers, application of their theory to international transboundary waters very quickly led to yet more probing questions, which pointed us to yet more interesting theorists. Our streams of research split and intersected like those of a mighty river's delta.

In retrospect, that *something* about water which makes it so special has to do with its very material factors: its fluidity, its multiple uses (for drinking, for growing food, for generating electricity), its cycling qualities (through solid, liquid and gaseous phases), narratives developed about water, and conceptions of our need for it. Each of these was found to be relevant in different admixtures to particular transboundary water arrangements. At the broadest level, the institutional arrangements were seen to be shaped by numerous processes that are active within very political economies and which are driven by global forces that are largely ignorant of and, yet, shaped by the use and movement of water around the world (see Schofield, 2022, in this issue).

The importance of a political–economic perspective is underscored by the fact that *hegemonic power* was just one of four aspects of state power that Tony identified in that table of determinants of international relations (Table 1); the others were hydrological capacity, economic capacity and access to global support – each of which may be seen as a different kind of power.

The path became clearer. Alongside *power*, other factors that had to be brought into the analysis included the food system, domestic and international policy-making processes, the construction of sanctioned discourses and ideas on how water should be managed and shared, the extent to which international cooperation can re-enforce both the norms and the arrangements, and the manner by which these arrangements change.

Power and hydro-hegemony

In situating hegemonic power, hydrological capacity, economic capacity and access to global support within the work of Dahl (1957), Bacharach and Lawler (1981), Frey (1993), and Lukes (1974, 1991), Daoudy (2005, 2009) emphasizes the importance of geographical position within the basin as well strategies in formal and informal negotiations. She notes how *structural* and *bargaining power* are drawn upon and countered, in this case by the Syrian state in its interaction in the Tigris and Euphrates arrangements it shares with Turkey and Iraq. Palestinian and Israeli water negotiators also used bargaining power during their negotiations over transboundary aquifers in the early 2000s,

Table 1. Factors affecting control of transboundary flows.

Riparians	Hydrological capacity (approximate water self-sufficiency, %)	Economic capacity	Hegemonic power	Access to global support
Nile River basin				
Ethiopia and others	100%	Weak and very weak	Very weak	Very little
Sudan	100%	Very weak	Very weak	Very little
Egypt *(downstream)*	70%	Moderate	Moderate	Significant
Jordan River basin				
Syria	70%	Moderate	Weak	Little
Lebanon	100%	Moderate	Very weak	Little
Israel *(mid-stream)*	25%	Strong, diverse	Strong	Very significant
Palestine	20%	Very weak	Very weak	Very little
Jordan	25%	Weak	Weak	Little
Tigris and Euphrates basins				
Turkey *(upstream)*	100%	Strengthening	Strong	Significant
Syria	90%	Moderate	Moderate	Very little
Iraq	100%	Weak (temporarily)	Weak	None

Note: The state controlling the bulk of the flow is highlighted. On different rivers, those dominant states are located at different points: up-, mid- and downstream.
Source: Adapted from Allan (2001, p. 224).

specifically through *active stalling* during negotiations (Zeitoun & Warner, 2006), just as Afghan negotiators did in relation to Iran over the Harirud (Thomas & Warner, 2015). Because these tactics seem to favour the side with greater military and political clout, Warner (2004) reached for Lukes' *third dimension of power* to explain how some outcomes may appear more-or-less predetermined.

The group redubbed this form of power *ideational*, and noted how it keeps a firm grip on what people think or believe is achievable so that strategies for control of water may be developed and executed. Tactics matter less than they might otherwise in such circumstances. An example of ideational power comes from the Palestinian negotiators' inability to confront the Israeli proposal of water desalination as a solution to the water issues the two sides were negotiating, or to get their preferred alternative onto the formal negotiations table (Zeitoun, 2008).[1] A further example comes from the use of International Water Law, which can be used to set the agenda and counter structural power – as reflected during the deliberations of the International Law Commission where down- and upstream riparian countries debated new legal framings in the codification of the UN Convention on the Non-Navigational Uses of International Watercourses (Daoudy, 2008).

The application of power theory to transboundary water arrangements does not end in the third dimension. If a transboundary water arrangement is observed to be resistant to change, in roughly equal parts because of the threat of force (military or otherwise) and because those involved generally agree with it, one possible explanation is some form of hegemony at the basin scale (Warner, 1992, 2004); that is, Gramsci's version of hegemony where social control is achieved through force-plus-consent. Achieving this degree of control is, of course, facilitated by a state's riparian position (Frey, 1993); downstream states are in a less favourable position to shape the arrangement as they might prefer, but hegemonic power – to return to Tony's phrase – could trump the disadvantage, as the case of downstream Egypt on the Nile (Cascão, 2008, 2009).

The analytical Framework of Hydro-Hegemony attempts to capture these various aspects. It incorporates very material aspects of power, such as riparian position and the ability to implement water projects, alongside more abstract dimensions of power, such as the ability to set agendas or influence external mediators towards a favoured position (Zeitoun & Warner, 2006). The framework has been refined through worldwide application (Hanasz, 2014; Bowen-Williams, 2011; Hensengerth, 2015, Kehl 2011; Tavakoli-Nabavi, 2017) with analysts questioning the state-centricity (e.g., Furlong, 2008), the suggested immutability of power relations (Farnum, 2014), and underplaying of influence of non-hegemonic actors (Tawfik, 2015). The framework has also led to further thinking on ways to counter hydro-hegemony (Cascão, 2008, 2009), and innovations which stress the relatively greater importance of ideational power (e.g., see the Circle of Hydro-hegemony in Menga, 2016) and soft power (Zeitoun et al., 2011).

Hegemonic discourses and the upstream/downstream dynamic

A key feature of the consent of the weaker side to a transboundary water arrangement is *sanctioned discourse* in which what is said and thought about the issue is that which is created or allowed or aligns with the interests of the more powerful actor (see Tripp, 2022, in this issue). Alternatives are downplayed in these cases – as explored in the case of the Nile, below. Borrowing largely from Hajer (1997), Tony Allan consistently highlighted the importance of discourse in all aspects of water use and abuse. He was aware that discourse can become globally hegemonic without remaining attached to a specific actor – and would cite integrated water resources management (IWRM), the Nexus, and many others (Allan, 2001, 2003; Allan & Mirumachi, 2010; Trottier, 2003). Precisely to shift discourses that were not questioned enough, he encouraged us to develop pithy observations that convey concepts to a managerially minded water world; for instance, his well-known 'water pessimists are wrong but useful, water optimists are right but dangerous' continues to guide those who straddle academic research and policy advice. Another one-liner originated from Warner (2004) asserting, 'upstreamers use water to get more power, downstreamers use power to get more water'.

The influence of a state's position on transboundary rivers has engaged minds since at least the 1990s. According to Lowi (1993), hegemonic upstreamers can ignore the interests of downstreamers without peril, a sentiment echoed throughout the more realist international relations literature and in practice – for example, in a 2018 strategy workshop bringing together high-level downstreamers' representatives from six European nations.[2]

Downstream states dread the development of large-scale infrastructure upstream because it may be used as a lever to get or keep them in line with upstream preferences – or to coerce their consent. For their part, upstream states may claim that such infrastructure buffers downstreamers against droughts and floods but, even unintentionally, an upstream calamity such as the collapse of a hydropower or tailings dam can do a lot of harm downstream. Downstream states can mobilize various forms of bargaining power to influence the discourse established by upstream states; this might be an alliance with a strong external power as Egypt was able to do for a century until Ethiopia became too powerful for its claims to be ignored (Cascão et al., 2020). The upstream state might also be undermined by downstream support for insurgents, as when the government of Syria gave shelter to the Kurdish PKK to gain a stronger hand in water negotiations (Conde, 2016). And in the 1990s, the government of the Netherlands tacitly supported litigation by substate (the city of Rotterdam) and non-state players (environmental non-governmental organizations (NGOs) and rainwater and horticultural operators depending on clean water) to get upstream Germany and France to cut pollution in the Rhine (Dieperink, 2011; Warner, 2011).

However, upon closer inspection, basin relations are not so clear-cut, even geographically. States can be upstream on one river and downstream on another, with different claims and interests in each case; Mexico is both up- and downstream of the United States on the Colorado River. Transboundary water arrangements became even more complex on the Euphrates and Tigris when the Islamic State/Da'esh laid siege to important dams to assert the statehood they aspired to and to inculcate loyalty from those dependent on the water (Daoudy, 2020; Von Lossow, 2016).

Of course, transboundary water arrangements do not have to be antagonistic. *Hegemon* stems from the Greek *to lead, to guide* – so, beyond the Gramscian understanding of hegemony as social order, hegemony can also be considered as first-among-equals leadership buttressed by authority. Hegemons tend to ascribe special benevolent roles to themselves as regional leaders, and may accept the consequences by taking a disproportionate burden upon themselves and taking on board non-hegemonic interests (Prys, 2008).

Coexisting conflict and cooperation

Where hard power is supported by soft power, relations under a hegemonic aegis do not have to be conflictive. Water diplomacy is often and widely promoted; the accompanying discourse is all about *cooperation*. But what *is* cooperation?

It was not long before Tony's insights into the politics and strategic aspects of transboundary water arrangements clashed with the unquestioned, ubiquitous and never-properly defined claims of cooperation. Much of the prevailing hydro-political analysis was found to be blind to the more subtle power plays and hegemony that is active even within water-cooperation initiatives. The group uncovered examples not only from the Nile Basin Initiative (NBI) (Cascão et al., 2020) and the Palestinian–Israeli Joint Water Committee (Selby, 2007, 2013) but also along the Rio Grande (Kistin, 2006) and the Orange–Senqu and other waters in the Southern African Development Community (Daoudy & Kistin, 2008, Kistin, 2010).

Viewed through the lens of hydro-hegemony, some of the conflictual processes looked quite useful – inasmuch as they brought issues into an arena where they could be dealt with discursively; and some of the so-called cooperative processes looked more like bullying – such as the conflict over dams in the La Plata river reflecting struggles to counter Brazil's overt control over the basin (Zeitoun et al., 2020a) or 'cooperation on India's terms' along the Ganges (Thomas, 2017). By our thinking, if water diplomacy were to be effective, it would need to rest on analysis that identified and interpreted the interaction between states much more clearly. Still more incisive would be to acknowledge that every transboundary water arrangement was characterized by cooperation *and* conflict, existing side by side.

The foundational hydro-political literature had established the lack of conflict over transboundary waters, relative to cooperation (Wolf et al., 2003; Yoffe et al., 2003). Studies had begun to extend analysis beyond the degree of water scarcity as a determinant of cooperation (Tir & Ackerman, 2009) or the degree of compliance to agreements (Bernauer & Siegfried, 2008) – thereby qualifying any state's consent to a transboundary water arrangement. Cooperation was also gauged by examining institutional characteristics that strengthen cooperative outcomes (Dombrowski, 2007) or tensions between agreement principles that guide state actions (Conca et al., 2006). However, such insights overlooked much of the messy, complex process of state engagement over water. These limitations were addressed by the conceptual advance on coexisting conflict and cooperation over transboundary waters which developed into the analytical framework of the Transboundary Waters Interaction Nexus (TWINS) (Mirumachi, 2015; Mirumachi & Allan, 2007). By identifying and interpreting the degrees of conflict and cooperation that make up any and all transboundary water arrangements, the concept sheds light on how political and economic factors shape the inter-state relations over time.

Importantly, TWINS also shifts the analytical focus from water-related events (and their conflict/cooperation propensity) to transboundary water *interactions*. This fresh analytical approach proved particularly beneficial in re-examining basins that had been considered to be relatively stable and even *cooperative*, such as the Mekong. Various river basin committees, agreements and development projects had characterized this basin as having a special 'Mekong Spirit' that could even overcome geopolitical tensions, especially during the Cold War (Mirumachi, 2015). However, power-informed analysis showed that the spirit was part of a construct by the donor community keen to see political stability through poverty reduction. *Cooperation* under the Mekong Spirit meant that the more powerful basin states could engage in international political dialogue and even actively implement international river basin projects with donor funding, but only to the extent it advanced their national interests (Mirumachi, 2015). The Mekong River basin has changed since the early days of the post-Cold War water governance, most dramatically with a series of dams and increased hydropower production. Recent years have seen cases of *shallow cooperation* where, while protocols exist for inclusive decision-making, they are incapable of dealing with the geopolitical, economic and financial drivers of dam-building (Zeitoun et al., 2020a).

TWINS analysis has extended thinking on water diplomacy on two fronts in particular (Warner et al., 2017). First, the approach asserts that water diplomacy needs to move beyond simply addressing disputes towards transforming existing transboundary water arrangements (Zeitoun et al., 2020a). Transformative water diplomacy is a platform to consider alternatives to existing decision-making, which requires a sharper focus on the harm to individuals, communities and the environment which are born out of coexisting conflict and cooperation. Second, the approach highlights how counter-hegemonic strategies must address mechanisms of coercive cooperation. Compliance with hegemonic arrangements can be achieved through seemingly cooperative measures such as awarding political pay-offs, financial rewards or establishing legitimacy that would seem to benefit the non-hegemonic states. Identifying the ways such coercive cooperation works enables new thinking on challenging and breaking consent to hydro-hegemony (Zeitoun et al., 2016).

Virtual hydro-hegemony

Another benefit of bringing hegemony into the analysis of transboundary water arrangements is that the line of enquiry turns quickly to what is excluded. Amongst many other aspects of any arrangement (such as the role of non-state actors, the legacy of relations, etc.) we can count soil water and virtual water. Incorporating these factors into the analysis shines the spotlight on food trade and soil water balances – both are highly relevant to transboundary water diplomacy. For instance, within the ongoing negotiations over the Grand Ethiopian Renaissance Dam (GERD), Egyptian negotiators are calling for precipitation and soil water to be incorporated into the Nile's overall water balance (FAO & IHE Delft, 2020). This would transform the dynamic of the negotiations because rainfall throughout Egypt is a negligible fraction of the rain that falls across the catchment in Ethiopia.

The fact that soil water is normally overlooked begs the question: Why? In consistently drawing attention to the water taken up by the roots of plants, Tony Allan emphasized that soil water cannot be pumped, priced or denied to anyone. Furthermore, when food is imported from another country, the water used to produce it – and which can be counted as embedded in the produce (Allan, 2011) – relieves the pressure on the importer's own water resources. But trade in virtual water embedded in a crop irrigated from a non-renewable or overexploited aquifer (Elhadj, 2022, in this issue) is what Greco (2012) (and Greco, forthcoming in the next part of this double issue) refers to as *bad virtual water*. It creates what Hoekstra (2014) calls 'the virtual water scarcity paradox': one state's thirst is satiated while another's is increased. Apart from ethical questions such as 'What gives someone the right to eat someone else's water?' (Greco, 2014), it also opens the research field to very relevant econometrics (Hanasaki et al., 2010; Oki et al., 2017; Tamea et al., 2014). In the last 10 years, scholars have calculated global volumetric quantities and major traders of virtual water, notably in terms of how virtual water transfers are unlikely to redress inequalities (D'Odorico et al., 2010; Dalin et al., 2012; Seekell et al., 2011). Dalin et al. (2017) considered the volumes coming from highly vulnerable groundwater sources to produce a global map of 'bad virtual water'; Pakistan, the United States and Mexico appear amongst the top five exporters of vulnerable groundwater exporters in terms of virtual water *per caput* for 2010.

In linking food trade and international transboundary water arrangements, the very physical contiguity of states is challenged: Sojamo et al. (2012) posed the question: 'Who manages virtual water, and subsequently, who impacts water security in the global political economy?'. The theoretical detachment from the earlier state-centred hydro-hegemony scholarship is only apparent; water embedded in food and, thus, traded among states is still considered *transboundary* and follows patterns expressed through the classic hydro-hegemony (and coexisting conflict and cooperation) analysis. When global food trade is considered, no water basin can be considered to be *closed* in the sense that there is not enough water in the system (Swatuk et al., 2015). All basins are now *open* and connected to every other water basin on Earth through *virtual water rivers* (Greco, 2012).[3] At the same time, outgoing or ingoing patterns of virtual water can greatly influence local transboundary water interactions among river basins and co-riparian states towards conflict or towards cooperation.

Along the Nile, Zeitoun et al. (2010) have considered the relevance of the water embedded in food trade for regional water security, and Cascão (2012) noted the potential destabilizing and counter-hegemonic role of foreign direct investments in water and land in Ethiopia and Sudan, including export-oriented irrigation schemes. Virtual water exports have had an opposite effect between Jordan and Saudi Arabia: their interaction over the transboundary Disi Aquifer accelerated following the activation of export-led irrigation schemes in Jordan. Using these volumes of water as a prior-use argument, the Jordanian side was able to initiate informal talks leading to a bilateral agreement over the shared aquifer (Greco, 2020).

Discourses and counter-hegemony on the Nile

The Nile is a paradoxical case when it comes to the role of power relations. For millennia, of all the Nile's now 11 riparians, Egypt has been by far the main user of the river. From its downstream position and although it contributes nil to the Nile's flows, Egypt remains the hegemon in basin-wide arrangements. Some of this is due to structural power – particularly the fertile delta and Egypt's location at the trade gateway for the Rea Sea between Africa, the Middle East and the rest of the Mediterranean. Egypt has also proven very able to exploit its bargaining power, which is partly a result of its early construction of the Aswan Dam (in 1904) and the bilateral agreement it forged in its favour with Sudan in 1959. Egypt's establishment of a hegemonic position came after completion of the High Aswan Dam in 1971, which provided a massive over-year storage capacity and boosted the country's agricultural output.

This may be considered a downstream triumph in many ways. However, the river comes from the rainy upstream catchments of the Blue Nile and Tekezze-Atbara in Ethiopia, and the White Nile. These cross a lot of potentially fertile land in midstream Sudan, which is second only to Egypt in terms of irrigation potential and, therefore, could become another major consumer of Nile waters. Constrained by the terms of the 1959 Agreement, successive Sudanese governments have often aligned their interests with those of Egypt, but this position has drastically shifted in recent years as Sudan has realigned its interests closer to upstream countries.

Contestation of Egypt's position from upstream states has remained rhetorical and without any major impacts on the water flows themselves – particularly through the decades-long Nile Basin Initiative (NBI). However, the transboundary water arrangement was severely challenged in 2010 when upstream states allied to push through the Cooperative Framework Agreement (Cascão, 2014) and, in 2011, when Ethiopia moved ahead with the GERD on the Blue Nile. Whether the GERD proves to be a counter-hydro-hegemonic project remains to be seen. On the one hand, Ethiopia has increased its structural power since the 1990s; its economy remains small relative to Egypt but execution of the GERD displays a new level of economic capacity and, in the mid-1990s, Ethiopia embarked on a counter-hegemonic strategy to increase its Nile knowledge base and to develop alternatives to the discourse that had been sanctioned by the most powerful actors, investing in studies and the capacity of key institutions and advanced a narrative on how multilateral cooperation on the Nile would lead to the de facto development of infrastructure projects upstream. This meant collaboration with external actors for political cover and additional funds to build the GERD and other Ethiopian dams.

The alternative counter-hegemonic scenario in the Nile Basin could have been multilateral cooperation. The basin hegemon would not allow it during the early years of the NBI, but Egypt imploded politically during the revolution in early 2011 and the NBI was aborted during negotiations of the Cooperative Framework Agreement soon afterwards. Ethiopia continued its counter-hegemonic tactics unilaterally – it announced the GERD in April 2011, and following 10 years of dam construction, pre-negotiations, negotiations, draft agreements and even external mediation, the GERD is in its second filling and produced electricity for the first time in February 2022. At the time of writing, Ethiopia is wracked by civil war; no legal or institutional agreement over water is in place, but also no impacts on water flows downstream and, ultimately, trilateral relations in one of the most observed but apparently unstoppable spirals in the modern history of the region.

Pursuing the research

Tony Allan's incessant probing of boundaries, and encouragement of others to do the same, clarified both the limits of research methodology and the future research agenda. Apart from the previously discussed benefits of better data on global food trade, analysis of the hidden politics of international transboundary water arrangements would also improve with greater access to state decision-making structures. This is doubly so when the power plays occur deep in the *shadow state* (i.e., the informal structures pulling the strings of or bypassing the formal state structures). He was later to remark that the secrecy of interstate negotiations coupled with untraceable decisions made in the shadow state made it all 'unresearchable'. Perhaps he was unaware of advances made in critical and transformative research methodologies or power research on non-action and non-decisions (Crenson, 1971; McConnell & 't Hart, 2019). The point taken from the polemic assertion is to sharpen research methods and check assumptions (which of course all researchers should do as a matter of course). In contexts where the researcher has little insight or trusted and diverse network, their analysis may be superficial: good data on transboundary projects are easier to acquire where the researcher is an active participant in diplomacy or is trusted by people within various influential groups. In such cases, first-hand interviews with key informants can be quite reflective and engaged with theory, but protecting the reputation or security of the informants means holding back data and substantiating points somewhat less forcefully. However, given that a hydro-hegemonic /transformative analysis actively seeks out alternative voices and ideas which, by definition, are less frequently heard, it is unsurprising that many people are keen to share their perspective of the story – and often push for sharper and deeper analysis.

The same applies to just about any contentious topic. The lesson learned from two decades of research is that long-term engagement with these key actors is crucial. Understanding how and why conflict and cooperation dynamics change requires collection of detailed data and constant conversations with informants, so research in this mode often cannot fit standard expectations of research project timelines or the mould of classic outputs of journal publications. Nevertheless, this *slow research* yields significant insights that simply cannot be derived through rapid policy analysis or narrow report-writing: as our research has shown, such expedient analysis of water diplomacy has been deeply flawed and even ruinous (Zeitoun et al., 2020b).

Future research will have to bear this in mind, and innovative analysis below the state level has since advanced. The work that Tony Allan undertook and inspired was focused on the political economy, which meant that while watersheds were disregarded, the policymaking problemsheds were either global or national. However, as subsequent research has shown, attentive detail to the harms on individuals, community and the environment brings out more details on the real, everyday impacts of power asymmetry (Zeitoun et al., 2020b). Tony Allan's later work focused on the role of relatively poor farmers who are subject to the raw side of gross power asymmetry in their relations with agribusiness. He noted how these farmers are left with few options and limited capacity to be good stewards of the land and water (Allan, 2019; Allan & Dent, 2021). This focus on the farmer can be applied to other stakeholders such as indigenous communities and riverine fisherfolk, for example, to demonstrate how their livelihoods are tied up with and subject to political and economic drivers spawned by decisions in distant office towers and central governments, often of the Global North.

A further opportunity to extend the analysis lies with the fact that much of Tony Allan's work remained silent on issues of ownership and access of such individuals. For example, the recent analysis of the impacts of large-scale agriculture and industry on the human right to drinking water is effective in bringing together virtual water flows and underpinning power relations that are blind to the local ramifications of water use (Mirumachi et al., 2021). Tony Allan's contribution to the analysis of transboundary waters will endure through all these and much more creative research paths.

Notes

1. The influence of ideational power is the only way to explain how the Palestinian negotiators consented to continue the so-called *cooperation* through the Israeli-Palestinian Joint Water Committee, which was strongly resisted by much of the Palestinian population and the negotiations' advisors (Selby, 2003a, 2003b), and which led to their own colonization (Selby, 2013).
2. See *From the Delta looking up.*
3. And quite distinct from *atmospheric rivers* of which we are now becoming aware (Waliser & Guan, 2017).

Disclosure statement

No potential conflict of interest was reported by the authors.

ORCID

Mark Zeitoun http://orcid.org/0000-0003-1791-0628
Ana Elisa Cascão http://orcid.org/0000-0002-5771-9951
Marwa Daoudy http://orcid.org/0000-0002-2223-847X
Francesca Greco http://orcid.org/0000-0002-1696-3472
Naho Mirumachi http://orcid.org/0000-0002-4923-2553
Jeroen Warner http://orcid.org/0000-0003-2847-8770

References

Allan, J. A. (2001). *The Middle East Water Question: Hydropolitics and the global economy*. I.B. Tauris.

Allan, J. A. (2003). IWRM/IWRAM: Integrated water resources allocation and management: A new sanctioned discourse? In P. P. Mollinga, A. Dixit, & K. Sthukorala (Eds.), *Integrated water resource management in South Asia: Global theory, emerging practice and local needs* (pp. 38–63). SaciWATERs.

Allan, J. A. (2019). Food, water and the consequences of society not valuing the environment. In J. A. Allan, B. Bromwich, M. Keulertz, & A. Colman (Eds.), *The Oxford handbook of food, water and society*. Oxford University Press.

Allan, J. A., & Dent, D. L. (2021). The cost of food: Consequences of not valuing soil and water and the people who manage them. In D. L. Dent & B. P. Boincean (Eds.), *Regenerative agriculture. What's missing? What else do we need to know?* (pp. 3–20). Springer Nature Switzerland.

Allan, J. A. T., & Mirumachi, N. (2010). Why negotiate? Asymmetric endowments, asymmetric power and the invisible nexus of water, trade and power that brings apparent water security. In A. Earle, A. Jägerskog, & J. Öjendal (Eds.), *Transboundary water management: Policy and practice* (pp. 13–26). Earthscan.

Allan, J. A. (2011). *Virtual water: Tackling the threat to our planet's most precious resource*. I.B. Tauris.

Bacharach, S. B., & Lawler, E. J. (1981). *Bargaining: Power, tactics, and outcomes*. Jossey-Bass.

Bernauer, T., & Siegfried, T. (2008). Compliance and performance in international water agreements: The case of the Naryn/Syr Darya Basin. *Global Governance, 14*(4), 479–501. https://doi.org/10.1163/19426720-01404006

Bowen-Williams, K. J. (2011). Tensions over hydroelectric developments in Central Asia: Regional interdependence and energy security. *Asia Pacific: Perspectives, 10*(2), 133–160.

Cascão, A. E. (2008). Ethiopia–Challenges to Egyptian hegemony in the Nile Basin. *Water Policy, 10*(S2), 13–28. https://doi.org/10.2166/wp.2008.206

Cascão, A. E. (2009). *Political economy of water resources management and allocation in the Eastern Nile River Basin* [PhD]. King's College London, UK.

Cascão, A. E. (2012) Nile water governance. *The Nile River Basin*. Routledge, 251–274.

Cascão, A. E. (2014). *Countering hydro-hegemony in the Nile Basin presentation given at seventh international workshop on hydro-hegemony*. London UEA Water Security Research Centre/ London Water Research Group. 12/13 May 2014.

Cascão, A. E., Feakade, W., Grossmann, M., Seid, A. (2020). Water diplomacy and conflict transformation in the *Nile* River Basin – the key role of the *Nile* Basin Initiative (NBI) over the past 20 years. In A. Kittikhoun & S. Schmeier (Eds.), *Water diplomacy and conflict management: The role of River Basin Organizations* (pp. 141–162). Routledge.

Conca, K., Wu, F., & Mei, C. (2006). Global regime formation or complex institution building? The principled content of international river agreements. *International Studies Quarterly, 50*(2), 263–285. https://doi.org/10.1111/j.1468-2478.2006.00402.x

Conde, G. (2016). Water and counter-hegemony: Kurdish struggle in the Tigris and Euphrates in Turkey. *Revista de Paz y Conflictos, 9*(2), 43–58.

Crenson, M. A. (1971). *The un-politics of air pollution: A study of non-decisionmaking in the cities*. The John Hopkins Press.

Dahl, R. (1957). The concept of power. *Behavioural Science, 2*(3), 201–215. https://doi.org/10.1002/bs.3830020303

Dalin, C., Konar, M., Hanasaki, N., Rinaldo, A., & Rodriguez-Iturbe, I. (2012). Evolution of the global virtual water trade network. *Proceedings of the National Academy of Sciences, 109*(16), 5989–5994. https://doi.org/10.1073/pnas.1203176109

Dalin, C., Wada, Y., Kastner, T., & Puma, M. J. (2017). Groundwater depletion embedded in international food trade. *Nature, 543*(7647), 700–704. https://doi.org/10.1038/nature21403

Daoudy, M. (2005). *The water divide between Syria, Turkey and Iraq: Negotiation, security and power asymmetry*. CNRS Editions (French).

Daoudy, M. (2008). Hydro-hegemony and international water law: Laying claims to water rights. *Water Policy, 10*(S2), 89–102. https://doi.org/10.2166/wp.2008.204

Daoudy, M. (2009). Asymmetric power: Negotiating water in the Euphrates and Tigris. *International Negotiation, 14*(2), 361–391. https://doi.org/10.1163/157180609X432860

Daoudy, M. (2020). Water weaponization in the Syrian conflict: Strategies of domination and cooperation. *International Affairs, 96*(5), 1347–1366. https://doi.org/10.1093/ia/iiaa131

Daoudy, M., & Kistin, E. (2008, March 26–29). *Beyond water conflict: Evaluating the effects of international water cooperation* Paper prepared for the 49th annual Conference of the International Studies Association, San Francisco.

Dieperink, C. (2011). International water negotiations under asymmetry: Lessons from the Rhine chlorides dispute settlement (1931–2004). *International Environmental Agreements, 11*(2), 139–157. https://doi.org/10.1007/s10784-010-9129-3

D'Odorico, P., Laio, F., & Ridolfi, L. (2010). Does globalization of water reduce societal resilience to drought? *Geophysical Research Letters, 37*(13), L13403. https://doi.org/10.1029/2010GL043167

Dombrowski, I. (2007). *Conflict, cooperation and institutions in international water management: An economic analysis*. Edward Elgar.

Elhadj, E. (2022). Ozymandias in the desert: irrigation in Saudi Arabia. *Water International, 47*(6), 969-978. https://doi.org/10.1080/02508060.2022.2123612

FAO & IHE Delft. (2020) *Water accounting in the Nile River Basin*. Remote Sensing for Water Productivity: Water Accounting Series

Farnum, R. (2014) *A critique of the London Water Research Group considering academic hegemony and traps in social justice research* (Discussion paper prepared for the Seventh International Workshop on Hydro-Hegemony). University of East Anglia

Frey, F. W. (1993). The political context of conflict and cooperation over international river basins. *Water International, 18*(1), 54–68. https://doi.org/10.1080/02508069308686151

Furlong, K. (2008). Hidden theories, troubled waters: Response to critics. *Political Geography, 27* (7), 811–814. https://doi.org/10.1016/j.polgeo.2008.08.005

Gramsci, A. (20031935). *Selections from the prison notebooks*. Lawrence and Wishart.

Greco, F. (2012, November 14–16). Introducing a historical perspective to virtual water flows: How hydro-metabolism explains the 'bad' virtual water 'flows' from non renewable/overexploited groundwater. EGU Leonardo Topical Conference Series (pp. 53–54), Turin. EGU

Greco, F. (2014, May 12/13). *Invisible 'virtual' water rivers: Blinding transboundary water interactions and consumers to the need for counter-hegemony*. Presentation to Seventh International Workshop on Hydro-Hegemony, London UEA Water Security Research Centre/London Water Research Group.

Greco, F. (2020). Coexisting cooperation and conflict on the Disi aquifer: The importance of considering virtual water. In M. Zeitoun, N. Mirumachi, & J. Warner (Eds.), *Water conflicts: Analysis for transformation*. Oxford University Press.

Greco, F. (2022, forthcoming). Tony Allan: a magic toolbox of theoretical frameworks, a never-ending story. *Water International*.

Hajer, M. A. (1997). *The politics of environmental discourse: Ecological modernization and the policy process*. Oxford University Press, Oxford Scholarship Online.

Hanasaki, N., Inuzuka, T., Kanae, S., & Oki, T. (2010). An estimation of global virtual water flow and sources of water withdrawal for major crops and livestock products using a global hydrological model. *Journal of Hydrology, 384*(3–4), 232–244. https://doi.org/10.1016/j.jhydrol.2009.09.028

Hanasz, P. (2014). Power flows: Hydro-hegemony and water conflicts in South Asia. *Security Challenges, 10*(3), 95–112. https://www.jstor.org/stable/26465447

Hensengerth, O. (2015). Where is the power? Transnational networks, authority and the dispute over the Xayaburi Dam on the lower Mekong mainstream. *Water International, 40*(5–6), 911–928. https://doi.org/10.1080/02508060.2015.1088334

Hoekstra, A. (2014). *The water footprint of modern consumer society*. Routledge.

Kehl, J. R. (2011). Hydropolitical complexes and asymmetrical power: Conflict, cooperation, and governance of international river systems. *Journal of World-Systems Research, XVII*(1), 218–235. https://doi.org/10.5195/jwsr.2011.429

Kistin, E. (2006) Qualifying cooperation over transboundary waters *Paper for Challenging the consensus: Water governance for Africa.* Bradford Centre for International Development, 1 December 2006.

Kistin, E. J. (2010) *Critiquing cooperation: The dynamic effects of transboundary water regimes* [PhD]. University of Oxford

Lowi, M. (1993). *Water and power – the politics of a scarce resource in the Jordan River Basin.* Cambridge University Press.

Lukes, S. (1974). *Power: A radical view* (2nd ed.). Macmillan. Basingstoke: Palgrave Macmillan.

Lukes, S. (1991). *Moral conflict and politics.* Clarendon Press.

Lukes, S. (2005). Power and the battle for hearts and minds. *Millennium: Journal of International Studies, 33*(3), 477–493. https://doi.org/10.1177/03058298050330031201

McConnell, A., & 't Hart, P. (2019). Inaction and public policy: Understanding why policymakers 'do nothing'. *Policy Sciences, 52*(4), 645–661. https://doi.org/10.1007/s11077-019-09362-2

Menga, F. (2016). Reconceptualizing hegemony: The circle of hydro-hegemony. *Water Policy, 18* (2), 401–418. https://doi.org/10.2166/wp.2015.063

Mirumachi, N. (2015). *Transboundary water politics in the developing world.* Routledge.

Mirumachi, N., & Allan, J. A. (2007, November 12–15). *Revisiting transboundary water governance: Power, conflict cooperation and the political economy* Proceedings from CAIWA International Conference on Adaptive and Integrated Water Management: Coping with Scarcity, Basel, Switzerland. CAIWA.

Mirumachi, N., Duda, A., Gregulsk, J., & Smetek, J. (2021). *Human right to drinking water: Impacts of large-scale agriculture and industry.* Directorate General for External Policies of the Union, European Parliament.

Nicol, A., Debevec, L., & Okene Ayaru, S. (2022). Water and complex problemsheds in Karamoja, Uganda. *Water International, 47*(6), 952-968. https://doi.org/10.1080/02508060.2022.2119525

Oki, T., Yano, S., & Hanasaki, N. (2017). Economic aspects of virtual water trade. *Environmental Research Letters, 12*(4), 4. https://doi.org/10.1088/1748-9326/aa625f

Prys, M. (2008). *Regions, hegemony and power: India and South Africa compared* [Unpublished D. Phil thesis]. University of Oxford, UK.

Schofield, R. (2022). Locating the channel and other tales from the river bank: constants and change in river boundary delimitation. *Water International, 47*(6), 979-1000. https://doi.org/10.1080/02508060.2022.2123613

Seekell, D. A., D'Odorico, P., & Pace, M. L. (2011). Virtual water transfers unlikely to redress inequality in global water use. *Environmental Research Letters, 6*(2), 024017. https://doi.org/10.1088/1748-9326/1086/1082/024017

Selby, J. (2003a). Dressing up domination as 'co-operation': The case of Israeli-Palestinian water relations. *Review of International Studies, 29*(1), 121–138. https://doi.org/10.1017/S026021050300007X

Selby, J. (2003b). *Water, power and politics in the Middle East – the other Israeli-Palestinian conflict.* I.B. Tauris.

Selby, J. (2007, May 12/13). *Beyond hydro-hegemony: Transnational hegemonic structures and national hegemonic projects* Presentation at the Third International Workshop on Hydro-Hegemony, London, London School of Economics/London Water Research Group

Selby, J. (2013). Cooperation, domination and colonisation: The Israeli-Palestinian Joint Water Committee. *Water Alternatives, 6*(1), 1–24. https://www.water-alternatives.org/index.php/volume6/v6issue1/196-a6-1-1/file

Sojamo, S., Keulertz, M., Warner, J., & Allan, J. A. (2012). Virtual water hegemony: The role of agribusiness in global water governance. *Water International, 37*(2), 169–182. https://doi.org/10.1080/02508060.2012.662734

Swatuk, L., McMorris, M., Leung, C., & Zu, Y. (2015). Seeing 'invisible water': Challenging conceptions of water for agriculture, food and human security. *Canadian Journal of Development Studies*, *36*(1), 24–37. https://doi.org/10.1080/02255189.2015.1011609

Tamea, S., Carr, J.-A., Laio, F., & Ridolfi, L. (2014). Drivers of the virtual water trade. *Water Resources Research*, *50*(1), 17–28. https://doi.org/10.1002/2013WR014707

Tavakoli-Nabavi, S. E. (2017). *More than water, more than human: A transdisciplinary sociology of water conflict in Central Iran* [Doctor of Philosophy]. Australian National University.

Tawfik, R. (2015). *Revisiting hydro-hegemony from a benefit-sharing perspective: The case of the Grand Ethiopian Renaissance Dam*. German Development Institute DIE.

Thomas, K. A. (2017). The Ganges water treaty: 20 years of cooperation, on India's terms. *Water Policy*, *19*(4), 724–740. https://doi.org/10.2166/wp.2017.109

Thomas, V., & Warner, J. (2015). Hydropolitics in the Harirud/Tejen River Basin: Afghanistan as hydro-hegemon? *Water International*, *40*(4), 593–613. https://doi.org/10.1080/02508060.2015.1059164

Tir, J., & Ackerman, J. (2009). Politics of formalized river cooperation. *Journal of Peace Research*, *46*(5), 623–640. https://doi.org/10.1177/0022343309336800

Tripp, C. (2022) Sanctioned discourse and the power of hegemonic imaginings. *Water International*, *47*(6), 896-900. https://doi.org/10.1080/02508060.2022.2124018

Trottier, J. (2003). *Water wars: The rise of a hegemonic concept – exploring the making of the water war and water peace belief within the Israeli-Palestinian Conflict*. Paper prepared for the green cross international project from potential conflict to co-operation potential, Geneva, Switzerland. Green Cross International.

von Lossow, T. (2016). The rebirth of water as a weapon: IS in Syria and Iraq. *The International Spectator*, *51*(3), 82–99. https://doi.org/10.1080/03932729.2016.1213063

Waliser, D., & Guan, B. (2017). Extreme winds and precipitation during landfall of atmospheric rivers. *Nature Geoscience*, *10*(3), 179–183. https://doi.org/10.1038/ngeo2894

Warner, J. (1992). *The politics of diversion – bridging troubled water in the Middle East* [Master's thesis]. University of Amsterdam

Warner, J. (2004) Mind the GAP – working with Buzan: The Illisu Dam as a security issue (Occasional Paper 67). SOAS Water Issues Study Group, School of Oriental and African Studies/King's College.

Warner, J. (2011). *Flood planning: The politics of water security*. I.B. Tauris.

Warner, J., Mirumachi, N., Farnum, R. L. Grandi, M., Menga, F., & Zeitoun, M. (2017). Transboundary 'hydro-hegemony': 10 years later. *WIREs Water*, e01242. https://doi.org/10.1002/wat2.1242.

Wolf, A. T., Yoffe, S., & Giordano, M. (2003). International waters: Identifying basins at risk. *Water Policy*, *5*(1), 29–60. https://doi.org/10.2166/wp.2003.0002

Yoffe, S., Wolf, A. T., & Giordano, M. (2003). Conflict and cooperation over international freshwater resources: indicator of basins at risk. *Journal of the American Water Resources Association*, *39*(5), 1109–1126. https://doi.org/10.1111/j.1752-1688.2003.tb03696.x

Zeitoun, M. (2008). *Power and water: The hidden politics of the Palestinian-Israeli conflict*. I.B. Tauris.

Zeitoun, M., Allan, J. A. and Mohieldeen, Y. (2010). Virtual water 'flows' of the Nile Basin, 1998–2004: A first approximation and implications for water security. *Global Environmental Change*, *20*(2), 229–242. https://doi.org/10.1016/j.gloenvcha.2009.11.003

Zeitoun, M., Cascão, A., Warner, J., Mirumachi N., Matthews N., Menga F., & Farnum R. (2016, May 11). Transboundary water interaction III: Contesting hegemonic arrangements. *International Environmental Agreements*, *17*, 271–294. https://doi.org/10.1007/s10784-016-9325-x

Zeitoun, M., Mirumachi, N., & Warner, J. (2011). Transboundary water interaction II: Soft power underlying conflict and cooperation. *International Environmental Agreements*, *11*(2), 159–178. https://doi.org/10.1007/s10784-010-9134-6

Zeitoun, M., Mirumachi, N., & Warner, J. (2020a). *Water conflicts: Analysis for transformation*. Oxford University Press.

Zeitoun, M., Mirumachi, N., Warner, J., Kirkegaard, M., & Cascão, A. (2020b). Analysis for water conflict transformation. *Water International*, *45*(4), 365–384. https://doi.org/10.1080/02508060.2019.1607479

Zeitoun, M., & Warner, J. (2006). Hydro-hegemony: A framework for analysis of transboundary water conflicts. *Water Policy*, *8*(5), 435–460. https://doi.org/10.2166/wp.2006.054

Part II

Beyond water: private enterprise, food security, and systemic change

I remember Tony

Carl Hausmann

We first met in Salzburg, Austria, at a conference by the International Policy Council for Food and Agriculture (IPC) in the mid-1990s. I was from the private sector and chief executive officer (CEO) of Cereol, a Paris-based, agriculture commodity trading and processing company. Tony gave a presentation about water and I was first impressed by his gentle courtesy and rather quaint appearance. However, very quickly I said to myself, 'he knows something that I must learn more about'.

'Virtual water' was a concept that I had not heard before, but anyone who works in agriculture knows that 'rain makes grain'. Rain (or lack thereof) in one growing area of the world impacts prices in all growing areas because crop yields and global grain supplies are impacted directly by rain. However, virtual water is a concept that takes some thinking to fully understand. When one sells a ton of wheat one is also trading in all the inputs including water that went into producing that wheat. There are thus many consequences of trading grain crops around the world. Once in China a government official said to a colleague that our price of soybeans seemed high, but China really appreciated all the water servicing that it took to grow the soybeans which China did not have available. Tony's presentation in Salzburg highlighted that Egypt had enough land, labour and technology to produce wheat for its population, but there was not enough rain or water servicing available to irrigate the crops, so Egypt was essentially importing their water servicing needs in the finished wheat product.

As today's food production for the world's growing population increases many agricultural areas are finding that local rainfall is insufficient, or water tables are dropping or river waters are being over-exploited. Indeed, the global grain trade intermediates the trade in water servicing requirements for food production around the world as much as it does in the grain itself. And I needed to better understand all aspects of that trade, not just grain prices. Where should global grain crops be produced and what trade routes are the most sustainable?

Over the next 25 years Tony and I met several times and I continued to deepen my understanding and respect for his perspective on the competing needs for limited water servicing that both nature and political systems are dealing with. In the long run nature will always win out, but in the short run political decisions about crop subsidies and how available water resources can be used will impact both water availability and the agricultural crops that are produced. Agricultural commodity trading houses need to understand this interplay and hopefully influence the outcomes for the benefit of all communities that they serve. I was honoured when Tony asked me to contribute a chapter to his *Handbook of Food Water and Society* (Oxford University Press, 2019). He felt that politicians and businesses need to be involved together with the academics so that society best allocates what nature offers for mankind – if we are smart we can all benefit.

I will miss Tony, both for his erudition and his gentleness. The rough world of global grain traders teaches many lessons through the results of your business, but I learned just as much from Tony and I will miss him more than my fierce competitors. How the world uses our limited water resources still needs study and negotiation amongst the different segments of society – for the benefit not just of grain traders or rich consumers but also the local populations whose very existence depends on the smart usage of these water resources.

Disclosure statement

No potential conflict of interest was reported by the author.

The role of the private sector in sustainable development

Rabi H. Mohtar

To Dr Tony Allan: scholar, colleague and friend. Tony and I served on the Water Security Council of the World Economic Forum (2008–10) – he was my mentor in engaging the private sector. He influenced my own thinking and that of so many others.

The private sector supplies most of our basic needs in the water, energy and food chains. These supply chains are our lifelines, but they are also responsible for most of the air, land, and sea emissions and pollution. Achieving a new business model for the sustainable development of these primary resources cannot be achieved unless concerned scientists and politicians engage and work with the private sector. Doing so requires a better understanding of the challenges and constraints on commodity flow along these supply chains. This is how we might do it.

Issues for discussion

The global challenge of sustainable development and the gaps in the water, energy and food chains that constrain access to primary resources

These have long been the focus of serious study, most recently by Aldaco-Manner et al. (2019), Carmona-Moreno et al. (2021), Daher et al. (2019), Dare and Mohtar (2018), Gardezi et al. (2020), International Renewable Energy Agency (IRENA) (2015), Keulertz and Mohtar (2019), Mohtar (2016), Mohtar et al. (2015), Pahl-Wostl et al. (2021), Portney et al. (2017), Purwanto et al. (2021), Walker (2020) and Weitz et al. (2017). The gaps are exacerbated by the interlinkages along and between these systems, and by external stresses including demographic change, climate change, poor governance, and lack of coherence among policymakers.

Shortfalls in primary resources have uneven, but always negative implications for availability and access to these necessities, particularly for the most vulnerable. To achieve sustainable development and universal access requires a good understanding of the entire supply chain from production, processing and transport to distribution. Identifying the challenges and enablers of these processes is the first step to improving their reliability and resilience.

The role of climate change in shortfalls of primary resources and its impact on water, energy and food production and use

Climate change has and will continue to constrain food production through extremes of storm, flood and drought. Global heating and seasonal shifts have also impacted the viability of crops, times of planting and harvest, and irrigation requirements. Temperature extremes have increased energy requirements for heating and cooling. Disruptions of supply chains by climate change are reported by Baskin (2020), Daher et al. (2017), Goodarzi et al. (2020), Harris and Wei (2018), and Huang et al. (2009), and International Water Resources Association (IWRA) International Scientific Committee (2021), Mohtar (2011, 2016, 2017), Mohtar and Daher (2015), Slay and Dooley (2020) and UN-ESCWA (2016).

Challenges to achieving the Sustainable Development Goals (SDGs) in the face of scarcity of human resources, capital and technology, and competing claims

In 2015, world leaders agreed on 17 SDGs that straddle sustainability, interconnectedness and global concerns. Each country has its own plan to achieve these goals, but it remains a daunting task to deal with a system that is complex, highly interdependent and involves every sector of society. Different countries are at different stages and levels of readiness and it is already clear that several of the goals will not be achieved by the target date of 2030. Scarcity of human and financial capital and the lack of coherent policies pose big challenges, and countries face painful trade-offs. Nevertheless, the SDGs have given momentum to the sustainability agenda and serve as a platform from which to achieve societal security. Apart from technological advances, policies are needed to incentivize good practice and change consumers' behaviour. In short, what is needed is a new business model that involves all sectors of society.

The Global Compact: the promise, what has been done and its further potential

In 2000, the United Nations Global Compact (UNGC, 2017) was formed to promote responsible business practices and UN values among the global business community and the UN System. UNGC has been active on projects that support national agendas in many countries, but it has been criticized for not doing enough, indeed for providing a *blue wash* for participating companies. The question remains as to how accountable these companies are, and whether the UNGC mandate can be expanded to include measurable outcomes and impacts beyond project-based investments.

The private sector has a key role in sustainability. Indeed, it is the dominant player in water, energy and food. A governance structure could facilitate public engagement but the nature of the structure needs further discussion and dialogue between the private and public sectors.

This commentary proposes that *attaining sustainable development is a process that requires collective effort with the private sector and the active engagement of the private sector*. Its objectives are to explore the roles that the private sector can play in sustainable development, with a focus on water, energy and food. It presents the key players that produce and manage these resources, the current and possible future role of the private sector, and a road map for further engagement.

Who produces and manages the primary resources?

The water supply chain

Who owns the shares?

In the United States, 88% of the population is served by a public community water system and the remaining 12% by private systems. Overall, in Organisation for Economic Co-operation and Development (OECD) countries, 200–300 million people (17–25% of the population) are served by the private sector (Perard, 2017). Worldwide and in major cities, 85% of water services are publicly owned and run (GWI, 2020). The jury is still out on which system is better in terms of efficiency and cost to consumers. Kopaskie (2016) discusses ways in which the private sector can assist water security in high- and middle-income economies, and presents various governance models for private-sector involvement and investment.

While there is growing interest in increased private-sector involvement and investment, without public regulation that holds the private sector responsible for public services such as water there is a risk of reduced access to these services (Lloyd-Owen, 2022, in this issue). Clearly, engaging the private sector in research, development, technology and innovation requires close cooperation and coordination among all stakeholders (Debaere & Kapral, 2021).

Food supply

Who owns the shares?

In the mid-20th century, crop yields were boosted by mechanization, crop breeding, fertilizers, pesticides and irrigation. Today, we are seeing a new agricultural revolution based on precision and digital agriculture that has the potential to further transform the global food system (Nature Food, 2020). However, over the last 20 years, public funding for agricultural research has declined: US annual public agricultural research spending dropped from US$6 billion to US$4.5 billion (Clancy et al., 2016; USDA-ERS, 2019), cutting core agricultural research including soil science and crop breeding, and cutting staff by 21% (Coe et al., 2020). Global private-sector spending on agricultural research increased from US$5.1 billion to US$15.6 billion per year between 1990 and 2014 (Fuglie, 2016), and private-sector agricultural research spending in the United States increased by over 60% between 2008 and 2013 (King et al., 2012). The food system is now controlled by a small pool of companies: a handful of corporations control 90% of the grain traded globally; six companies provide most the seeds, fertilizers and pesticides; Wal-Mart, which feeds one in five American shoppers, is the 10th largest economic entity in the world; big industrial farms comprise only 1% of farmers, but control 65% of farmland.

Although there is increased recognition of the importance of private-sector development for food and nutrition security, illustrated in the UN 2030 Agenda for Sustainable Development (UNSDG, 2015) and the Malabo Declaration on Accelerated Agricultural Growth and Transformation (African Union, 2014), the 2017 UN Global Compact Progress Report found that one-third of its signatories had yet to develop any measurable SDG targets; and 60% of the companies assessed by

PricewaterhouseCoopers were not engaging meaningfully (PWC, 2019). In its statement to the World Summit on Food Security, the European Centre for Development Policy Management (ECDPM, 2013) noted:

- Smallholders are important to food supply and security.
- There is a huge need to improve current agricultural infrastructure and institutions.
- Food supply chains are very inefficient.
- The growing water scarcity and competition for water challenges future expansion of food production and puts at risk the supply for present world population.
- Attention must also be paid to biodiversity, soil depletion and land availability and use.

Energy supply chain

Who owns the shares?

For oil and gas, the 13 biggest energy companies, as measured by their oil reserves, are government owned and operated. Saudi Aramco, Gazprom (Russia), China National Petroleum Corp., National Iranian Oil Co., Petróleos de Venezuela, Petrobras (Brazil) and Petronas (Malaysia) are each bigger than ExxonMobil, the largest of the multinationals. State-owned companies now control more than 75% of all crude oil production (Bremmer, 2010), but the private sector will play a crucial role in the transition to renewable energy.

The US special presidential envoy for climate, John Kerry, said during the Clean Power Conference organized by the American Clean Power Association: ‘Probably something like $20 trillion will be allocated to the market over the next years’ (Anderson, 2021). After stagnating in 2020, global power-sector investment is set to increase by around 5% to more than US$820 billion in 2021. Renewables dominate investment in new power generation and are expected to account for 70% of the US$530 billion spent on new generation capacity in 2021, with the remainder to be spent on grids and storage. Thanks to rapid technological improvements, a dollar spent on wind and solar photovoltaic deployment today results in four times more electricity than a dollar spent on the same technologies 10 years ago (IEA, 2021a).

Renewable investment has thrived in markets with well-established supply chains, and where lower costs are accompanied by regulatory frameworks that provide cash flow visibility and lenders and financiers understand these sectors well and seek to support sustainable projects. Corporate-sector demand for clean electricity to meet sustainability targets has also played a role. Much of the spending in 2020 was concentrated in a handful of markets, most notably the People’s Republic of China, the United States and Europe; for the sixth consecutive year, 2020 saw capital spending in the power sector higher than for oil and gas supply (IEA, 2021b).

As for power production, the US Energy Information Administration (EIA) reports that investor-owned utilities serve three of every four utility customers nationwide. That is, for 70% of the country, electricity is delivered by private, for-profit, investor-owned companies (EIA, 2019); the remaining 30% are served by utilities owned by municipalities. Worldwide however, publicly owned distribution utilities outnumber private ones (Alkhuzam et al., 2013).

Understanding the business model

CSR is not enough

Private-sector companies are sustained by their profits so their activities must be self-supporting. Many of these companies report to a board of investors who expect growth in their portfolio, so executive goals and strategic plans prioritize financial margins; their environmental, ecological and other footprints rarely explicitly enter their business models. Water, energy and food providers operate on very small profit margins: their survival relies on economies of scale, precision and perfection of the supply chain operation. The *green investment* of some new-trend companies is not mainstream to these sectors.

Corporate social responsibility (CSR) is practiced by many big and multinational corporations and involves investment in social welfare projects. In most cases, CSR operations are not linked to the company's business operation but are viewed as a public relations campaign by the company. Irrespective of where one stands on these issues, CSR is clearly not enough to transform company operations: real change is not going to happen unless social responsibility is linked to the company's core business model.

The role of investors

Meaningful change will happen when the company business model changes. A more sustainable model requires a change in mindset, not only of company executives but also in the strategic targets set by the investors (Allan & Dent, 2021). These targets may clash with the board's financial targets, in which case long-term gains in resources efficiencies, human and ecological health must be valued rather than short-term profits. This is where analysis of these trade-offs should be communicated and discussed with the board to broaden their perspectives and change the business-as-usual operational model.

The future and the role of the private sector

It is evident that a large portion of our vital resources of water, food and energy is provided by the private sector. It is also evident that long-term sustainability of water, food and energy supply requires engagement, support and leadership from the private sector. To better engage with the private sector, society needs:

- To better understand the business model(s) of the private sector, their operation, clients, business goals and limitations.
- To co-identify points of entry for a new business model driven by common, social and ecological values. This includes identifying the roles of the private sector in transitioning to a sustainable business model.
- To engage investors in transformation of the business model and quantitatively demonstrate the long-term sustainability gains over short-term profits. This requires targeted trade-off analysis and analytics that recognize that one size does not fit all. Good practices suited to one supply chain do not necessarily work for others.

- To co-design a sustainability campaign that engages all stakeholders – including consumers – regarding sustainable production, consumption and pricing of key commodities. There is a price to be paid for a more sustainable commodity and the consumers must be willing to pay it.
- To ensure equitable access to commodities and services: the business model must not deny access to these basic commodities.
- To develop new policies to facilitate the transition to a new, transformational model (incentives or taxes on unsustainable practices).
- To build capacity through formal and informal workforce training, considering the supply chains as systems that can be improved and offering innovative solutions.
- To target consumption and behavioural challenges to the exploitation and conservation of primary resources.

Conclusions

The private sector is a key player in the management of water, energy and food. Its role in sustainable development cannot be ignored. Engaging the private sector requires understanding its operations and limitations. Co-design of transformational business models is needed and must involve all stakeholders, including government, private-sector decision-makers and consumers.

Disclosure statement

No potential conflict of interest was reported by the author.

References

African Union. (2014). Malabo declaration on accelerated agricultural growth and transformation for shared prosperity and improved livelihoods. Doc. Assembly/Au/2 (Xxiii). Retrieved January 23, 2022, from https://au.int/sites/default/files/documents/31247-doc-malabo_declaration_2014_11_26.pdf

Aldaco-Manner, L., Mohtar, R. H., & Portney, K. (2019). Analysis of four governance factors on efforts of water governing agencies to increase water reuse in the San Antonio Region. *Science of the Total Environment*, *647*, 1498–1507. https://doi.org/10.1016/j.scitotenv.2018.07.366

Alkhuzam, A. F., Arlet, J., & Lopez-Rocha, S. (2013). *Private versus public electricity distribution utilities: Are outcomes different for end-users?* World Bank Blogs. Retrieved January 24, 2022, from https://blogs.worldbank.org/developmenttalk/private-versus-public-electricity-distribution-utilities-are-outcomes-different-end-users

Allan, J. A., & Dent, D. L. (2021). The cost of food: Consequences of not valuing soil and water and the people who manage them. In D. L. Dent & B. P. Boincean (Eds.), *Regenerative agriculture. What's missing? What else do we need to know* (pp. 3–20). Springer Nature Switzerland.

Anderson, J. (2021) Private sector investments called key to driving energy transition. S&P Global Platts. Retrieved January 24, 2022, from https://www.spglobal.com/platts/en/market-insights/latest-news/electric-power/060821-private-sector-investments-called-key-to-driving-energy-transition

Baskin, K. (2020). *Supply chain resilience in the era of climate change.* Sloan School, MIT Management. Retrieved January 23, 2022, from https://mitsloan.mit.edu/ideas-made-to-matter/supply-chain-resilience-era-climate-change

Bremmer, I. (2010, May 22). The long shadow of the visible hand: Government-owned firms control most of the world's oil reserves. *Wall Street Journal*, Retrieved January 24, 2022, from https://www.wsj.com/articles/SB10001424052748704852004575258541875590852

Carmona-Moreno, C., Crestaz, E., Cimmarrusti, Y., Farinosi, F., Biedler, M., Amani, A., Mishra, A., & Carmona-Gutierrez, A., eds. (2021). *Implementing the water-energy-food-ecosystems nexus and achieving the sustainable development goals.* UNESCO, European Commission, IWA.

Clancy, M., Fuglie, K., & Heisey, P. (2016, November 10) US agricultural R&D in an era of falling public funding. *Amber Waves*, USDA Economic Research Service. https://go.nature.com/377eRB2

Coe, M. T., Evans, K. M., Gasic, K., & Main, D. (2020). Plant breeding capacity in U.S. public institutions. *Crop Science*, *60*(5), 2373–2385. https://doi.org/10.1002/csc2.20227

Daher, B., Lee, S., Kaushik, V., Blake, J., Askariyeh, M. H., Shafiezadeh, H., Zamaripa, S., & Mohtar, R. H. (2019). Towards bridging the water gap in Texas: A water–energy–food nexus approach. *Science of the Total Environment*, *647*, 449–463. https://doi.org/10.1016/j.scitotenv.2018.07.398

Daher, B., Lee, S., Mohtar, R. H., Asaka, J. O., & VanDeveer, S. D. (2017). Security, climate change, and the resource nexus. In R. Bleischwitz, H. Hoff, C. Spartaru, E. van der Voet, & S. van Deveer (Eds.), *Routledge handbook of the resource nexus* (pp. 45–63). Routledge.

Dare, A., & Mohtar, R. H. (2018). Farmer perceptions regarding irrigation with treated wastewater in the West Bank, Tunisia, and Qatar. *Water International Wicked Problems of Water Quality Governance*, *43*(3), 460–471. https://doi.org/10.1080/02508060.2018.1453012

Debaere, P., & Kapral, A. (2021). The potential of the private sector in combating water scarcity: The economics. *Water Security*, *13*, 100090. https://doi.org/10.1016/j.wasec.2021.100090

ECDPM. (2013). The role of the private sector in agriculture and food security. European Centre for Development Policy Management. Retrieved January 23, 2022, from https://ecdpm.org/topics/role-private-sector-food-security/

EIA. (2019). Investor-owned utilities served 72% of U.S. electricity customers in 2017. US Energy Information Administration. Retrieved January 24, 2022, from https://www.eia.gov/todayinenergy/detail.php?id=40913

Fuglie, K. O. (2016). The growing role of the private sector in agricultural research and development world-wide. *Global Food Security*, *10*, 29–38. https://doi.org/10.1016/j.gfs.2016.07.005

Gardezi, N. Z., Steel, B. S., & Lavado, A. (2020). The impact of efficacy, values, and knowledge on public preferences concerning food–water–energy policy tradeoffs. *International Journal of Environmental Research and Public Health*, *17*(22), 8345. https://doi.org/10.3390/ijerph17228345

Goodarzi, M. R., Vaghee, H., & Mohtar, R. H. (2020). The impact of climate change on water and energy security. *Water Supply*, *20*(7), 2530–2546. https://doi.org/10.2166/ws.2020.150

GWI. (2020). Global water tariff survey. Global Water Intelligence, Oxford. Retrieved January 23, 2022, from https://www.globalwaterintel.com/products-and-services/market-research-reports/tariff-survey-2020

Harris, S., & Wei, D. (2018). Why climate resilience and supply chains go hand in hand. BSR. Retrieved January 23, 2022, from https://www.bsr.org/en/our-insights/blog-view/climate-change-supply-chains-go-hand-in-hand

Huang, C., Kim, S., Song, K., Townshend, J. R. G., Davis, P., Altstatt, A., Rodas, O., Yanosky, A., Clay, R., Tucker, C. J., & Musinsky, J. (2009). Assessment of Paraguay's forest cover change using Landsat observations. *Global and Planetary Change*, *67*(1–2), 1–12. https://doi.org/10.1016/j.gloplacha.2008.12.009

IEA. (2021a). Global energy supply investment by sector, 2019–2021. International Energy Agency. Retrieved January 24, 2022, from https://www.iea.org/data-and-statistics/charts/global-energy-supply-investment-by-sector-2019-2021-2

IEA. (2021b). World energy investment 2021, executive summary. International Energy Agency. Retrieved January 24, 2022, from https://www.iea.org/reports/world-energy-investment-2021/executive-summary

International Renewable Energy Agency (IRENA). (2015). *Renewable energy in the water, energy & food nexus* (Eds., R. Ferroukhi, D. Nagpal, A. Lopez-Peña, T. Hodges, R. H. Mohtar, B. Daher, S. Mohtar, & M. Keulertz). UAE.

International Water Resources Association (IWRA) International Scientific Committee. (2021). One water, one health: Water, food and public health in a changing world. S Koo-Oshima, Y Li, RH Mohtar (eds.). *Final report of IWRA Online Conference, in collaboration with FAO.* 7–9 June, Paris. Retrieved January 23, 2022, from https://iwraonlineconference.org/

Keulertz, M., & Mohtar, R. H. (2019). *Governance and cooperation over food and SDG2 in the Arab region.* EDA insight research and analysis. Part of the New Governance for the Environment in the Arab Region Series. Emirates Diplomatic Academy.

King, J., Toole, A., & Fuglie, K. O. (2012). *The complementary roles of the public and private sectors in US agricultural research and development.* USDA-ERS Economic Brief 19.

Kopaskie, A. (2016). *Public vs private: A national overview of water systems.* University of North Carolina School of Governance, Environmental Finance Center. Retrieved January 23, 2022, from https://efc.web.unc.edu/2016/10/19/public-vs-private-a-national-overview-of-water-systems/

Lloyd Owen, D. (2022). The private sector and water services: A reflection. *Water International, 47*(7), 1032–1036. https://doi.org/10.1080/02508060.2022.2133416

Mohtar, R. H. (2011). An integrated sustainability index for effective water policy, and D. Waughray (Ed.), *Water security: The water–food–energy–climate nexus* (pp. 240–248). Island Press/Covelo. World Economic Forum Water Initiative.

Mohtar, R. H. (2016). Integrated water, energy, and food governance: A Qatari perspective. In M. Tok, L. Alkhater, & L. Pal (Eds.), *Policy-making in a transformative state* (pp. 295–307). Palgrave Macmillan. https://doi.org/10.107/978-1-137-46639-6_11

Mohtar, R. H. (2017). Climate change and the water–energy–food nexus in the MENA region. *OCP Policy Center. PB-17/39.* Texas A&M University

Mohtar, R. H., Assi, A. T., & Daher, B. T. (2015). Bridging the water and food gap: The role of the water–energy–food nexus. *UNU-FLORES Working Paper Series 5* UN University

Mohtar, R. H., & Daher, B. (2015, March 4–5). Policy briefs for the water, energy and food nexus in the Arab Region. Prepared for GIZ ACCWaM Regional Programme for Adaptation to Climate Change in the Water Sector in the MENA Sector. In cooperation with the League of Arab States and the Arabian Gulf University inception workshop, Cairo, Egypt. Arabic.

Nature Food. (2020). Editorial: Public–private roles beyond crop yields. *Nature Food, 1*(6), 311. https://doi.org/10.1038/s43016-020-0109-7

Pahl-Wostl, C., Gorris, P., Jager, N., Koch, L., Lebel, L., Stein, C., Venghaus, S., & Withanachchi, S. (2021). Scale-related governance challenges in the water–energy–food nexus: Toward a diagnostic approach. *Sustainability Science, 16*(2), 615–629. https://doi.org/10.1007/s11625-020-00888-6

Perard, E. (2017). Water supply: Public or private? An approach based on cost of funds, transaction costs, efficiency, and political costs. *Policy and Society, 12*(3), 193–219. https://doi.org/10.1016/j.polsoc.2008.10.004

Portney, K. E., Vedlitz, A., Sansom, G., Berke, P., & Daher, B. T. (2017). Governance of the water–energy–food nexus: The conceptual and methodological foundations for the San Antonio Region Case Study. *Current Sustainable/Renewable Energy Reports, 4*(3), 160–167. https://doi.org/10.1007/s40518-017-0077-1

Purwanto, A., Sušnik, J., Suryadi, F. X., & de Fraiture, C. (2021). Water–energy–food nexus: Critical review, practical applications, and prospects for future research. *Sustainability, 13*(4), 1919. https://doi.org/10.3390/su13041919

PWC. (2019). *SDG challenge 2019: Creating a strategy for a better world.* PricewaterhouseCoopers Retrieved January 23, 2022, from https://www.pwc.com/gx/en/sustainability/SDG/sdg-2019.pdf

Slay, C., & Dooley, K. (2020). Improving supply chain resilience to manage climate change risks. HSBC Center Paper. The Sustainability consortium, Arizona State University and University of

Arkansas. Retrieved January 23, 2022, from https://sustainabilityconsortium.org/download/35952/

UN-ESCWA. (2016). *Developing the capacity of ESCWA member countries to address the water and energy nexus for achieving SDGs: Regional policy toolkit* (R. H. Mohtar, ed). UN Economic and Social Commission for Western Asia.

United Nations Global Compact. (2017). Progress report: Business solutions to sustainable development. Retrieved January 23, 2022, from https://d306pr3pise04h.cloudfront.net/docs/publications%2FUN+Impact+Brochure_Concept-FINAL.pdf

United Nations Sustainable Development Goals (UNSDG). (2015). Retrieved January 23, 2022, from https://www.un.org/sustainabledevelopment/development-agenda/

USDA-ERS. (2019). Agricultural research funding in the public and private sectors. US Dept Agriculture, Economic Research Service. Retrieved January 23, 2022, from https://go.nature.com/304fzNT

Walker, S. (2020). Triple threat: Water, energy and food insecurity. World Resources Institute. Retrieved January 23, 2022, from https://www.wri.org/insights/triple-threat-water-energy-and-food-insecurity

Weitz, N., Strambo, C., Kemp-Benedict, E., & Nilsson, M. (2017). Closing the governance gaps in the water–energy–food nexus: Insights from integrative governance. *Global Environmental Change*, *45*, 165–173. https://doi.org/10.1016/j.gloenvcha.2017.06.006

The private sector and water services: a reflection

David Lloyd Owen

I enjoyed many discussions with Tony Allan about the pros and cons of private sector participation (PSP) in water and wastewater. He was, by nature, sceptical about PSP while I was and remain a critical friend. His principal concerns were that asset ownership ought to remain in public hands, and the scale and nature of the information and management asymmetries between the public and private sectors. This viewpoint considers how his observations have tallied over the past two decades, and outlines some pertinent developments in PSP during that time.

Tony Allan's reservations

Certainly, Tony Allan's reservations about asset ownership resonate. When the ten10 English and Welsh water and sewerage companies were privatized in 1989, it was assumed they would remain as public limited companies. Today, just three retain a market listing: United Utilities (North West), Pennon (South West) and Severn Trent, while Glas Cymru (Dŵr Cymru Welsh Water) is a not-for-profit company. Anglian, Southern, Thames and Yorkshire are owned by private equity companies, and Northumbrian and Wessex are subsidiaries of international conglomerates. If these companies had, instead, been granted concession contracts, say for 30 years (as Tony would have preferred), the fates of most would have been different. They would have had a powerful incentive to deliver the best public health, customer service and environmental performance, especially with their concessions subject to periodic renewal. At the same time, their incumbent assets would have remained in public ownership, while new assets developed during each concession period would be handed over at their expiry.

The information asymmetries are more nuanced. It is not necessarily the fault of major water companies that they employ a large number of competent managers, engineers and scientists; Vivendi, the former Générale des Eaux, is far more than a utility – it is a centre for water research and innovation, especially in integrating disparate assets into systems designed to deliver a better service at a lower cost and with zero carbon in mind. One of the most important outcomes of the Sustainable Development Goal process (SDG 6, universal safe drinking water and safe sanitation by 2020) has been the UN-Water's Global Analysis and Assessment of Sanitation and Drinking-Water (GLAAS) assessments of financial and managerial capacity across the world. For the first time, we have an idea about just how worryingly incompetent many governments are when it comes to their water utilities and how indifferent they are about public health.

For example, in 2018–19, GLAAS found that 60–85% of national water, sanitation and hygiene (WASH) plans were adequately costed, with 9–15% having sufficient finance (at least 75% of what is needed) to carry them out; 41–54% had assessed the human resources needed to reach SDG 6 but only 6–14% had sufficient human resources; 52% of national plans considered making safely managed water access the norm, but only 30% considered safe urban sanitation. The rest still assume delivery of, essentially, non-potable water and unhygienic sanitation in 2030 and beyond. A total of Twenty20 countries were able to assess their funding needs and the gap between these and the funds currently being generated. Collection of government spending data rose from 5% in 2009–10 to 55% by 2016–17, and for household spending (including tariffs), from 5% in 2011–12 to 33% by 2016–17. Revisions of the coverage data by the WHOWorld Health Organization (WHO) and UNUnited Nations's (UN) Joint Monitoring Programme in 2015 and 2021 also highlight the need to know where we are before future goals can be met.

No private-sector company ought to countenance such an information vacuum. In the early years of global PSP (1989–2004), information and management asymmetries were used by some contract bidders to alter contract terms after gaining the concession. This hardly formed the basis of a constructive relationship between the company and the municipality and many of these contracts were terminated early. In some cases, these asymmetries resulted in hubris: in West Manila, West Jakarta and Buenos Aries, US dollar-denominated project debt was financed by local currency revenues. During a 30-year contract in countries such as the Philippines, Indonesia and Argentina, there is a high probability that the local currency will depreciate sharply against the dollar, making their debt service costs unsustainable, given the implications of tariff rises. In each case, the inevitable happened and the contract was terminated early.

Two decades ago, PSP was controversial: when Michel Camdessus broached the topic of water finance at the Kyoto World Water Forum in 2003, the meeting was broken up by demonstrators; many contracts were cancelled well ahead of their intended time; many books and articles punctuated by exclamation marks were published. These days, the debate has faded into the background, while the water sector is preoccupied with adapting to climate change and aiming to meet the SDG 6 targets. What has happened?

How PSP has developed in recent years

Between 1999 and 2016, I prepared fifteen15 editions of a water yearbook (published by Masons, Pinsent Masons and Arup, respectively) which provided a global survey of PSP from both a country and a company perspectives, so I was able to monitor in some detail the development of PSP over a period of time and to see it at its best – and worst.

My definition of PSP covers companies which own and operate their assets outright (the USUnited StatesA, England and Wales), or have a minority share of the assets along with those who operate and manage assets on behalf of a municipality (operation and maintenance (O + &M) contracts) for contracts of at least five years in length, and companies which are responsible for developing and running new assets (concessions) either on their own or while managing the assets already in place.

Some leases and concessions have been renewed several times, with examples in France, Germany, Italy and Spain having been renewed for appreciably more than a century. In terms of population served, there were 138 million people served through long-standing contracts, along with the asset-owning statutory water companies in England and Wales (13.6 million people covered in 1989) and investor-owned utilities in the USUnited StatesA. New contract awards have been identified dating back to 1987.

The current period of PSP dates from a group of contract awards made in 1987. Contracts dating back to 1585 have been identified, with Thames Water, for example, tracing its origins to a contract (the New River Company) awarded in 1609. Between 1987 and 2015, 1903 awards have been identified, covering 2080 separate contracts, serving a total of 1052 million people.

Today, international players such as Veolia (France) and Acciona Agua (Spain) are active wherever they believe they can operate. Regional players such as VA Tech Wabag (India) and SAUR (France) have focussed on a limited number of countries, while local players operate in one country. Local players can be very large as in the case of SaoSão Paulo's SABESP (28 million) or Shanghai Industrial Holdings (66 million) while, at the other end, Cholderton Water serves 2,500 people in rural England.

The nature of PSP has changed beyond expectation. In 2003, it appeared that a group of five global players were to dominate the market, then accounting for 73% of the total population served by PSP awards. By 2015, their share had fallen to 24% with the leading five players at the time accounting for 35% of the total. Instead, the shift has been towards nationally based companies, especially in China and, more recently, in India. PSP has not seen global behemoths straddle the world, rather it has encouraged a large number of local companies to become involved in providing these services.

Considering the overall expansion of PSP, looking at five-year periods between 1995–99 and 2020–14, awards of water-only contracts have ranged from 94 million to 106 million people per period. In contrast, the number served for sewerage-only rose consistently from 16 million in 1995–99 to 153 million for 2010–14, while combined water- and- sewerage contracts fell from 55 million to 16 million over the same period. The water- and- sewage contracts typify the all-embracing construction of customer-service concessional contracts favoured by the World Bank's International Finance Corporation (IFC) from the 1990s. They have fallen from favour, especially due to the more contentious nature of these contracts arising from the direct customer contact involved. In contrast, developing a sewage treatment works does not involve contact with customers, although delivering benefit to stakeholders. The IFC model assumed that international players would provide a significant source of the debt funding needed for developing new infrastructure – but finance and politics turned out unfavourably.

Population growth and service extension has been another growth driver, with 78 million people being covered in 112 contracts between 1989 and 2015 (where comparative service coverage data isare available). Growth comes through serving the previously unserved, adjacent utilities being absorbed into a larger player, or by population growth within the service area.

Project cancellations

Between 1995 and 2014, 120 contract endings covering 119.1 million people were identified, which is equivalent to 12% of all identified contract awards in population terms. These could be where the contract was unilaterally cancelled (48.2 million), the municipality bought out the concession (26.4 million), or the contract had reached its agreed expiry date (44.5 million). Contract losses in 1995–99 were 2.9 million, rising to 44.1 million in 2000–04, 42.5 million in 2005–09 and 29.6 million in 2010–14. While 21% of all contracts awarded to international players have ended, just 5% of local company contracts had, falling to 2% in 2005–14. Cancelled contracts accounted for 86% of endings in 1995–99, falling to 43% in 2000–04 and 55% in 2005–09 and easing further to 11% in 2010–14.

While some contracts should never have been contemplated (most notoriously, Bolivia's Cochabamba PSP in 2000, which in effect lasted four days) or read as case studies in how not to run a contract (Azurix's British management team arriving in Argentina on the anniversary of the victory in the Falkland Islands showed a limited appreciation of client sensitivities), international players have become both more selective and effective in recent years.

It is evident that PSP should not be contemplated unless there is a transparent tender process based upon enforceable performance targets, regular corporate reporting, appropriate affordability measures, and tariff rises being dependent on the actual delivery of improved services.

PSP today and tomorrow

The greatest paradox about PSP is its innate pragmatism, even in spite of itself. A municipality may have no motive to serve the unserved unless there is political capital to be gained by doing this, effectively excluding those living in informal settlements. The operator of a concession ought to answer to regulatory rather than political scrutiny; contractual obligations to extend water access in, for example, Manila and Jakarta has made connecting informal settlements the norm rather than the exception.

PSP in water will never be a panacea. Neither should it invariably be ruled out. It can be justified when it brings about tangible gains in affordability, service quality and sustainability. This depends on effective economic, environmental and public health regulators who are free from political interference and are adequately resourced to monitor utility performance.

The emasculation of the Environment Agency (EA) in England since 2010 is a case in point. The EA's reliance on self-reporting by privatized utilities gave rise to unacceptable sewage discharges by Thames and Southern, whereby they appear to have decided that the possibility of penalties was easily outweighed by the cost savings in discharging untreated sewage. In 2021, Southern was fined £90 million for offences carried out between 2010 and 2015, while in 2019 Ofwat ordered it to return £123 million to its customers along with a £3 million fine for offences between 2010 and 2017. Prior to 2010, the company had been convicted or cautioned for a further 168 offences, demonstrating that lesser fines were no deterrent.

What has been missing from almost all of the debates for and against PSP is an evidence-based approach with beneficial outcomes in mind. If the private sector is truly capable of offering win–wins for all stakeholders, its case studies should be well-suited to critical scrutiny. Likewise, outright opponents of PSP ought to consider greater priorities – such how the water sector is going to rise to the challenges of climate change and meeting SDG6. Universal access to safe, sustainable, and affordable water and sanitation ought to override ideology.

Disclosure statement

No potential conflict of interest was reported by the author(s).

ORCID

David Lloyd Owen http://orcid.org/0000-0002-2094-9008

Water governance and system coordination across diverse risk-management cultures

Brendan Bromwich, Damian Crilly and Jyoti Banerjee

Adoption of a systems approach is a current theme in UK environmental policy but implementation is challenged by differing interpretations. We draw on Cultural Theory to categorize different kinds of organization, their assumptions about system control, and comparative contributions to managing water and environment. For the private sector to engage with nature at its true value, we need to bridge global business and environmental systems – a challenge of scale, resolution, and culture. We argue that understanding organizational cultures around risk will enhance layered, collaborative governance at regional and catchment levels, and draw out the comparative strengths of public, private and voluntary organizations.

Introduction: a missing middle or a need to bridge separate systems?

The call for a systems perspective to create more integrated and transformative approaches to environment, water and infrastructure is loud and clear (e.g., Defra, 2018; Ofwat, 2017). The ambition to bridge silos, tackle causes rather than symptoms of problems, and identify multi-benefit interventions for which costs can be shared is intuitively attractive. But, with just a few pilot projects to draw upon, the practicalities remain unclear.

Major steps towards an integrated approach at the catchment level began with the Water Framework Directive (WFD) that laid down a scientific basis for water quality, but implementation focused on the outputs rather than the transformative approach to which it aspired (Giakoumis & Voulvoulis, 2019). Whereas WFD was top-down and government-led, Defra's Catchment Based Approach (CaBA; Defra, 2013) has been bottom-up, coordinating local collaboration. It has encouraged wider participation but has not brought big changes in catchment characteristics, and private-sector engagement has been patchy at best. From the catchment perspective, there is perceived to be a *missing middle* – a lack of engagement across agriculture, business and local government. Substantial change remains elusive. There are initiatives exploring this territory – such as the landscape enterprise networks (LENs) convened by Nestlé, 3Keel and others,[1] United Utilities' Catchment Systems Thinking[2] and Wessex Water's EnTrade[3] – but these are pioneers rather than game changers.

Framing the problem of private sector engagement as a *missing middle* from catchment systems reflects an environmental perspective but does not engage business. The framing is important because it creates the rationale for the response. If we think of environmental actions as primarily driven by public-sector resources with some additional private funding, then it might be valid to talk about a top-down regulatory and policy model that needs to connect with bottom-up actions. If there is a disconnect, we might argue that there is a missing middle, but businesses are more likely to talk of climate- and nature-related externalities in their business model. Their model has global rather than catchment boundaries – and they have little or no awareness of either WFD or CaBA.

If, by contrast, we frame the problem as a global business system that is failing to manage its impact on and interface with the natural environment, then a different scale of system re-integration is needed. We argue that this perspective is required. And the opportunity exists – the climate crisis has attracted huge commitments to net zero actions from corporate chief executive officers (CEOs) – but when we compare the modest scale of catchment management programmes (routinely less than £1 million) and the hundreds of millions of pounds in fund management, there is clearly a difference of scale to be bridged between environmental and business systems management.

Moreover, while private-sector engagement is very necessary, models of catchment governance that depend solely on market mechanisms fail to reflect the complexity of socio-ecological systems such as catchments. For example, Helm (2020, 2015) draws on operational models in the power sector to propose a catchment system operator/regulator who would procure services from the private sector to implement an overarching catchment plan. Comparisons between the management of power and water as infrastructure systems are important. But parallels between management of infrastructure systems and management of the natural environment are not comparing like with like. Our alternative considers organizational culture and risk management of systems, drawing on different approaches to coordination, and the creation of outcomes in catchments and landscapes.

Policy drivers

Droughts in 2018, 2020 and 2022, storms in 2021 and 2022, the Covid-19 pandemic, Brexit, and a government commitment to 'net zero' – what Tony Allan called 'emblematic events' – set the scene for a vigorous discourse on the need for a change of policy. Several initiatives might combine to implement substantial change:

- The government's 25-Year Environment Plan (Defra, 2018) and Encouraging a Natural Capital Approach[4] offer a vision and suitable metrics for system integration.
- Engineering institutions are driving a more holistic, integrated approach, referred to as Industry 4.0 (World Economic Forum, 2021), in which infrastructure is perceived as a system, drawing attention to the outcomes of that system rather than an exclusive focus on the infrastructure itself.

- Business is being driven to action on its climate,[5] nature[6] and social[7] impacts; and increasing uptake of multi-capital accounting (Banerjee, 2019) creates a framework for this reporting.[8]

The list of policy initiatives calling for a systems perspective reflects broad ambitions and, consequently, a range of sectoral and organizational cultures that have to be navigated to achieve collaboration. Defra's Participatory System Mapping at the catchment level (Mott MacDonald & CECAN, 2020) highlighted the cultural disconnect between different organizations working in catchments: one workshop participant perceived a 'lack of level playing field' when their environmental non-governmental organization (NGO) interacted with an infrastructure-focused government agency. Notwithstanding the challenges, collaboration across a range of organizational cultures offers different insights and can respond to problems with a wider set of tools.

Theoretical contributions

Three theoretical constructs inform the discussion:

- *Complexity*: a system may be described as *complicated* if it has a wide range of different, interacting components; it is described as *complex* if these interactions exhibit non-linearity, non-proportionality, unpredictability and there is a process of learning by the components in response to change (Bicket et al., 2020). Complicated systems may be modelled: complex systems are hard to model.
- The categories of bias in the perception of systems described as *Cultural Theory* (Douglas, 1986; Thompson et al., 1990).
- The concept of the *wicked problem* (Rittel & Webber, 1973).

Cultural Theory categorizes four approaches to managing systems: *risk-controlling*, which relies on rules or infrastructure to constrain system outcomes; *collaboration* or *pooling risk*, which relies on collective effort; *capitalization* and *trading* of risk, which enables transfer of risk to those best placed to produce cost-effective outcomes; and *acceptance of risk*, whereby variable outcomes are tolerated. These strategies reflect organizational cultures that it describes as *bureaucratic (or hierarchical)*, *egalitarian*, *individualist (or entrepreneurial)* and *fatalist (or risk-accepting)*. Decisions relating to the environment are socially embedded – that is, people's choices are influenced by collective, cultural norms rather than simply a rational choice to maximize personal benefit. Cultural Theory observes that different groups of people may be more (or less) influenced or constrained by cultural norms and may be more (or less) integrated in social networks that shape or contest cultural norms. The four approaches to risk reflect cultural bias, illustrated in Figure 1 by a matrix of strong and weak compliance with social norms and strong or weak engagement with social relations (and consequently policy discourse):

- A *bureaucratic* or *hierarchical* culture seeks to control risk and favour authoritarian management based on the assumptions that they have authority to create rules and that, in general, people are inclined to comply. This approach prevails in

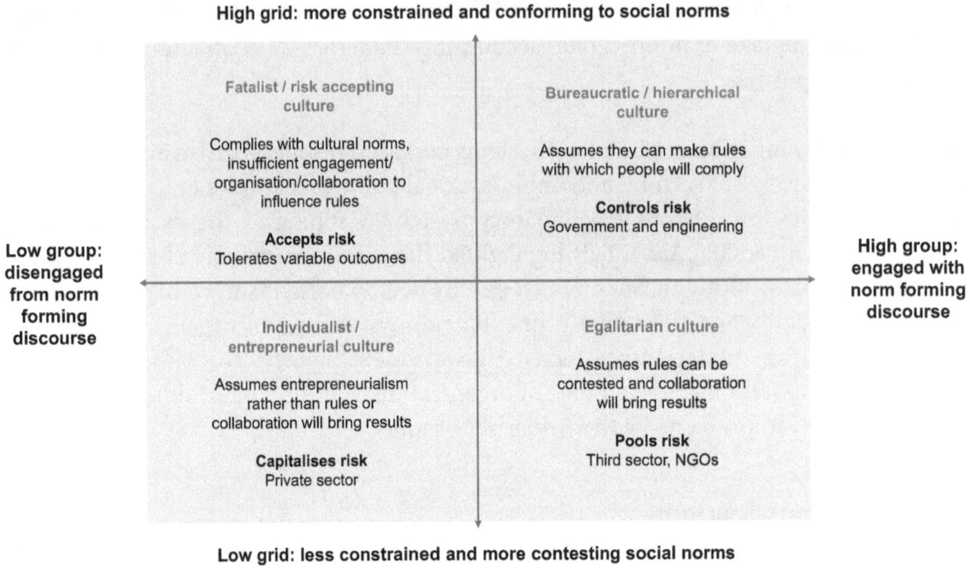

Figure 1. Categories of organizational culture and risk management – grid-group model.

government (which creates policy) and among engineers (who see technology and infrastructure as the means of controlling complex systems).

- An *egalitarian* culture pools risk and favours collaborative management, assuming that people will work together for a common objective. Coordination and collective action are strengths of NGOs or coordinating bodies.
- An *individualist* or *entrepreneurial* culture that capitalizes risk (seeing it as an opportunity) and favours competitive strategies – assuming that lower regulation enables innovation and creativity to solve problems. Capitalizing risk provides a reward for those willing to profit from taking on problems and creating a more efficient and effective solution than others – the strength of the private sector.
- A *fatalist* culture that accepts risk – assuming that tolerating variable outcomes is preferable to making the effort to control, coordinate or capitalize risk. This may either be an approach taken by those operating in relative isolation (e.g., family farms) or accepting residual risk as part of a wider risk strategy.

These categories provide a language to describe different kinds of organizations and their comparative strengths and weaknesses for different approaches to complex systems (Allan, 2002).

The third relevant theoretical construct is the dichotomy of *wicked* and *tame* problems. A wicked problem is one that cannot be easily defined, and cannot be resolved by a single analytical approach. By contrast, a tame problem can be defined and resolved. The optimization of an engineering problem such as a pipeline design is a tame problem – the variables of cost, energy and flow can be defined and an optimal solution defined. Conversely, the optimization of a river catchment is a wicked problem, in which optimizing one element (say, water resources) may have complex links with other outcomes (such as biodiversity, social amenity, and water quality) that make it hard to articulate the problem as a whole. Wicked problems display the following features:

- There is no definitive formulation: formulation is part of the problem.
- There is no clear point at which a wicked problem is solved, and there are no right or wrong solutions. Wicked problems can be managed in more or less effective ways with a combination of methods creating more or less beneficial results over time.
- An observer's perception of how a problem may be tackled creates a bias in how they are likely to articulate a problem.

Cultural Theory and wicked-problem theory come together with observations that each of the groups identified in Cultural Theory deals with wicked problems in ways that reflect their own bias: government documents often refer to *policy levers* that could be used to control a system; likewise, engineers emphasize the potential for digital methods and infrastructure to control systems; environmental NGOs underline the need for collective action and collaboration; whereas the private sector looks for opportunities to add value and create profit.[9] In short, each group may present a different, simplified solution that tackles a problem in a way that is logical within their own world-view but fails to address complex problems in their entirety.

Solutions to problems that are framed in just one of these ways may have a deceptive elegance and appear to solve the problem. This is misleading – as may become clear if other perspectives are considered or, as H.L. Mencken quipped: 'For every complex problem there is an answer that is clear, simple, and wrong.' Thompson et al. (1990) argue that it is better to address wicked problems with a blend of solutions that dynamically balance different emphases over time, producing better ongoing management of systems rather than deceptively neat resolutions.

Evolution of catchment system management

The WFD framework, a bureaucratic approach, identifies the health of river catchments according to their chemical, biological and geomorphological states. Its implicit assumptions, principally the potential of regulatory compliance to achieve integrated system change, have proven to be overoptimistic. The CaBA has added a collaborative, egalitarian approach that has mobilized collective action on the environment at the catchment level. It can convene a coalition of the willing but it struggles to bring in actors with other motivations – such as the profit-seeking private sector. Where NGOs are essentially collaborative in culture, the private sector is essentially competitive – creating a particular challenge for collaborative governance.

Helm, by contrast, assumes actors will engage with entrepreneurial motivation and fails to value the contribution of collaboration motivated by collective endeavour for the common good. The problem with Helm's proposition is highlighted by contrasting the power system, from which his framework originates, and the water system. There are relatively few variables in a power system (infrastructure, energy and finance), and so it is possible to set up a regulated market to optimize system performance. The same holds for the engineered elements of the water system – the operation of infrastructure for the delivery of water at the right time, place and quality for a fair price.

In both cases, the limited number of variables means that markets can be constructed and regulated to deliver efficient results. They are both tame problems. The difficulty comes with the social and ecological nature of catchments, which have numerous significant system outcomes. Their complexity means that there is no single optimized solution; e.g., optimization of a catchment for water resource availability would undermine its biodiversity or its social outcomes, and vice versa. Operation of a catchment is a wicked problem. Plural Rationality (Thompson et al., 1990) suggests that reliance on any one or any pair of these three system cultures will not bring the desired results and, consequently, a blend of all three active cultures is needed.

By adopting a local, participatory approach, CaBA tries to balance the diverse objectives inherent in operation of a complex system. It draws on motivation of the public good; participants motivated by a collective notion of the public good can make compromises to achieve a balance of overall benefits. In this light, we may draw an analogy with the health system. It is complex and diverse; some problems are tame (e.g., hip replacements, cleaning wards) and can be optimized by efficient use of markets and good management, other elements are complex and, in some cases, a collaborative approach incentivized by common good yields results; non-profit organizations whose role is enhanced by motivation to collaborate for the common good include cancer research and nursing, children's hospitals, and others. This motivation and collaborative approach is similar to the Royal Society for the Protection of Birds (RSPB), the World Wide Fund for Nature (WWF), rivers and wildlife trusts and smaller organizations relevant to water and environment.

Current development of water policy is drawing on collaboration through the establishment of regional planning groups. What is emerging, and what we are advocating, is the evolution of the water management system to blend all three active system cultures – risk regulation of government, collaboration and risk pooling of the voluntary sector, as well as a market-based approach by those who have capacity to capitalize risk and deliver results at scale.

Organizational culture and respective strengths in system management

Environment agency engagement in catchments

A comparison of three different ways that the Environment Agency (EA) engages with catchments demonstrates the comparative advantages of different organizational arrangements:

- EA catchment coordinators operate in an egalitarian style to mobilize collective action in catchment partnerships with environmental NGOs.
- The asset management team engages with a bureaucratic framework to design and implement engineering works. It controls risk by defining infrastructure outcomes (e.g., level of flood protection) and setting contracts with private-sector organizations (who capitalize that risk).

Table 1. Cultural comparison of the catchment partnership and flood partnership – observations by interviewees.

Some of the benefits of the egalitarian convening organization
1) It was positive that the host was a third-sector organization. The personal qualities of the host were also mentioned like being a non-domineering person that manages to move the partnership along.
2) There have been some successes on working together on small projects in the partnership. One . . . potential area for partnership working was getting rid of invasive species – an issue that is non-political and everyone can agree that it's the right thing to do.
3) The partnership aims for collaborative working, meaning consensus building.
Some challenges of being egalitarian – and looking at the formal organizations (hierarchy/bureaucracy) to bring benefits of their approaches
1) Partners have ideas for what they would like to do but the mechanisms for making it happen are not there. This may be because the representatives on the partnership don't have the right level of seniority to make changes in their organization.
2) Partnership plan isn't driving delivery. No action plan that states what the priorities are and who is delivering on them.
Looking across at the flood partnership and seeing some of the contrasting benefits that a more hierarchical/ bureaucratic approach brings
1) The strategic group in the flood partnership has a high level of personal commitment by senior representatives. Representatives at CEO level, elected members and senior directors attend the strategic group. The group oversees a practitioner's group that deals with day-to-day delivery of action.
2) The flood partnership has a published plan with clear responsibilities for who delivers what. Members have a specific role, are held accountable for delivering it, and report back on activities in the strategic group.
Comments that highlight the benefits of hybridity/complementarity
1) It would be helpful to have a list of partnership projects that was endorsed by the EA and clarified what status the partnership projects have in relation to the EA's own projects and advice.
2) For organizations like the EA with a regulatory role, partnership with third-sector organizations facilitated a different conversation with local communities and the public compared to what the EA could do by itself.

Source: Mott MacDonald and CECAN (2020).

- Flood partnerships combine the strengths of bureaucratic/hierarchical and egalitarian cultures by mobilizing collaboration, but retain an overall programme control.

An EA study interviewed members of a catchment partnership who articulated the relative advantages and disadvantages of being an egalitarian network in contrast to the hierarchical strengths of the flood partnership (Table 1).

Multi-sector collaboration in regional planning

The categories of Cultural Theory informed the approach of Water Resources South East (Mott MacDonald, 2021) in their analysis of multi-sector engagement with their regional resilience plan:

- Power system operators are entrepreneurial. To attract finance for new power installations they have to demonstrate the value proposition in terms of risk and reward. Power stakeholders seek to control their own risk by having secure rights to water licences; they are unwilling to accept the variability inherent in procuring water in a market. However, having ensured their own security of supply, they would be happy to trade potential surplus water in markets to organizations prepared to accept a greater variability in their own supplies. The sector operates with individualist/entrepreneurial logic, balancing outcome risks and finance on a clear monetized basis.

Table 2. Wyre catchment project: allocation of risks among stakeholders.

	Buyers	Investors	Landowners	Rivers Trust(s)
NFM construction/delivery risk	£	£££	Nil	Reputational
NFM performance risk	££	££	Nil	Reputational
Contractual/counterparty risk	£	££		
External risks – policy environment	£	£		

Note: NFM, natural flood management. Source: Hird (2022).

- Agricultural users exhibit both entrepreneurial and fatalist approaches to managing water risk. Farmers with multi-year supply contracts that could be jeopardized by a single harvest failure (such as soft fruits for supermarkets) are entrepreneurial in securing water resources, for example, willing to pay for connection to a public water supply as security against drought. Farmers operating in an open market tolerate variable weather and variable outcomes – they are risk accepting.
- The Canal and Rivers Trust is an NGO with a corresponding egalitarian approach to risk. It has a 1-in-20-year service aspiration – keeping the canals going routinely but accepting that for the 1-in-20 dry year there are likely to be others with more pressing water demands and, therefore, a degree of service failure is acceptable for the common good.

It is, therefore, important that planners at the regional level understand the diversity of risk cultures amongst water users, and create planning processes that engage all sectors. The question of scale is important in understanding business engagement; the paper industry is important in Kent, but was notably absent from Defra's catchment systems project, whereas it engaged actively in Water Resources South East.

Procurement of catchment-level outcomes – River Wyre, Lancashire

The Wyre Natural Flood Management Project encompasses approximately 70 ha of natural flood management features that could reduce the frequency of flooding to some 120 properties, with significant cost savings to United Utilities, EA, local authorities, the insurance industry, locally based businesses and homeowners. These cost savings could generate a long-term revenue stream to repay investment in reduced flood risk.

A debt-based, commercial transactional model was developed using a *special purpose vehicle*: a commercial interest company that will draw upon external finance for the capital delivery, which will be repaid over nine years through contracts with buyers (those who will benefit from the intervention) and sellers (farmers and landowners who will host and manage the interventions). The company has secured £1.25 million operating capital from a range of investors. The scheme structure reflects the relative capacity of different organizations to carry risk, as shown in Table 2.

The Rivers Trust, an environmental NGO, carries no financial risk, just reputational risk. Landowners are not required to take on risk. The risk is held by investors and buyers, representing organizations with the best capacity to handle the capitalization of risk. Investors in the scheme are rewarded for their willingness to take on the risk with a 6% return on their investment.

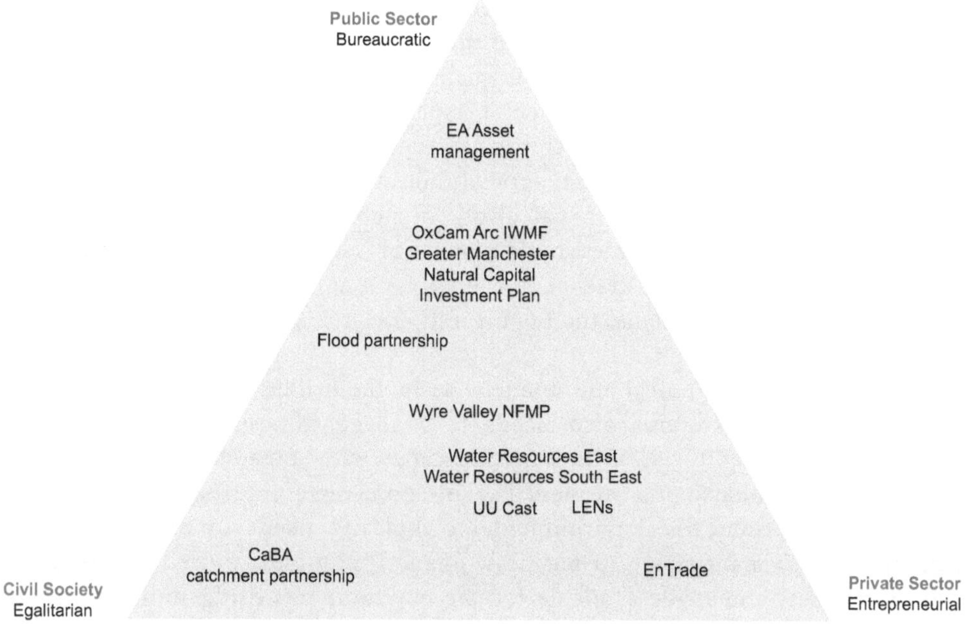

Figure 2. Collaborative management of water resources and environment.

Conclusions: plural governance and bridging systems

In light of taking this systemic perspective on water governance, we propose:

- That the water sector adopts a plural approach to governance that embraces different organizational cultures, with their associated approaches to risk management.
- A collaborative approach to bridging environmental and business systems that produces a radical transformation in environmental outcomes.

Figure 2 plots a number of water management initiatives according to their approaches to water governance. The WFD remains the bureaucratic framework for water quality in catchments. CaBA represents the egalitarian, collaborative approach at the catchment level. The regional planning groups such as Water Resources East and Water Resources South East are private-sector initiatives (but taking a coordinating role), as are the LENs initiatives. The OxCam Arc and the Greater Manchester Natural Capital Fund are collaborative efforts led by government. We are not arguing that all initiatives need to combine the three cultural approaches to risk. On the contrary, we argue that different approaches are fit for different problems, and a diverse, targeted and mutually complementary approach is needed.

The Wyre project is important in that its institutional arrangement is explicitly designed around the capability of different types of organization to manage risk – an approach we are advocating here. However, the investment at this stage is small and a lot more work is needed to create financial instruments that are attractive to fund managers seeking to invest at scales greater by orders of magnitude.

Scale is a critical dimension of the challenge. Business and water resource systems operate at scales larger than the catchment, although the catchment is the key system boundary for flooding and water quality. In France, for example, an ability to combine the regional and catchment perspectives is achieved by water parliaments for each of the six main river basins. They provide a coordinating body at basin level that reflects government, agricultural, municipal and business concerns. Beneath this structure, there is a patchwork of public and private arrangements. A regional perspective – informed by local nuance – will also help integrate other planning frameworks such as transport, energy, food and health. By seeking synergies across these wider systems, the opportunity for creating transformative change is increased.

In addition to geographical and financial scale, the bridge between the environment and business systems also needs to manage differences in resolution. Environmentalists rightly seek mosaic landscapes with a patchwork of different land uses and habitats, but business has no immediate interest or capacity to engage at this resolution. Environmental complexity needs to be packaged in a manner that is manageable to business: financial complexity needs to be managed by business and made available for the environment. Bridge building needs work on both sides.[10]

Notes

1. See https://landscapeenterprisenetworks.com/.
2. See https://www.unitedutilities.com/corporate/responsibility/stakeholders/catchment-systems-thinking/.
3. See https://www.entrade.co.uk/.
4. See https://www.gov.uk/guidance/enabling-a-natural-capital-approach-enca/.
5. See https://www.fsb-tcfd.org/.
6. See https://tnfd.global/.
7. See https://www.gov.uk/government/groups/social-impact-investment-taskforce/.
8. See https://www.integratedreporting.org/resource/international-ir-framework/.
9. See, for example, the Integrated Reporting Framework that describes systems in terms of multiple capitals (IIRC, 2013).
10. The views discussed in this article are those of the authors and not (necessarily) of their organizations.

Disclosure statement

No potential conflict of interest was reported by the authors.

ORCID

Brendan Bromwich http://orcid.org/0000-0003-0204-2928
Damian Crilly http://orcid.org/0000-0003-4622-5400
Jyoti Banerjee http://orcid.org/0000-0002-3645-3576

References

Allan, J. A. (2002). *The Middle East water question. Hydropolitics and the global economy.* IB Tauris.

Banerjee, J. (2019). Integrating multi-capital thinking into business decisions. In J. A. Allan, B. Bromwich, M. Keulertz, & A. J. Coleman (Eds.), *The Oxford handbook of food water and society* (pp. 760–776). Oxford University Press.

Bicket, M., Christie, I., Gilbert, N., Penn, A., Hills, D., & Wilkinson, H. (2020). *Complexity and what it means for policy design, implementation and evaluation.* CECAN. https://www.cecan.ac.uk/wp-content/uploads/2020/08/EPPN-No-16-Complexity-and-what-it-means-for-policy-design-implementation-and-evaluation-.pdf

Defra. 2013 *Catchment based approach: Improving the quality of our water environment.* https://assets.publishing.service.gov.uk/government/uploads/system/uploads/attachment_data/file/204231/pb13934-water-environment-catchment-based-approach.pdf

Defra. 2018 *A green future: Our 25-year plan to improve the environment.* https://assets.publishing.service.gov.uk/government/uploads/system/uploads/attachment_data/file/693158/25-year-environment-plan.pdf

Douglas, M. (1986). *How institutions think.* Syracuse University Press.

Giakoumis, T., & Voulvoulis, N. (2019). Water framework directive programmes of measures: Lessons from the 1st planning cycle of a catchment in England. *Science of the Total Environment, 668*, 903–916. https://doi.org/10.1016/j.scitotenv.2019.01.405

Helm, D. (2015). *Catchment management, abstraction and flooding: The case for a catchment system operator and coordinated competition.* New College.

Helm, D. (2020). Thirty years after water privatization—is the English model the envy of the world? *Oxford Review of Economic Policy, 36*(1), 69–85. https://doi.org/10.1093/oxrep/grz031

Hird, D. (2022, May13). *Showcase of the Wyre River Natural Flood Management Project.* Green Finance Institute Webinar. https://www.youtube.com/watch?v=Iductkx8EU4

IIRC. (2013). *International IR framework.* Value Reporting Foundation, The International Integrated Reporting Council. https://www.integratedreporting.org/wp-content/uploads/2015/03/13-12-08-THE-INTERNATIONAL-IR-FRAMEWORK-2-1.pdf

Mott MacDonald & CECAN (2020). *Systems analysis for water resources.* [Final report]. Defra. http://randd.defra.gov.uk

Mott MacDonald. (2021). *WRSE Resilience Phase 2: Multi-sector resilience and systems approaches.* [Final report]. https://www.wrse.org.uk/media/bowilcvo/wrse-multi-sector-resilience-phase-2.pdf

Ofwat. 2017 *Resilience in the round.* https://www.ofwat.gov.uk/wp-content/uploads/2017/09/Resilience-in-the-Round-report.pdf

Rittel, H. W., & Webber, M. M. (1973). Dilemmas in a general theory of planning. *Policy Sciences, 4* (2), 155–169. https://doi.org/10.1007/BF01405730

Thompson, M., Ellis, R., Wildavsky, A., & Wildavsky, M. (1990). *Cultural theory.* Routledge.

World Economic Forum. (2021). *Infrastructure 4.0: Achieving better outcomes with technology and systems thinking.* https://www3.weforum.org/docs/WEF_Infrastructure_Technology_Adoption_2021.pdf

Chronic crisis: 30 years on from the Dublin Principles and still no market to value water

Martin Keulertz and Phil Riddell

ABSTRACT

Solving the global water crisis requires water to be economically mobile. Well-regulated water markets that can allocate scarce water resources to their best economic, social and environmental uses will require rigorous water accounting; access to the resource by legal, transparent and enforced permits; and a market where permitted water can be sold by users that have more than enough to other potential users that can make better use of it. This will involve exclusion of the weaponization of water for the exercise of power and dismantling market distortions such as agricultural subsidies that are rarely aimed at better water management.

Tony Allan spent his final years of service from 2015 to 2021 emphasizing: 'Water is managed by farmers, who all work in the private sector' and that approximately 90% of all water withdrawn from nature is consumed in global food systems where it is subject to a 'failed market' because it is nowhere valued in supply chains and by consumers. Standing on his shoulders, we will examine why there is still no proper valuation of water and go on to suggest a framework for introducing markets to alleviate this most neglected crisis.

Introduction

The global water crisis is not a distant threat: it is here and now. Water resources are heavily depleted in every densely populated place and in every major agricultural region; extreme examples cry out in India, the Middle East, Southern Europe, Central Asia and North America. Soil water and, thereby, groundwater recharge is badly managed almost everywhere, whether by neglect or ignorance. And water pollution is on the rise everywhere, and not only in areas with low water management capacities. In essence, water management is much overlooked by society.

This is not new. Thirty years ago, the Dublin Principles declared water to be *an economic good*, purposely highlighting its crucial role in the present and future well-being of economies and societies. Yet, there is still confusion about how these principles may apply in practice, not least because of the dichotomy between the private actions of farmers in the front line who control the fate of effective rainfall in its entirety and, thereby, recharge of streams and aquifers and allocation of abstracted water to various water users, which lies predominantly in the domain of the public sector, especially in emerging and developing countries. This means that scarce water is generally not allocated by markets but rather by political interests and, in some such examples, water becomes an instrument of political power.

Despite a lot of talk about increasing the role of the private sector in water management, few changes have taken place to date and those that have taken place have been less than convincing – although billions of dollars have been spent on purportedly equitable water management to help farmers and other sections of society in the name of social equity, economic advantage and environment sustainability. Is there a different way to manage and allocate water that increases not only physical efficiency but also economic efficiency? Might it include markets where appropriate? And could it reduce public sector expenditure?

The Dublin Principles did not gain immediate traction among water managers – especially the fourth principle: 'Water has an economic value in all its competing uses and should be recognised as an economic good.' However, within a few years – possibly as a result of donor pressure in the case of client states – policy-level statements began to appear along the lines of: 'we now acknowledge that water is an economic good, therefore we are going to charge for it volumetrically'. Such notions are nonsense: they confound or concatenate volumetric charges (which have no allocative characteristics) with resource prices (which do). So, instead of seizing the beneficial opportunities arising from a better understanding of water as an economic good, public-sector water managers (and in many cases their development partners) have continued to skirt around the innate problems that water imposes on society. These problems arise directly from water's very fluidity and its awkward juncture with every aspect of life, the wider economy, society and the environment. By analogy with the concept of a *wicked problem* (see Bromwich et al., 2022, in this issue), water has also been called a *wicked resource* (Grafton, 2017). Here we take a fresh look at how water is managed and allocated in the real world, and at the options for establishing market mechanisms to allocate scarce water resources.

The value of water

At the conclusion of the International Conference on Water and the Environment in Dublin in 1992, water experts from around world promulgated the Dublin Statement on Water and Sustainable Development, also known as the Dublin Principles. They recognized: (1) that water is a finite and essential resource; (2) the importance of women in water management; (3) the need to take a participatory approach to water management; and (4) that water has an economic value and should be treated as an economic good (Lane & Jarman, 1998).

Thirty years on, although there are frequent calls to treat water as an essential resource that poses significant risks to global prosperity and security (OECD, 2017; World Economic Forum, 2015), little has changed: several of the most significant river basins such as the Nile, Mekong, Euphrates and Tigris are approaching the abyss (Zeitoun & Mirumachi, 2010) and depletion of global groundwater resources has more than doubled (Siebert et al., 2010; Wada et al., 2010). Recently, water emergencies have been declared in various predominantly agricultural areas due to prolonged droughts (in California,

Australia, the Mediterranean region, Africa, South, Central and West Asia). Some water analysts have warned that 4 billion people may face water scarcity by mid-century (Mekonnen & Hoekstra, 2016). We observe that the first Dublin Principle has become a self-fulfilling prophecy.

The other three principles have no less urgency. The discourse around women features mostly in discussions around water, health and sanitation in which women play a key role fetching water for their households and to ensure community health (Sinharoy & Caruso, 2019): yet, water policy and planning remains the domain of mostly male, cornucopian engineers, who believe in technological solutions to confront the crisis without the need to induce institutional change first (Molle et al., 2009). And with respect to the third principle, despite several frameworks such as Integrated Water Resources Management (IWRM), the water–energy–food and other nexus constructs, or the water stewardship concept, no real breakthrough has been achieved. In his final years, Tony used every opportunity to contrast the fact that 'farmers manage water' with the pitiful involvement of crucial stakeholders such as farmers in water management (Allan, 2019; Allan & Dent, 2021). However, the fourth principle could be the game changer: if water were properly valued as an economic good it could, at last, be treated as a finite and essential resource whilst giving stakeholders the role they require to manage and allocate it effectively. Again, Tony used his wit to distinguish between *small water* and *big water*, to highlight the overwhelming role of agriculture in water consumption. Big water, or *food water*, accounts for more than 90% of all water abstracted from nature: small or *wash water* has a much smaller footprint and it can be treated and reused (Keulertz & Allan, 2017); the same goes for industrial water.

Several debates emerged in the 1990s and thereafter on how to interpret the principle, that water has an economic value. Perry's (1998) insight on potential impacts on different uses of water such as irrigated agriculture, water markets or financial instruments such as tradable water options, concludes that complex institutional arrangements need to be in place for a successful economic valuation of water resources. He also pointed out the complex relationship of farmers in society, exemplified through deliberate market distortions such as agricultural subsidies that are rarely aimed at better water management but, rather, at societal needs such as food security. Thirty years on, the fourth Dublin Principle remains the most difficult to implement.

At this point it is essential to understand that, notwithstanding possible issues of food security and social safety nets, the fourth Dublin Principle means that agriculture has to compete for its share of the available water, and hence that water markets provide the arena in which that competition takes place. But note also that, as explained below, the kind of market suggested here is for water that although originally allocated to a particular user is no longer needed by them as a result of technical and institutional measures that: (1) reduce their demand for water; and while, (2) increasing the value of its use.

Facts **and** ***trends*** **to instigate better water management**

What, exactly, is *effective* and *good* water management? Perry (2013) noted that *effective* water management is objectively measurable when policy outcomes achieve the desired output but *good* water management is far more subjective. It often depends on the advocate, as witnessed in several initiatives by companies in the food and beverage sector whose objectives may not necessarily be in line with those of the rest of us. Similarly, Giordano and Shah (2014) went as far as to call for a complete abandonment of coarse-grained concepts such as IWRM in favour of more-fine-grained, pragmatic concepts that may work in a given catchment.

Certain germane facts can be observed since the Dublin Principles were promulgated. For instance, in the period from 2000 to 2020, globally traded agricultural commodities have nearly doubled in terms of weight (890 million to 1.6 billion tonnes) and tripled in value (US$405 billion to US$1.2 trillion; Chatham House, 2022). The volatility of global food markets in 2022 after the Russian invasion of Ukraine underscores how much has changed in recent decades; trade has been globalized with a lesser proportion of exports from food bowls that cater for many export markets (*The Economist*, 2022). Climate change has brought greater water-resource challenges, whether floods or droughts, that curtail crop yields (IPPC, 2021). At the same time, demographic change and urbanization have intensified dramatically (Riddell & Gujja, 2019); since 1992, the world population has grown by from 5.4 billion to 7.7 billion (UNDESA, 2017).

In addition to changing facts, come trends. Water has become a vital resource in the supply of energy, exemplified in the water–energy–food nexus (Mohtar & Daher, 2012; see Matthews et al., 2022, in this issue). Diets have changed from traditional patterns to more meat-based Western patterns in developing and developed countries alike – in fact, what used to be a Western diet has become a global phenomenon (Willett et al., 2019). Such diets are generally moreover more water intensive, which compounds the challenge. Against a background of such challenges, digitization of technology can be especially helpful in quantifying available water resources and determining water use, for example, China's water conservation project making use of geographical information system (GIS) data to inform decision-making (Cho, 2018).

These facts and trends inspire or, even, require a fresh look at the fourth principle in order finally to achieve progress on how to value water as an economic good. Any such discussion should include a market as a basis for improved water management and allocation. Hirshleifer et al. (1969) were amongst the first to examine how water could be managed by economic principles based on market conditions. They cautioned decision-makers that: 'water markets effect a major change, rather than a merely marginal modification, in the trend of current practice and thought on water supply problems' (p. 69). An immediate issue is that water is characterized by various opportunity costs; as Rougé et al. (2017) put it: 'the marginal opportunity cost of water refers to benefits forgone by not allocating an additional unit of water to its most economically productive use at a specific location in a river basin at a specific moment in time' (p. 1).

This alerted modellers who sought to compute water to its most economically productive use in, for example, river basins. An example from the Murray–Darling basin in Australia shows that such attempts were flawed given political considerations along with simple human greed – factors which do not adhere to strictly rational decision-making. Beasley (2021), himself a senior counsel, describes how the law of the land and international agreements that stipulated prior partitioning of water resources to the environment have been circumvented for private gain with the connivance of the government of the day. The fact remains that water is a common good, subject to political considerations that have often nothing to do with *effective* or *good* management principles; water has to deliver *inter alia* food, drinking water, sanitation, ecosystem services, industrial uses, tourist uses, energy, farmer welfare and urban prosperity; it is required to be all things to all people. And focusing on economic matters, it would be best if it can be allocated to its economic value; that is, it could become an economically mobile resource, being used for services that provide the best returns in an economy, which is best ensured in a market. Hence, we aim to establish what a market entails.

Accounting as the foundation for water markets

Before establishing a water market, there needs to be a rigorous assessment of the water resources in any given catchment. This must include historical data to understand the lowest inflows into water storage (lakes, rivers, streams, groundwater) over a long assessment period of, for example, 100 years. This routine accounting of water storage must be continuously followed up by water authorities to provide an allocation announcement on a given day of the year – for instance, it could be made after the rainy season – and must be regularly updated according to actual rainfall and inflows.

Most importantly, authorities responsible for allocating water resources must acknowledge the different needs of different water users. First, there are the systemic high-security entitlement users that generally include municipalities for drinking water, industry, livestock and agriculture; these priorities are usually specified in a country's water policy. There may also be environmental flow regulations as in the case of the Murray–Darling but environmental flows also vary from catchment to catchment. Moreover, there must be possibilities for individual water users to manage their water through mechanisms that better meet their needs and plans, for example, water users may be allowed to keep a certain volume of water from the previous year's allocation in storage for use in the next year (but, if the water storage facility breaks, the accounts must set to zero and there must a deduction of the total volume of water carried over from the pool for general allocation in the next year). What is crucial is continuously accounting for water, which can be achieved by technology (for instance, smart water metres to track water use in storage from month to month and year to year including an adjustment for evaporative loss). Technology can also be applied to provide information for a centrally managed public registry of entitlements, seasonal allocations, actual deliveries, temporary and permanent trades (how much to whom and where, from whom and where, at what price – all three components of licence, entitlement and delivery share), what subsystem the water is sourced from, and any conditions attached.

Beyond accounting, the preconditions for success and effectiveness of a water market powerfully include a basic allocation process that sets limits on the volumes of unused water that can be traded through right-in-use and on the licensing process for permanent sale of entitlements. This is made more complicated by a permit to use a given amount water in a certain location over a certain season, a water entitlement (m^3 as a percentage of the permitted amount) and a service (in reticulated systems) that enables the water to be delivered to the permitted location. Trading rules – a form of allocation procedure – are also implemented where, for example, it is not physically possible to move water from the source of the seller's location to the prospective buyer's property. To take a specific example from the mid and lower Lachlan River in New South Wales, where seepage and effluent creek flows are so large that moving 1 m^3 downstream might require a release of

7 m^3 to ensure delivery (Thurtell & Wetti, 2011), so downstream trades are not allowed. Salt content and accumulation can be another constraint. There are also capacity constraints in parts of the main rivers – which are the primary conduit of bulk supply – and may, therefore, restrict the ability to deliver extra volumes from certain upstream reaches to reaches downstream.

Other regulatory requirements may also need to include measures such as annual use limits (AUL) for irrigation water – in order to prevent environmental externalities such as salinization and waterlogging from rice production or pasture which could arise where farmers are able to access unlimited amounts of additional water from their water market. There are also processes to track annual crop water demand (ET),[1] so that if rainfall is low and ET high, the AUL can be seasonally adjusted to allow full crop water demand to be satisfied providing that a farmer can access enough water (market again). This procedure also has to be publicly accessible, generic and clearly communicated.

Establishing the water market

First, it must be understood that although volumetric water charges are a perfectly rational way by which to recover the operational and capital costs of water resources management and service delivery, they do not represent the *value* of water. Neither do volumetric charges have the intrinsic allocative characteristics needed to move surplus to shortage. The value of water, sometimes called the *resource price* can be simply defined as the price a willing buyer will pay to a willing seller for a given quantity of water, of a particular quality, at a specific location, at a specific time. It is important to understand, however, that if there is no competition for the water as in the case of water allocated by the regulator under a permit system, then it has no intrinsic resource price. But this does not mean there may be no buyer for a portion of the permitted water that is not needed by a permit holder. If such a transaction reduces the opportunity cost of the water involved, then that water has been allocated in a way that increases the economic efficiency of its use. As such, the water can be said to be economically mobile. However counter-intuitive this may appear to be, under the right conditions and with a suitable institutional landscape, economic water use efficiency increases both environmental flows and access to water by the poor (Cai et al., 2001).

However, this introduces a fundamental and knotty problem, as well as an institutional challenge.

The fundamental problem is that unless the state itself becomes one of the willing buyers, then it is faced with an insurmountable challenge when trying to value water using resource prices – because these are determined by factors beyond a state's realistic ability to monitor. The institutional challenge can be summarized by the question: If the state is not institutionally equipped to allocate water using its resource price, is there an institutional concept that could? Here, we argue that a *well-regulated water market* could, indeed, allocate water according to its resource price. But, first, we need a brief diversion to argue that prioritizing increased economic water-use efficiency (agglomerated at an appropriate scale) is the right way to solve the problem of inefficient physical water-use efficiency.

In the world of water, there is a persistent obsession with fixing *physical water-use efficiency* without any consideration as to how the water involved could be reallocated in a way that allows the benefits of increased physical water-use efficiency (which by definition are local) to be transformed into *economic water-use efficiency* (which should more broadly benefit the economy, society and environment). Physical water-use efficiency can be thought of as the ratio of water actually used to the net amount of water withdrawn from source. In water supply, for instance, physical water-use efficiency is compromised by non-revenue flows, while in irrigation it is compromised by low return flows and badly regulated withdrawals from source. With this in mind, it is important to understand two more key principles.

The first involves Jevons' paradox that shows that demand for a resource is directly proportional to the efficiency with which it is used (Grafton et al., 2018). In the case of irrigated agriculture, Jevons' is manifested by reduced return flows – because irrigators use the 'saved' water to expand their irrigated area, or to change their farming systems to thirstier but more profitable crops. The reader will note the inverted commas framing the word *saved*. This brings us to the second principle. A commodity can only be saved by its owner. In most polities, ownership of water is vested in the state, which manages it on behalf of its people and environment. Thus, a farmer who has increased his or her physical water-use efficiency and who cooperates with the state to avoid Jevons' paradox has not saved any water; they have simply not used some of the water they had hitherto been using or were permitted to use. Invoking Jevons' paradox does nothing in the absence of regulated withdrawals and incentives that reduce the amount used.

The first matter to note with respect to regulated withdrawals is that the term *not-used* is meaningless unless there is a specified amount of water that is available for use. In a well-regulated water sector, this is defined by a water-use permit. This is not the same as a water right, which is something that the state must guarantee, as in the United Nations (UN) right to water through Resolution 64/292 (UN Water, 2020). Rather, it is an indicative maximum amount of water that the state has decided is available for use by a particular user during a specific period of time and from a specific body of water. But, in some years it may be necessary to reduce the amount actually allocated to a permit holder. This could be due to inadequate refilling of storage infrastructure as a result of a poor preceding rainy season, or it could be a precautionary measure if long-term forecasts suggest that an unusually dry season lies ahead. Either way, whether it is the full amount covered by the permit or merely an allocated subset thereof, the regulator has granted the water to the user on the assumption that there is no competition for it. This is important because any resource for which there is no competition does not have an intrinsic resource price. This is another invalidation of a state's claim that a volumetric charge recovering the capital and operating expenditures of water resource management and service delivery reflects the value of water. The only value that permitted water has, is what is sometimes called the *added value of water*. The added value accrues to water resource management, storage, treatment, service delivery and regulation etc., and has nothing to do with water's resource price, if any.

Which brings us, at last, to water markets. By providing a mechanism that facilitates a response to pricing signals, a market provides a way that a commodity not used by its permit holder can be sold to someone who values it more. In this way, a well-regulated market is the best way to distribute a scarce resource. And this entire article is predicated on the assumption that water is, indeed, scarce in relation to demand.

However, the concept is widely misunderstood by governments, many of which think water markets mean that they must sell water. It does not – although there is no reason that a government should not participate in a water market. It could make a great deal of sense if governments could buy permitted water that is not used by the permit holder. Another objection raised by governments, especially in Islamic countries, is that water is a gift from God and cannot be bought or sold. The same could be said of land. Nonetheless, water prices have to be overseen by a public regulator to ensure fairness – that is, that the price should be limited to that which a willing buyer will pay a willing seller at the applicable time and place. This is precisely the resource price that transfers water from someone who does not need it to someone who needs it more in such a way that reduces its opportunity cost. A water market intended to facilitate the economic mobility of water is all about the trading of permitted water that is not used but which may have a resource price because someone wants to buy it. This is not about a market for tradable water rights, or permits, it is about the water itself – sometimes called the *right-in-use*. As with any marketable commodity, someone with unused but permitted water simply informs the market, wherein someone that needs water offers a price. Because water is fungible, this means that the unused water remains in or returns to source and the buyer withdraws it elsewhere in the water resource. This is much easier to say than do but, if functional water markets are an appropriate way to allocate scarce water in a way that increases the economic efficiency of water used in a river basin, the difficulties must be faced and dealt with. The same has been true of, for example, many surgical procedures that went through a process of trial and error before becoming successful and widely applied. Necessity is the mother of invention.

What are the challenges?

Surely, there are many challenges – not least political objections and uncertainties that will be difficult in countries where water scarcity is an instrument of political power – but, essentially, a system of water permits has to be established. Permits, however, are meaningless in the absence of cumulative flow metres to measure both withdrawals and return flows. This will be costly and further political decisions may be required as to who pays. While it may be perfectly reasonable for a water-user association to pay, it would be unreasonable to expect a small-scale irrigator in a poor country to pay. This introduces another difficulty: scale.

While it may be possible for an individual farmer not to use his or her entire allocation, there is unlikely to be a market for the small amounts of water that would be involved. To solve this problem, it is necessary to aggregate individual surpluses into larger quantities that will be of interest to a market. There are two ways to do this, and they are not mutually exclusive. A group water permit could increase any surplus for trading purposes. Alternatively, any quantity of water could be consolidated in a water *bourse*. Users wanting water would make an offer to the bourse and, if their price is accepted, they could withdraw the amount of water they have purchased. With this in mind, and as argued by Beasley (2021), participation by the state as buyers in such a market would be far more effective than rolling out expensive programmes of so-called irrigation modernization with the embedded threat of Jevons' paradox. Proceeds accruing to the sales would then be passed back to the provider in proportion to their investment. Of course, the bourse would take a percentage and a sales tax could be levied from the sale and allocated to recovery of some of the added value costs of water service.

An example of a functional water market

One of the authors visited the Jati Lahur Basin in Indonesia while on mission for the Food and Agriculture Organization (FAO) in 2004. He observed that, after generating hydropower, water flowing through the Jati Lahur dam (situated some 70 km south-east of Jakarta) passes through seven water-user associations, each with a volumetric water use permit shared between their smallholder membership. By a combination of increased physical water-use efficiency and capacity-building, the farmers learned to grow more rice with less water. They could have used the balance of their right-in-use to expand rice production (Jevons' paradox refers) but opted for a different use of the water to which they were entitled but no longer needed. Together, they decided how much of their right-in-use could be spared and this was left in the river to flow downstream to Jakarta, where it is withdrawn for use by bottling enterprises who pay the water users.

As a result, the economic efficiency of water use increased in the small basin for the following reasons: there was no need to change the dam's operating rules so the energy generated by hydropower was not compromised (Brunner et al., 2019); and farmers' incomes increased, not only because they were growing more rice, but also because there was a market for the water they no longer needed (and the supply was very stable because of the dam and its operating rules). The market was, of course, the botting industry in Jakarta.

However, order to reach that market, the water had to flow from the rice fields to the bottling plants. This is important because one of the reasons increased economic efficiency also leads to increased access for the poor, the environment and other economically relevant uses (especially non-consumptive uses) is that economically mobile water is also physically mobile water. Accordingly, the increased environmental flows downstream of the irrigation offtakes resulted in increased capture fisheries. Thus, a simple market arrangement between the water-user associations and the bottling industry also produced economic and environmental benefits.

Three further lessons may be drawn from this example of a functional market. First, other than the water-use permits, the arrangement was informal yet highly successful. Second, the additional benefits (fish and bottled beverages) increased the benefits that previously accrued to the dam (which were otherwise limited to energy and irrigation). Third, when one of the authors of this paper reported back to the Ministry of Water, he was told that this was indeed a water market but that water markets were illegal in Indonesia – so the government decided to call it something else that could be thought of a de facto water market. Here, we are calling for more of the same pragmatism with respect to what is still considered off-limits by many planners and policymakers.

Conclusions

For all the talk about water as an economic good, not much progress has been made to value water according to the Dublin Principles, in particular economic principles. So, we have taken the liberty of thinking outside the box. The first prerequisite to set up a water market is to physically account for water in order to allocate it to different water-user groups which, itself, presents complex challenges.

An institutional overview of how water markets could be set up to increase economically efficient water management practices immediately highlights knotty problems such as Jevons' paradox and the cultural challenges proceeding from weak government understanding about the difference between service charges and resource prices. There is no denying that widespread introduction of water markets will be immensely challenging, with many moving parts: political, institutional, infrastructural, legal and regulatory. But as Tony Allan continuously reminded us, 'we don't live in economies but in political economies' (Allan, 2011). Cutting the Gordian knot of political economies may be one of the biggest challenges to management and allocation of water in line with its economic value.

Even more challenging will be the achievement of global water security without the means to allocate scarce water according to economic priorities. We posit that all the challenges can be solved if regulators and market-based allocation mechanisms work hand in hand for the common good. The time available to solve the chronic crisis is running out.

Note

1. ET is defined as a combination of evaporation and transpiration, which is measured in order to better understand crop water requirements (Perry, 2013).

Disclosure statement

No potential conflict of interest was reported by the authors.

ORCID

Martin Keulertz http://orcid.org/0000-0003-1226-1403

References

Allan, J. A. (2011). *Virtual water: Tackling the threat to our planet's most precious resource.* I.B. Tauris.

Allan, J. A. (2019). Food, water and the consequences of society not valuing the environment, and J. A. Allan, B. Bromwich, M. Keulertz, & A. J. Colman (Eds.), *The Oxford handbook of food, water and society* (pp. 858–879). Oxford University Press.

Allan, J. A., & Dent, D. L. (2021). The cost of food: Consequences of not valuing soil and water and the people who manage them. In D. L. Dent & B. P. Boincean (Eds.), *Regenerative agriculture. What's missing? What else do we still need to know?* (pp. 3–20). Springer Nature Switzerland.

Beasley, R. (2021). *Dead in the water: A very angry book about our greatest environmental catastrophe . . . the death of the Murray–Darling basin.* Allen & Unwin.

Bromwich, B., Crilly, D., & Banerjee, J. (2022). Water governance and system coordination across diverse risk-management cultures. *Water International, 47*(7), 1037–1047. https://doi.org/10.1080/02508060.2022.2133074

Brunner, J., Carew-Reid, J., Glemet, R., McCartney, M., & Riddell, P. (2019). *Measuring, understanding and adapting to nexus trade-offs in the Sekong, Sesan and Srepok Transboundary River Basins*. IUCN: Viet Nam Country Office.

Cai, X., Ringler, C., Rosegrant, M. (2001). Examining physical and economic efficiencies of water use through integrated economic–hydrologic water modelling. In M. Marino & S. Simonovic (Eds.), *Integrated water resources management* (pp. 75–81). IAHS Press.

Chatham House. (2022). *Data/resourceytrade.earth.*

Cho, S. (2018). *China's experience in tackling water scarcity through sustainable agricultural water management*. World Bank Water Blog. Retrieved July 18, 2022 https://blogs.worldbank.org/water/china-experience-tackling-water-scarcity

The Economist. (2022, May 21). The coming food catastrophe. *The Economist.*

Giordano, M., & Shah, T. (2014). From IWRM back to integrated water resources management. *International Journal of Water Resources Development, 30*(3), 364–376. https://doi.org/10.1080/07900627.2013.851521

Grafton, R. Q. (2017). Responding to the 'wicked problem' of water insecurity. *Water Resources Management, 31*(10), 3023–3041. https://doi.org/10.1007/s11269-017-1606-9

Grafton, R. Q., Williams, J., Perry, C. J., Molle, F., Ringler, C., Steduto, P., Udall, B., Wheeler, S. A., Wang, Y., Garrick, D., & Allen, R. G. (2018). The paradox of irrigation efficiency. *Science, 361* (6404), 748–750. https://doi.org/10.1126/science.aat9314

Hirshleifer, J., De Haven, J. C., & Milliman, J. W. (1969). *Water supply: Economics, technology, and policy*. RAND corporation. https://OSTI.gov/USDeptEnergy

IPPC. (2021). *Climate change 2021: The physical science basis.*

Keulertz, M., & Allan, J. A. (2017). *What is food–water and why do we not account for it? in: The Oxford handbook of water politics and policy.* Oxford University Press.

Lane, J., & Jarman, J. (1998). Six years on-what happened to the Dublin Principles? *Waterlines, 16* (3), 10–11. https://doi.org/10.3362/0262-8104.1998.004

Matthews, N., Schoonbaert, B., & Burlon, E. (2022). Unexpected bright spots: How the pandemic, climate change and biodiversity loss are shaping the evolution of the nexus. *Water International, 47*(7), 1140–1146. https://doi.org/10.1080/02508060.2022.2133485

Mekonnen, M. M., & Hoekstra, A. Y. (2016). Four billion people facing severe water scarcity. *Science Advances, 2*(2), e1500323. https://doi.org/10.1126/sciadv.1500323

Mohtar, R., & Daher, B. (2012). Water, energy, and food: The ultimate nexus. In *Encyclopedia of agricultural, food and biological engineering* (2nd ed., pp. 1–5). Taylor & Francis.

Molle, F., Mollinga, P. P., & Wester, P. (2009). Hydraulic bureaucracies and the hydraulic mission: Flows of water, flows of power. *Water Alternatives, 2*(3), 328–349. https://www.water-alternatives.org/index.php/alldoc/articles/vol2/v2issue3/65-a2-3-3/file

OECD. (2017). *Water risk hotspots for agriculture*. OECD Studies on Water. https://doi.org/10.1787/9789264279551-en

Perry, C. J. (1998). *Water as an economic good: A solution, or a problem.* International Water Management Institute.

Perry, C. J. (2013). ABCDE+F: A framework for thinking about water resources management. *Water International, 38*(1), 95–107. https://doi.org/10.1080/02508060.2013.754618

Riddell, P., Gujja, B. (2019). Water and rice production: Looking to the future, and J. A. Allan, B. Bomwich, M. Keulertz, & A. Colman (Eds.) *The Oxford handbook of food, water and society* (pp. 606–630). Oxford University Press.

Rougé, C., Harou, J. J., Pulido-Velazquez, M., & Matrosov, E. S. (2017). *Estimating the economic opportunity cost of water use with river basin simulators in a computationally efficient way. Geophysical Research Abstracts* 1941 EGU2017-1634.2017 EGU General Assembly. 16348.

Siebert, S., Burke, J., Faures, J. M., Frenken, K., Hoogeveen, J., Döll, P., & Portmann, F. T. (2010). Groundwater use for irrigation – A global inventory. *Hydrology and Earth System Sciences, 14* (10), 1863–1880. https://doi.org/10.5194/hess-14-1863-2010

Sinharoy, S. S., & Caruso, B. A. (2019). On World water day, gender equality and empowerment require attention. *Lancet Planetary Health, 3*(5), e202–e203. https://doi.org/10.1016/S2542-5196(19)

Thurtell, L., & Wetti, P. (2011). *Environmental water delivery: Lachlan River*. Commonwealth Environmental Water.

UNDESA. (2017). *World population prospects: The 2017 revision*.

UN Water. (2020). *10th anniversary of the UNGA resolution on the human rights to water and sanitation*.

Wada, Y., van Beek, L. P. H., van Kempen, C. M., Reckman, J. W. T. M., Vasak, S., & Bierkens, M. F. P. (2010). Global depletion of groundwater resources. *Geophysical Research Letters*, *37*(20), 1–5. https://doi.org/10.1029/2010GL044571

Willett, W., Rockström, J., Loken, B., Springmann, M., Lang, T., Vermeulen, W., Garnett, T., Tilman, D., DeClerck, F., Wood, A., Jonell, M., Clark, M., Gordon, L. J., Fanzo, J., Hawkes, C., Zurayk, R., Rivera, J. A., De Vries, W., Sibanda, L. M., ... Murray, C. J. L. (2019). Food in the Anthropocene: The EAT–lancet commission on healthy diets from sustainable food systems. *The Lancet*, *393*(10170), 447–492. https://doi.org/10.1016/S0140-6736(18)31788-4

World Economic Forum. (2015). *Global risks 2015 10th edition*. WEF.

Zeitoun, M., Mirumachi, N. (2010). Transboundary water interaction: Reconsidering conflict and co-operation. In K. Wegerich & J. Warner (Eds.), *The politics of water: A survey* (pp. 297–316). Routledge.

When the virtual water runs out: local and global responses to addressing unsustainable groundwater consumption

Iman Haqiqi, Chris J. Perry and Thomas W. Hertel

ABSTRACT

Given the growing importance of groundwater in irrigated crop production, policies aimed at restricting groundwater use create fears of intensified food insecurity. Yet, a comprehensive quantitative analysis is required to evaluate the impacts of groundwater sustainability restrictions on food system. Using a multi-scale multi-system framework integrating economic and biophysical determinants of sustainability, we find that the local economic impacts of a groundwater sustainability policy are often substantial. However, due to market-mediated responses, including surface water substitution, expansion of rainfed production, relocation and virtual trade in blue water, the final impact on global food prices and production is surprisingly modest.

Introduction

Tony Allan's seminal insight, captured in the phrase 'virtual water', was that trade in agricultural commodities was effectively the transfer of huge quantities of water from producing countries to consuming countries – for example, producing 1 kg of wheat or rice involves the transpiration of about 1000 kg of water. So the location of this production can have a significant impact on groundwater abstraction.

Allan's main interest was in the consequence of these transfers: the extent to which political stability was preserved in countries that rely on imported 'virtual water' to ensure food security. There are two dimensions to this linkage. First, the avoidance of 'water wars' as the import of virtual water made it unnecessary to 'capture' the underlying natural resource itself; and second, as a direct consequence of this trade, enhanced food security. Few rulers of fragile democracies – and non-democracies – could survive long when the shops have no bread (Dizard, 2022).

In this paper, we explore another perspective on virtual water: many surplus food-producing countries are 'exporting' large quantities of groundwater that is being 'mined' from their aquifers. Stated bluntly, they are exporting their environment, free of charge, to food-importing countries. There will be consequences to this simple insight, and we explore, within a multi-scale framework, through a set of models and calculations, the impacts of a 'best case' scenario whereby groundwater consumption is constrained to

Supplementary data for this article can be accessed at https://doi.org/10.1080/02508060.2023.2131272

a 'sustainable' equilibrium by prudent management. What changes are likely in the pattern of commodity production, locally and globally, and what changes in prices can be expected as a consequence of such 'good governance'?

Sustainable groundwater use – some definitions

For groundwater, what does sustainability mean? Groundwater systems are often highly complex, comprising different layers that are sometimes directly linked, sometimes partially linked and sometimes completely independent, affected by the nature of the medium(s) through which, and in which, groundwater is stored, surface topology, land cover, lakes, rivers and artificial interventions such as irrigation systems. Here, the objective is not to address these physical variables in detail – though in any specific case, they are fundamentally important.

The commonly adopted definition of sustainable groundwater use is that abstraction should not exceed recharge. This apparently logical formulation is at best misleading (Bredehoeft, 2002) and useful analysis of the status of a groundwater system, and the consequences of over-abstraction, require a rather more careful specification of the scenario being evaluated.

A groundwater system is in *equilibrium* when inflows are equal to outflows. Inflows include natural recharge from rainfall, additional infiltration of water imported via surface irrigation systems, return flows from local groundwater abstractions and lateral inflow from surrounding hills or aquifer systems. Outflows include evaporation from wet surfaces and capillary rise, transpiration by plants, lateral flows towards streams or surrounding aquifers, and pumping. When these two sets of flows are equal, the water table remains, when observed over a period of years, essentially constant, varying seasonally and annually due to actual rainfall patterns, associated pumping rates, etc. Yet even from this simple formulation we see immediately that any increase in net abstraction from an aquifer – for example, by introducing a new tubewell – must have implications for one or more of the outflows. The water table must reach a new equilibrium, at a lower level at which one or more outflow(s) is reduced – lateral outflows to streams, or water consumption by natural vegetation, flows to surrounding aquifers or evaporation from waterlogged soil. When this has happened, inflows and outflows again equate, and stability is re-established.

Already, the definition of 'sustainable' becomes more complicated than is implied by the simple rule that abstraction should be less than recharge (indeed, if abstraction is less than recharge, the water table will rise over time). In the example just described, *equilibrium* is restored after an increase in abstraction, and again, over a period of years, the observed status of the groundwater system will remain constant. Yet the impacts of the reduced outflows to vegetation and local streams may cause damage to a downstream wetland or aspects of the natural landscape which were previously supported. Environmentalists might well argue that these are not 'environmentally sustainable' outcomes as generally interpreted: rather they are a new, less desirable, equilibrium. Downstream users – irrigators, fishermen, ferry operators, etc. – will similarly have negative views of the new, 'sustainable' scenario.

As the process of increased groundwater demand for irrigation and other sectors unfolds, we pass through a continuum of equilibria, in each of which the stability test (inflows = outflows) is met (e.g., Chinnasamy & Agoramoorthy, 2016). Eventually all the natural outflows fall to zero, and incremental abstraction is supported by a continuous reduction in aquifer storage – a scenario that is literally 'unsustainable' because eventually the aquifer will be depleted, or salinized, or otherwise rendered unfit for use. During this stage, there are no points of equilibrium. The water table continues to fall until eventually the final equilibrium arrives when an aquifer is depleted to the point that use is restricted to whatever recharge still, sporadically and unpredictably, reaches the saturated zone. The 'mining' element, in this scenario, necessarily reduces to zero.

Why is this more complex perspective on 'sustainable' groundwater use important? It is commonplace to report that, because measured abstraction is less than recharge, there is 'net availability' of groundwater for development (e.g., Chatterjee & Ray, 2016) This position is apparently based on the assumption that the 'available' resource is currently disappearing somewhere, because such a conclusion is rarely (never?) associated with a reportedly *rising* water table. Alternatively, it might be assumed that the outflows that are impacted by further abstraction have no value. In this phase of development, the properly evaluated 'sustainable' yield is, or should be a *political* decision based on trade-offs among alternative water allocation regimes – as such it is an issue of governance, to which information (measurements, observations, modelling) contribute to political decisions that determine allocations. Here we include in the spectrum of political interventions the decision to do nothing and allow the tragedy of the commons (Hardin) to unfold. While to our knowledge, this policy is never made explicit, this scenario is not rare.

In fact, all water resources development (e.g., diversion from rivers, construction of dams, water harvesting, de- or re-forestation, conversion to new irrigation technologies) involves some degree of reallocation of water among users and uses. Food security, and rural employment are increased when irrigation water is diverted from estuaries and wetlands. While such developments have vastly increased food production, and hence have benefited society generally, they are not without costs, and it is the role of governments, through political processes, to evaluate the underlying trade-offs and intervene appropriately.

In sum, there are four stages of groundwater development: (1) the natural state of precipitation, vegetation and runoff; (2) progressive human interventions that expand use (agriculture, domestic water supply, etc.) at the expense of other outflows, while maintaining equilibrium between inflows and outflows; (3) groundwater mining, where outflows exceed inflows, and the aquifer is progressively depleted; and (4) effective depletion of the aquifer, often associated with: (a) subsidence of the land surface, which in many areas is already ongoing and substantial (Galloway & Burbey, 2011); (b) compaction of the soil, so that infiltration is restricted and storage of soil moisture in the profile is reduced; and (c) extended time for any infiltration to reach the saturated zone, due to a combination of (b), above, and the ever-increasing depth to the water table. Unfortunately, many aquifers around the world are at the third stage listed above, and in the absence of interventions by the relevant authorities, they will automatically progress to the fourth stage of irreparable damage.

In this paper, we assess the impacts of restricting groundwater abstraction to sustainable levels – halting 'mining' of groundwater, and returning to stage 2, above. This intervention results in the redistribution of production and trade, and associated local and global price changes. Achieving this goal would constitute improved governance, requiring interventions by the relevant authorities to reduce groundwater abstraction in many areas by defining allocation policies, introducing laws that reflect those policies and providing institutional arrangements to enforce the laws.

These interventions will be politically challenging, and some countries will fail. In the following analysis we attempt to evaluate the local and global economic costs associated with this move to 'good governance'. Such governance would allow for continued, controlled use of aquifers as temporary buffer storage and preserving a resource that can be assigned to priority uses during periods of severe drought. The alternative of allowing aquifer depletion and loss of the buffer function will undoubtedly be far more costly – an issue to be addressed in a further study.

The state of the world's groundwater

Various analysts have assembled data and models to assess groundwater status globally and for major regions (Famiglietti, 2014; Siebert & Döll, 2008, 2010; Wada et al., 2012, 2014). While estimates vary depending on methodologies and time periods analysed, there is consistency among all analysts that current irrigation, in major agricultural areas, is dependent on groundwater abstraction rates that are depleting the underlying aquifers.

Added to these concerns, the rate of increase is remarkable. According to Wada et al. (2014) global aquifer depletion almost tripled between 1960 and 2010, and is projected to almost double again by 2099 (though whether these quantities of water actually exist must be questioned). Among individual countries, the increased rate of depletion between 1960 and 2010 was most pronounced in India (almost tripling) and the United States, China and Pakistan (roughly doubling). In Saudi Arabia, fortunately an outlier, consumption increased sixfold – mostly due to the now-abandoned policy of pursuing self-sufficiency in wheat, but abstraction is still driven by substantial production of irrigated fodder.

The next sections address the likely impacts – locally and globally – of the reductions in water supply that would bring stability to aquifers. As background, the expectations of Wada et al. and of Famiglietti are quoted in full:

> The current degree of non-sustainable use may compromise the future livelihoods of millions of people and their living standards.(Wada et al., 2014)

> Vanishing groundwater will translate into major declines in agricultural productivity and energy production, with the potential for skyrocketing food prices and profound economic and political ramifications.(Famiglietti, 2014)

Methods

Experiment design

To permit a unified analysis of groundwater sustainability policies, we pair a hydrological model with an agro-economic model of global crop production at the level of 5--arc minute grid cells (Figure 1). These two models have been successfully linked (Woo et al., 2022) and have shed light on the impacts of surface water scarcity (Liu et al., 2017),

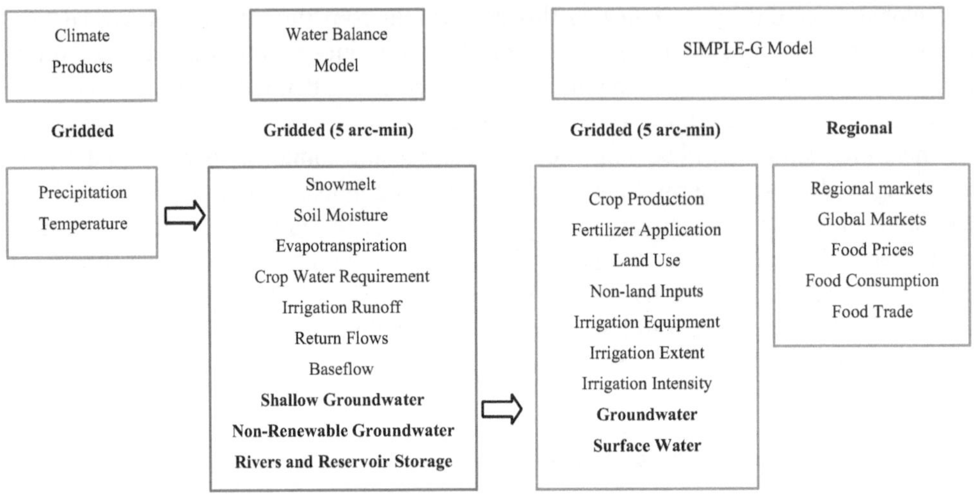

Figure 1. Employing hydroclimatic information to inform the economic model about irrigation water availability and non-renewable groundwater irrigation. The Water Balance Model (WBM) is a global gridded framework for modelling water mass balance at each grid cell. It simulates the vertical water exchange between atmosphere and land as well as the horizontal flow of water through river networks, baseflow and run-offs. Daily precipitation and temperature from climate products as well as information regarding agriculture and water demand at each location are the input to WBM. The simulated changes in surface water and groundwater volumes are fed into the SIMPLE-G model where decisions about land and water use are modelled. The regional outcomes are determined via the interaction of gridded crop outputs and regional as well as global market responses.

as well as the yield impacts of compound hydro-climatic extremes (Haqiqi et al., 2021). Here, we employ the outputs of the Water Balance Model (WBM) following Grogan et al. (2017). WBM is a validated and widely used macro-scale hydrological model (Wisser et al., 2010; Grogan, 2016; Grogan et al., 2022). Consumptive water requirements are calculated by crop and by growth stage, based on soil moisture, temperature and irrigation status. The sources of irrigation in WBM can be reservoirs, rivers, shallow groundwater and non-renewable groundwater if available. The model also considers non-beneficial consumption and return flows through irrigation runoff, baseflow between groundwater and surface water, and percolation of irrigation return flows to shallow groundwater. The multi-scale nature of this model, building from the grid cell to global hydrology, makes it an appropriate model for this study.

To assess the sustainable level of groundwater abstraction, we follow Grogan (2016, 2017), wherein the maximum allowed level of groundwater use is determined such that there is zero non-renewable groundwater use, thereby ensuring a stable equilibrium level of groundwater. For each grid cell, we calculate the required reduction in abstractions to achieve this objective. This level is calculated by running WBM over the period 1980–2009, while not allowing abstraction from non-renewable resources. The sustainable level of abstraction varies each year depending on weather conditions. To avoid excessive complexity and in order to focus on long-run conditions, we take the 30-year long-term average as the basis for determining the sustainable level of abstraction. We use historical global observational climate products based on ERA-interim (European Centre for Medium-Range Weather Forecasts-reanalysis; Dee et al., 2011), MERRA (Modern-Era

Retrospective Analysis for Research and Applications; Rienecker et al., 2011), NCEP (National Centers for Environmental Prediction; Saha et al., 2014) and UDEL (University of Delaware; Willmott & Matsuura, 2001). The 30-year average sustainable level of abstraction is used to inform the economic model about long-run groundwater availability. There are other possible definitions of sustainability and we will discuss the implications of this choice later.

One challenge we face is that SIMPLE-G-Global is constructed based on economic and agricultural information *c.*2017. In other words, the crop production, cropped area, yields and prices reflect a more recent market condition than that reflected in the historical weather data. In addition, the size of SIMPLE-G grid-cells is 5 arc-min, while the historic WBM runs are for global 30 arc-min grid cells. Having determined the long-run historic sustainable level of groundwater withdrawals, by grid cell, we take the current conditions of water resources based on recent WBM runs from 2012 to 2018 based on GLEAM v3 (Global Land Evaporation Amsterdam Model; Martens et al., 2017) at 5 arc-min grid cells. This allows for consistency with the 2017-based economic model. We assume all the smaller (5 arc-min) grid cells follow their underlying larger grid-cells in terms of non-renewable groundwater rates.

Table 1 summarizes the situation for major groundwater users in the world. Overall, 27% of irrigation water consumption is linked to non-renewable resources. However, this is substantially higher in some regions: 62% for Iran and 39% for India. Compared with total global crop water demand, the non-renewable groundwater contribution is around 6% (although this is 43% for Iran and 23% for India). The outputs of the WBM in terms of groundwater and surface water use in crop production are in line with other studies in the literature (Bierkens & Wada, 2019; Chapagain & Ysbert Hoekstra, 2011; Mekonnen & Hoekstra, 2020; Siebert & Döll, 2010; Gleeson et al., 2012).

Figure 2 shows the distribution of cropped area relying on irrigation and non-renewable groundwater around 2017. 'Green' water is that provided to crops by rainfall; 'blue' water comprises irrigation water from surface and groundwater. Figure 2 shows the hotspots of groundwater crisis are most prominent in the United States, China, India, Pakistan and the Middle East where crop water demand in some of the grid cells is obtained nearly entirely from non-renewable resources. As seen here, the problem of non-renewable groundwater abstraction tends to be quite concentrated.

Table 1. Total crop water demand and contribution of non-renewable resources for major groundwater users, *c.*2017.

	Total crop water demand	Green water contribution		Blue water contribution		Non-renewable groundwater	
	(km^3 yr^{-1})	(km^3 yr^{-1})	%	(km^3 yr^{-1})	%	(km^3 yr^{-1})	%
India	493	203	41%	289	59%	112	23%
China	438	223	51%	215	49%	64	15%
Pakistan	101	2	2%	98	98%	21	21%
USA	428	374	87%	55	13%	13	3%
Iran	6	2	31%	4	69%	3	43%
Mexico	48	36	76%	12	24%	3	5%
World	3848	2934	76%	915	24%	244	6%

Note: The percentage shows the share in total crop water consumption.

Source: Outputs of global Water Balance Model (WBM) based on GLEAM v3. Aggregated from 5 arc-min grid cells (authors' calculations based on Grogan, 2016, 2017, 2022).

(a)

(b)

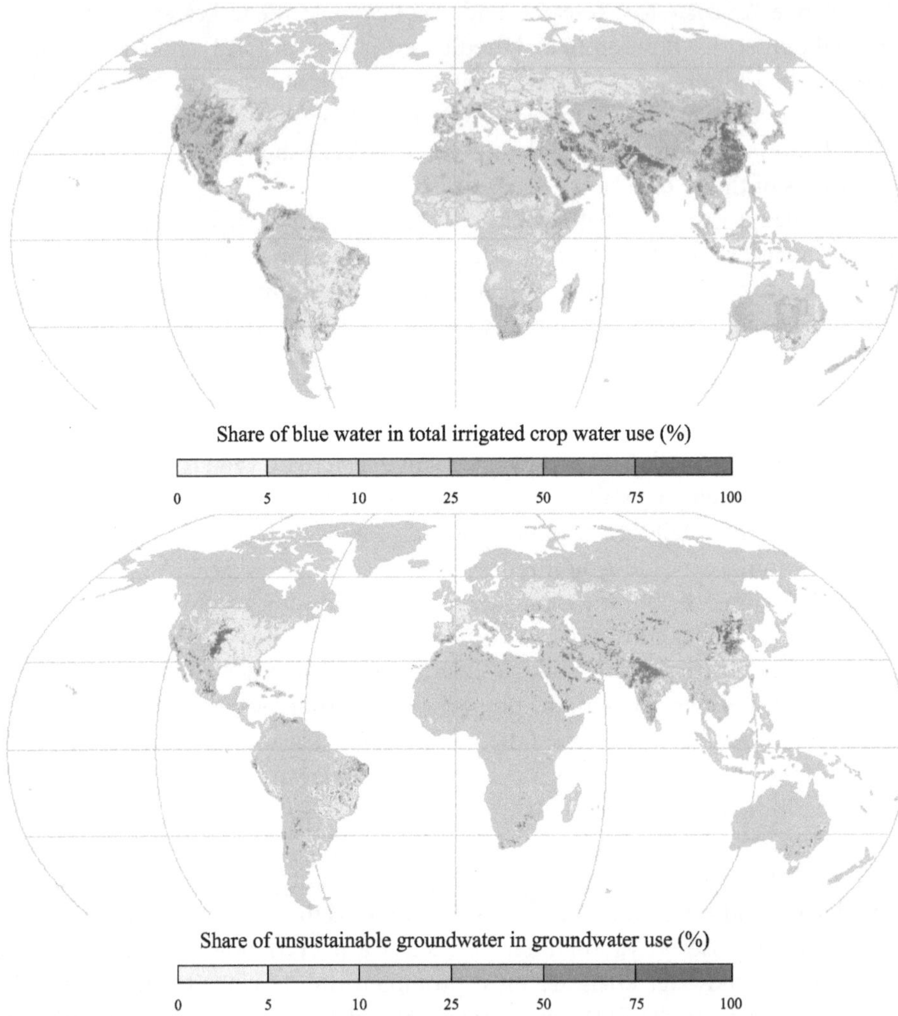

Figure 2. Spatial distribution and the significance of irrigation and unsustainable groundwater. (a) Share of blue water in total gridded crop water consumption around 2017. Red colouring means there is little green water available to crops. Therefore, if there is crop production in these locations, it must be irrigated. (b) Share of unsustainable groundwater in total gridded irrigation water consumption around 2017. Colour red means crop production mainly relies on unsustainable groundwater. Grey colouring refers to areas not irrigated.

To capture the local, national, and global impact and responses to sustainability groundwater restrictions, we employ a gridded economic model called SIMPLE-G (Haqiqi et al., 2020) with more than 1.3 million grid cells (5 arc-min, squares of width 9.26 km at the equator). Major components of regional food demand and gridded crop supply of the SIMPLE-G model are illustrated in Figure 3. Each grid cell represents a distinct unit of agricultural production on which competition for land and water resources play out, within and between rainfed and irrigated crop production. SIMPLE-G is a multiscale framework for the integration of economic and biophysical determinants of sustainability. It includes markets at regional, subregional and grid-cell

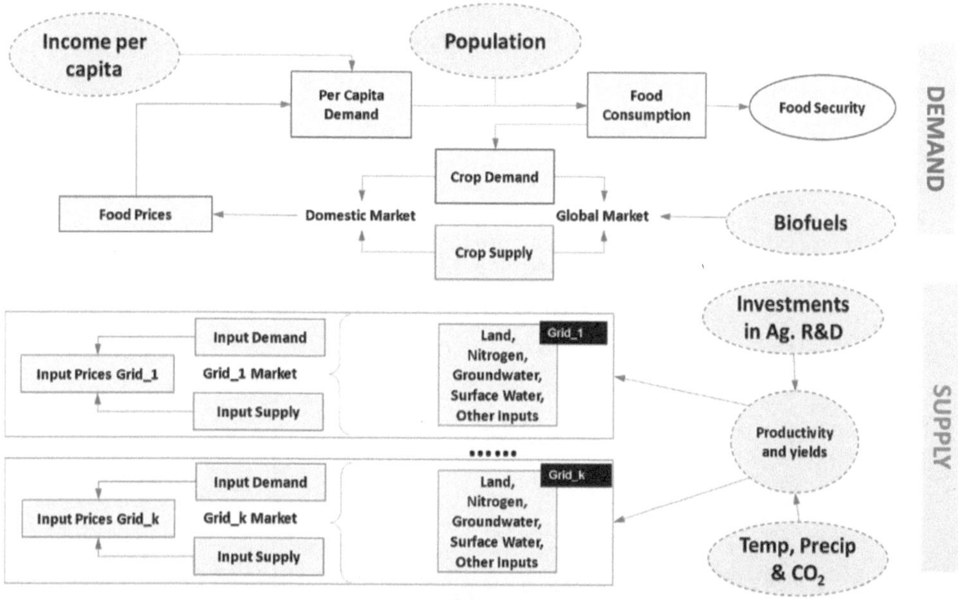

Figure 3. Overall structure of the SIMPLE-G model. The model determines the global and local prices of agricultural commodities in a supply–demand framework. The demand for agricultural products by global regions is considered as a function of population, per capita income, biofuel mandates and global prices. The demand endogenously determines the likely shifts in diets and the ensuing implications for food security. The supply side is modelled at the grid-cell level as a function of market prices, technological change and weather, although the latter two forces are unchanging in this study. The model determines the equilibrium local prices, land use, water use, fertilizer application and agricultural demand for other inputs.

levels. In the face of groundwater sustainability restrictions, the model solves for the gridded equilibrium level of irrigated and non-irrigated land, extraction of groundwater and surface water for irrigation and crop production. It also determines the new equilibrium local, national and global crop prices, along with domestic use and international trade in crops. Crops are consumed directly, as well as indirectly in the form of processed food, as well as livestock products, at the level of 17 world regions, with these consumer demands for foodstuffs responding to prices and per capita incomes.

To focus our analysis on the impact of groundwater sustainability restrictions, we assume no changes in regional population, income and biofuel demands. However, the demand responses to changes in food prices will be captured in the interaction of demand and supply at the regional markets. Also, we assume no change in local climate conditions to allow us to clearly assess the impact of groundwater restrictions alone. For a study considering compound socio-economic and environmental stressors within this framework, see Haqiqi et al. (2020).

The SIMPLE-G model does not model the behaviour of individual farmers, but the outcome at the aggregate level at each grid cell for all crops aggregated into a single composite commodity. The crop output of each grid cell is differentiated from other grid cells, reflecting these differences in crop composition. This product differentiation applies at both national and regional (17 world regions) levels. In any given grid cell,

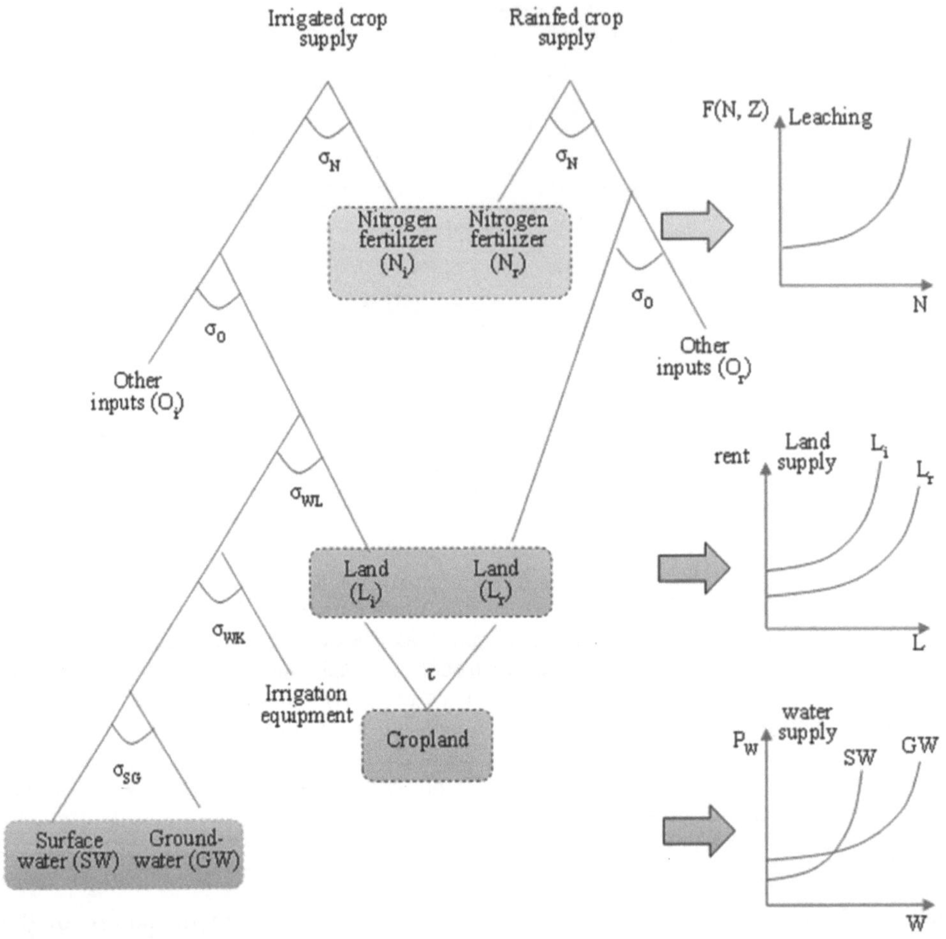

Figure 4. Schematic representation of grid-specific functional forms. SIMPLE-G assumes a supply–demand framework for production inputs at each grid-cell. (a) To calculate the demand for inputs, a nested constant elasticity of substitution (CES) function for irrigated and rainfed practices is used. This is a common function widely used in economics due to its flexibility in handling highly non-linear relationships working based on relative prices and production scale. Here σ shows the substitution parameter for each nest. The bottom nest (SG) determines the surface water (SW) and groundwater (GW) composite. Then the water-capital (WK) nest determines the combination of water and irrigation equipment. The water-land (WL) nest then determines water and land composite. Finally, the N and O nests determine how production combines WL with fertilizer and other inputs. (b) To calculate the supply of land, a constant elasticity of transformation is assumed with quantity-preserving characteristics. (c) To determine the supply schedule of groundwater and surface water, the supply elasticities are determined based on the volume of available water (Haqiqi, 2019).

yield, water use and nitrogen fertilizer use in crop production are weighted averages over all crops, but we distinguish between two farming practices within each grid cell: rainfed versus irrigated crop production. Figure 4 illustrates the gridded supply structure and agricultural production function with major inputs and parameters of the model for both irrigated and rainfed crop supply.

Virtual water trade is introduced in this version of SIMPLE-G to represent the water embedded in international trade of agricultural commodities. There is a rich literature around the concept, applications and modelling of virtual water trade (Allan, 1997, 2003, 2011; Dalin et al., 2017; Rosa et al., 2019). Here, changes in virtual water trade are distinguished by source and these changes follow the corresponding changes in local production and regional exports (for details, see the supplemental data online).

Implementation of groundwater restrictions in the SIMPLE-G model

As discussed above, the groundwater sustainability limits are calculated by the Water Balance Model (WBM) considering the hydrological relationships and connections within and across grid cells. To allow for a deeper understanding of the distinct forces at work when groundwater sustainability restrictions are imposed, we first run the SIMPLE-G model assuming no change in surface water use, irrigation intensity and irrigation extent. This restricted scenario generates the largest possible impacts on food prices, and food consumption.

While informative, this simplified scenario does not capture adaptation to the sustainability policy, nor does it capture the spillover effects. Therefore, we consider additional scenarios whereby the system responds to these direct impacts. A first line of response involves a change in the composition of irrigation water resources. Depending on availability and relative costs, the model measures the change in surface water withdrawals as cost-minimizing producers attempt to reduce the impacts of groundwater restrictions on production. Then, the model consider changes in irrigation intensity in terms of water suppled and used by crops. At the level of a grid cell, this can be due to a change in the composition of crops within the grid cell or it can reflect a shift to a different deficit irrigation strategy that is not explicitly modelled. All these changes can affect the rate of recharge to groundwater and thus changes the initial sustainability limit. They also have implications for downstream surface water availability. To capture the downstream effects, we include another sustainability scenario with iterations between the SIMPLE-G economic model and the WBM hydrological model. In this scenario the decisions regarding surface water, irrigation intensity, and irrigation extent are transferred to the WBM. Then WBM will give us the implications for surface water availability downstream and a revised required change in groundwater restrictions. We will show why these return flows are important for determining the spatial pattern of production at the local level.

Results

Restricting groundwater consumption alters global agricultural production. If there are no other sources of water, which is the case for many hotspots of unsustainable groundwater abstraction, the immediate direct impact is a fall in production. According to our calculations, the consequences of this ‘first round’ of responses to the groundwater sustainability restriction is a 12.3% reduction in global crop production (around 2 billion tons of corn-equivalent crops). Despite comprising only 6% of global water consumption, the proportionate reduction in production is twice as large due to the relatively high productivity of groundwater. However, as the food system responds to this groundwater constraint, the final impact on global crop production is greatly moderated and limited to only −0.2% corresponding to a change of −28.4 million tons in corn-equivalent crop production.

Figure 5 decomposes the change in global crop production due to groundwater sustainability restrictions. We label them as follows: 'no adaptation' (the direct effects), 'substitution' of surface water and other inputs, 'rainfed conversion', 'relocation' and 'global trade'. The initial margin of adjustment involves substituting surface water for groundwater, where feasible, and using other farm inputs more intensively (Figure 5, bar b). This adaptation results in a 3.5% moderation in the global crop output decline. The next farm-level response considered is the conversion of irrigated land to dryland crop production, where rainfall is adequate. Despite rainfed production having a lower yield (or simply not being possible in some locations), in the aggregate, these results suggest that rainfed substitution can offset one-third of the damage at the global level (3.18% increase in Figure 5, bar c). The consequences of conversion from irrigated to rainfed crop production depend on the relative yields, changes in local land rents and the demand for alternative land uses. The next margin of response in Figure 5 (bar d) is due to the national relocation of production. Rising prices in countries where groundwater is restricted motivates expansion of national production in other suitable locations, especially in regions with strong, price-inelastic, domestic demand (e.g., South Africa, India and China). The extent of expansion in other grid cells depends on biophysical and economic conditions in each grid cell, but at a global level, these adjustments offset about one-sixth of the direct impact (2.1% increase in Figure 5, bar d). Similarly, international trade in agricultural production offsets the initial fall in production through exports and imports and the ensuing trade in 'virtual water'. Figure 5 (bar e) offsets 3.32% of the global production decline. Table 5, columns 2 and 4, provides more detail on the change in virtual water trade due to the groundwater sustainability policy.

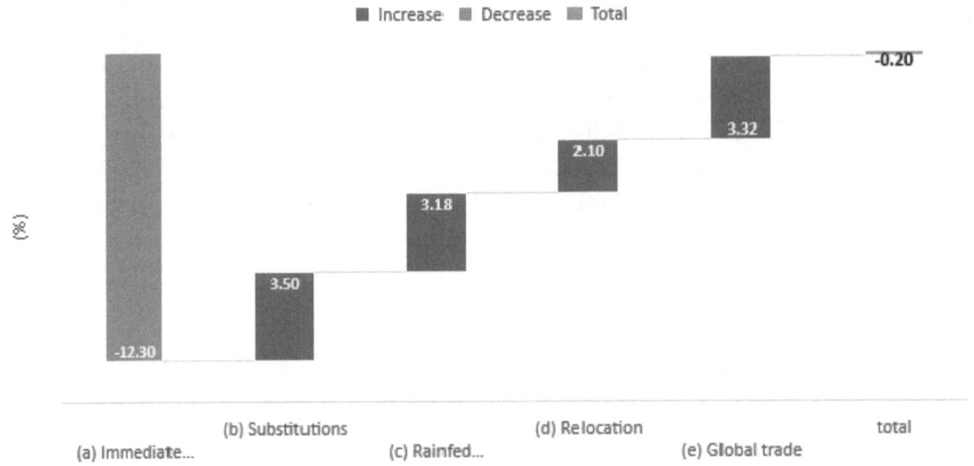

Figure 5. Decomposition of global crop response to global groundwater sustainability restrictions. The leftmost (orange) bar shows the immediate/direct impact before adaptation. The subsequent bars incorporate successive market-mediated adaptations; (b) substitution of surface for groundwater as well as other economic responses by farmers; (c) conversion of irrigated to rainfed area; (d) expansion of domestic production in response to higher prices; and (e) price-induced expansion in other regions.

At the global level, implementing the sustainability policy implies a 27% reduction in groundwater consumption (244 km^3), corresponding to a 6% fall in total water consumption in agriculture from all sources (rainfall, surface and groundwater). Due to multi-scale responses, the long-run impact on crop prices is quite modest (0.4%). However, the price increases are much larger in some local markets. The analysis projects an increase in global surface water withdrawal by 3.42% in the long run. On the other hand, despite the fact that irrigated production has declined by 1.03%, rainfed production has increased by 0.4%. This nearly offsets the reduction in irrigated output. This is also reflected in cropped area: −0.43% change in irrigated area and +0.21% change in the rainfed area, leading to a global cropped area increase of 0.21% (Table 4, columns 2–8).

At the country level, the long-run production impacts are small as similar adaptations can occur within a country. The findings suggest that in China, India, the United States, Pakistan and Iran, total annual production will decline by −12.5, −6.0, −3.0, −2.6 and −2.1 million tons of corn equivalent crops. Also, the overall long-run employment impacts are positive for many countries. According to the model output, in China, India, Indonesia, Bangladesh and Turkey there will be 296,600, 236,900, 66,700, 57,500 and 34,600 new agricultural jobs due to spillovers and expansions in croplands. The largest negative employment effects are in Pakistan and Vietnam by −36,000 and −24,800 jobs.

Spatial distribution of the impacts

This section illustrates the final changes in equilibrium levels of input demand and crop supply in response to enforcement of sustainable groundwater use, globally. These are equilibrium outcomes of the model and reflect interactions between local and global agricultural markets. The spatial heterogeneity of the impacts is due both to heterogeneity in the magnitude of the reduction in groundwater as well as biophysical differences across locations. The findings suggest a significant change in the pattern of irrigated production, water consumption, irrigated cropped area and employment in irrigated agriculture as a consequence of the sustainable groundwater policy.

Impact on cropped area

Figure 6 shows the spatial distribution of the impact on cropped area. The red and yellow colourings illustrate contraction and green colouring shows expansion in percentage change. The grid cells with a high dependency on unsustainable groundwater would face a sharp reduction in irrigated cropped area coupled with an increase in rainfed cropland. The net change for such areas is a decline in cropped area. These hotspots include California's Central Valley, the US Ogallala Aquifer, the Mississippi basin, parts of India and Pakistan, and parts of China, North Africa and the Middle East. The analysis projects a slight increase in both irrigated and non-irrigated area for the rest of the world.

(a)

(b)

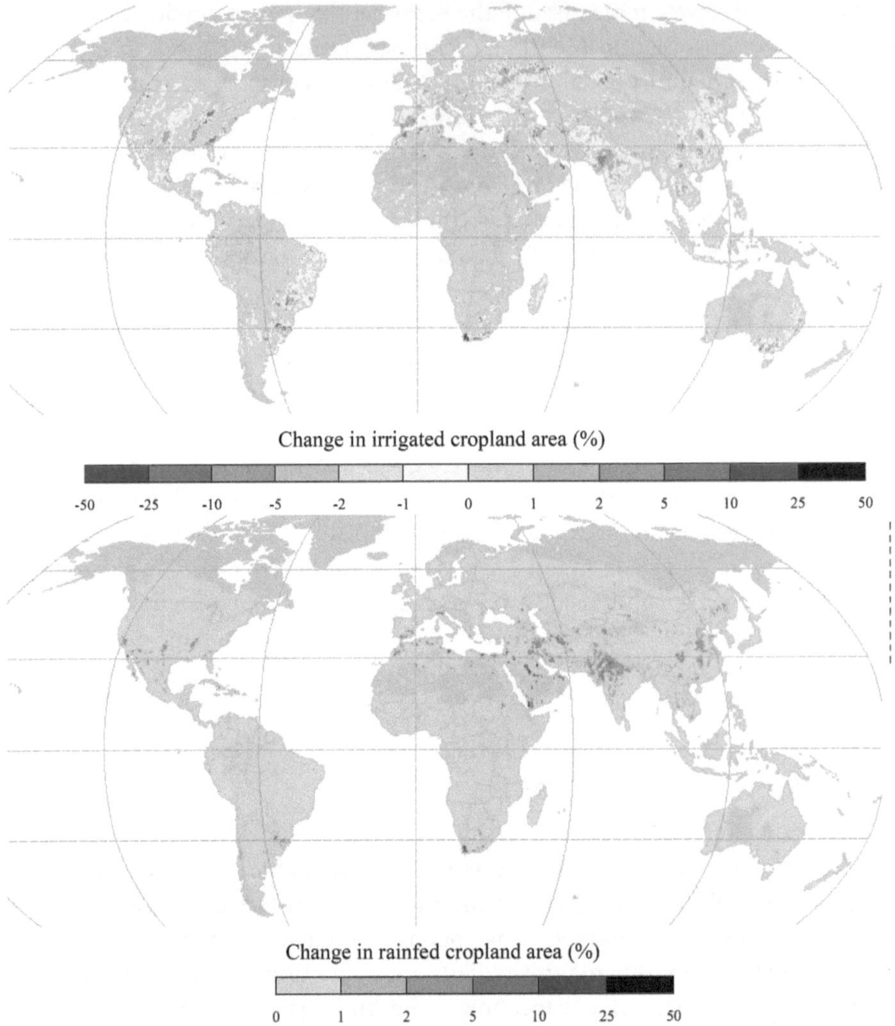

Figure 6. Global cropped area responses to global groundwater sustainability restrictions. (a) Percentage change in irrigated cropped area and (b) percentage change in non-irrigated (rainfed) practices as estimated by the SIMPLE-G model based on groundwater sustainability scenarios produced by the Water Balance Model (WBM). The maps illustrate the conversion to rainfed in targeted locations as for most of the grid-cells with large reduction in irrigated areas; a clear increase in rainfed areas is projected. Both maps show a moderate increase in irrigated and rainfed areas (light-green colouring). The expansion in global cropland area and conversion to rainfed are important channels of adjustment and economic responses to groundwater sustainability policy.

Impact on surface water

Figure 7 shows the spatial distribution of the impact of the groundwater conservation policy on surface water irrigation in percentage change. It shows an increase in surface water irrigation for almost all irrigated areas with the greatest increases in the hotspots of non-renewable groundwater use. As this is in percentage change, the magnitude of the change in surface water irrigation depends on availability and current uses. The increase in surface water irrigation reflects substitution of groundwater or a change in the location and scale of

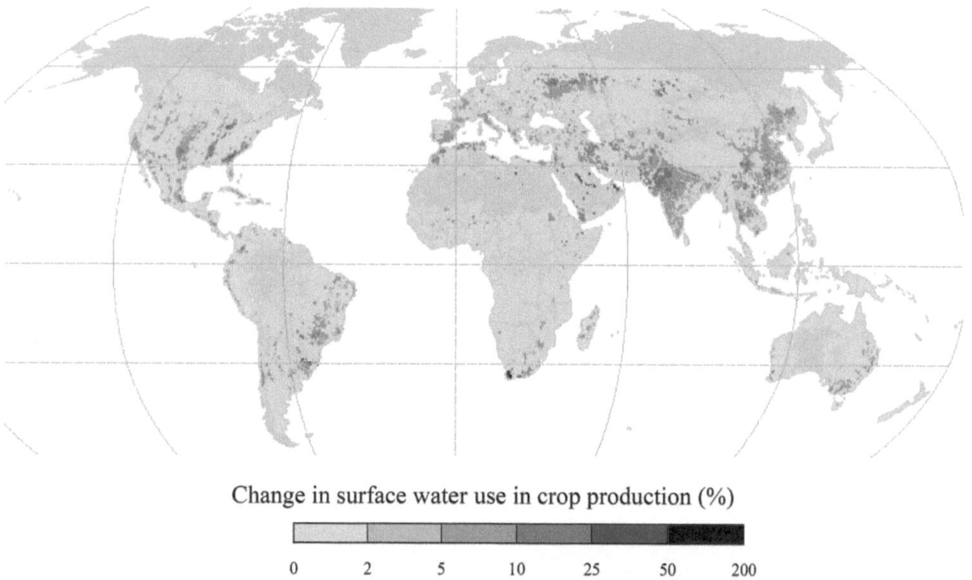

Figure 7. Global changes in surface water irrigation in response to global groundwater sustainability restrictions. The map shows the percentage change in surface water use in irrigation. The impact on surface water can be decomposed into a 'substitution effect' and a 'scale effect'. If the substitution effect is dominant, the change in surface water use in targeted areas is positive and compensates for the decline in irrigated areas. If the scale effect is dominant (less irrigated area), the change in surface water use in the targeted area is negative. Dark green means an increase in surface water use with a dominant substitution effect; and light green colouring shows the locations where there is an increase in surface water irrigation due to a positive scale effect, absent the substitution effect.

production. These changes can be decomposed into a 'substitution effect' and a 'scale effect'. If the substitution effect is dominant, the percentage change in surface water use in targeted areas is positive and more than compensates for the decline in irrigated areas. If the scale effect is dominant (less irrigated area overall), the change in surface water use in the targeted area is negative. (The Appendix in the supplemental data online explores the extreme case wherein the expansion of the surface water irrigation is restricted in targeted areas.)

Impact on production

Figure 8 illustrates the impact of enforcing groundwater sustainable use on equilibrium irrigated and rainfed crop production. These are the consequences of the changes in water use, land conversion, employment, local rents, crop prices, farm revenues, and other local and global changes. In general, irrigated production declines in hotspots of unsustainable groundwater while rainfed production increases. The increase in surface water use in some dry locations may cause an increase in irrigated production next to the targeted areas or may not be enough to recover the irrigation production (look at the Middle East, for example).

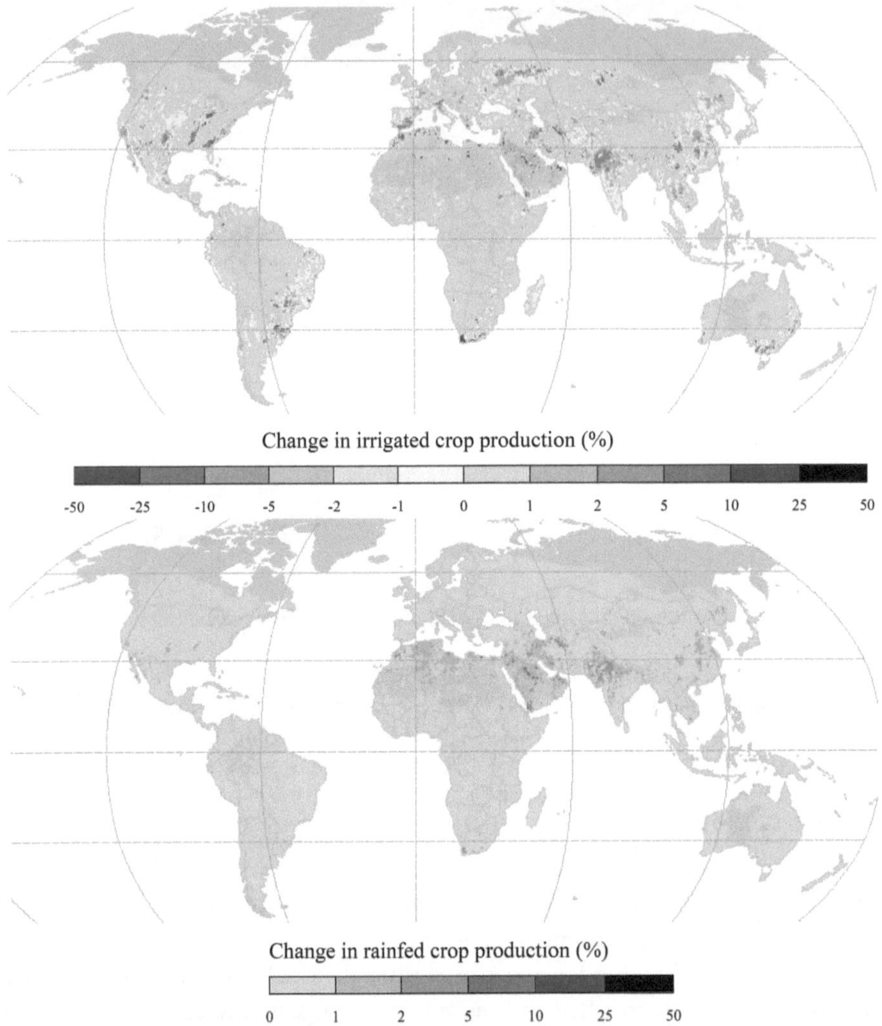

Figure 8. Global changes in crop production in response to global groundwater sustainability restrictions. The map shows the percentage change in (a) irrigated crop production and (b) rainfed crop production. Red and yellow colouring denotes a reduction in production and green colouring represents increases in production. While the irrigated production in the hotspots of unsustainable groundwater is substantially reduced, they show the biggest percentage increase in rainfed production. This conversion to rainfed agriculture compensates for some of the immediate impact on irrigated production. In addition, light green colouring on irrigated and rainfed maps indicate an expansion of crop production in non-targeted locations. While the patterns are similar to changes in cropped area, the percentage change in production is smaller, because rainfed crops have lower yields.

Regional outcomes

Restricting groundwater irrigation has diverse implications for regional crop production, land use and food prices. The long-run results at the regional scale are reported in Table 2 and the absolute changes are provided in Table 3. In terms of percentage changes, the largest price increases arise in the Middle East, North Africa and South Africa. In general, rainfed production increases while irrigated production declines. In those countries facing less severe restrictions, conversion to rainfed and surface water substitution (driven by higher local prices) can fully offset the decline in irrigated production.

In absolute terms, the biggest decline in crop production is predicted in China, India, the Middle East and North Africa by −12.5, −8.8, −3.6 and −2.5 million-ton change in crop production, respectively.

Table 2. Long-term regional impacts of restricting groundwater irrigation to renewable resources (percentage change in crop price, production and area, by region).

		Crop production (% change)			Cropped area (% change)%		
Region	Crop price (% change)	Total	Irrigated	Rainfed	Total	Irrigated	Rainfed
East Europe	0.23%	0.12%	−0.71%	0.21%	0.07%	−0.29%	0.09%
North Africa	1.10%	−0.72%	−1.21%	0.91%	0.19%	−0.41%	0.43%
Sub-Saharan Africa	0.13%	0.06%	−0.46%	0.11%	0.06%	−0.14%	0.06%
South America	0.29%	0.10%	−0.50%	0.28%	0.11%	−0.18%	0.14%
Brazil	0.28%	0.12%	−1.37%	0.33%	0.12%	−0.77%	0.20%
Australia and New Zealand	0.34%	0.09%	−0.90%	0.35%	0.11%	−0.37%	0.15%
Europe	0.34%	0%	−1.46%	0.35%	0.09%	−0.62%	0.18%
South Asia	0.80%	−0.40%	−1.20%	0.74%	0.04%	−0.54%	0.55%
Central America	0.38%	−0.15%	−0.94%	0.35%	0.09%	−0.36%	0.24%
South Africa	0.63%	−1.53%	−3.95%	0.58%	0.11%	−1.61%	0.32%
Southeast Asia	0.24%	−0.01%	−0.71%	0.21%	0.06%	−0.35%	0.16%
Canada	0.32%	0.36%	−0.24%	0.37%	0.14%	−0.13%	0.15%
United States	0.43%	−0.26%	−1.45%	0.43%	0.09%	−0.69%	0.25%
China	0.64%	−0.27%	−0.91%	0.65%	0.07%	−0.32%	0.46%
Middle East	1%	−0.58%	−1.19%	0.76%	0.14%	−0.30%	0.43%
Japan and Korea	0.10%	0.01%	−0.11%	0.11%	0.02%	−0.08%	0.09%
Central Asia	0.40%	−0.14%	−0.36%	0.29%	0.07%	−0.07%	0.11%
World	0.40%	−0.20%	−1.10%	0.40%	0.10%	−0.50%	0.20%

Source: Research findings based on SIMPLE-G.

Table 3. Long-term regional impacts of restricting groundwater irrigation to renewable resources (actual change in crop production and area, by region).

	Change in crop production (1000 MT)			Change in cropped area (1000 ha)		
Region	Total	Irrigated	Rainfed	Total	Irrigated	Rainfed
East Europe	889	−549	1390	146	−38	183
North Africa	−2497	−3241	738	53	−34	87
Sub-Saharan Africa	783	−568	1331	130	−13	143
South America	591	−747	1285	78	−16	95
Brazil	1108	−1628	2709	76	−40	117
Australia and New Zealand	129	−286	400	34	−9	43
Europe	39	−3303	3328	96	−80	176
South Asia	−8820	−15,781	6576	87	−541	628
Central America	−620	−1484	869	38	−37	74
South Africa	−1266	−1381	276	16	−25	40
Southeast Asia	−80	−2506	2381	69	−91	161
Canada	635	−20	617	55	−1	56
United States	−2983	−6100	3189	137	−190	327
China	−12,522	−24,971	11,874	98	−217	315
Middle East	−3581	−4992	1469	81	−67	148
Japan and Korea	11	−87	97	1	−3	4
Central Asia	−193	−326	128	13	−3	16
World	−28,374	−67,968	38,655	1209	−1404	2613

Source: Research findings based on SIMPLE-G.

Implications for virtual water trade

Groundwater restrictions are expected to alter the patterns of agricultural trade. The biggest percentage change in agricultural trade flows is expected for North Africa, South Africa and the Middle East where production is sharply reduced and imports rise. The model suggests that trade will play an important role in meeting the demand for food in these regions. (Country-level results are reported in Table A6 in Appendix A in the supplemental data online.) The largest percentage declines in production are estimated to be in Saudi Arabia, Jordan, South Africa, Oman and Iran where production drops by 19.0%, 10.5%, 7.6%, 4.8% and 2.8%, respectively (see Table A6 online). In these countries, there is little possibility to use surface water to substitute groundwater (indeed, rivers in these areas are often over-abstracted), and there is little room for expanding rainfed agriculture at competitive production costs. On the other hand, production is expected to increase in Turkey by 1.7% as access to more land and water is less challenging. In larger countries such as India, China and the United States, the relocation occurs predominantly within the country (Figure 6).

Table 4 shows the impacts of groundwater sustainability restrictions on 'virtual trade in blue water'. For all the regions, exports of virtual blue water decrease, with the biggest decline in the United States and South Asia. Note that this is largely due to the expansion of rainfed agriculture, both through conversion from irrigation as well as from cropped area expansion. Expanding rainfed crop production likely leads to a rise in virtual trade in green water, which we have not quantified in this paper. In summary, the groundwater and SW virtual trade will decline by 10.8 billion m^3.

Implications for employment

The employment is calculated assuming a gridded labour market with no labour mobility across grid-cells. Labour is part of the non-land composite input within a Leontief form

Table 4. Trade impacts of restricting groundwater irrigation (percentage change in supply to the domestic market, exports and imports).

	% Change in supply		% Change in demand
Region	Domestic	Exports	Imports
East Europe	0.02	0.54	−0.45
North Africa	−0.56	−2.83	1.96
Sub-Saharan Africa	0.01	0.78	−0.75
South America	−0.01	0.34	−0.28
Brazil	0.01	0.38	−0.29
Australia and New Zealand	−0.01	0.18	−0.1
Europe	−0.05	0.1	−0.11
South Asia	−0.35	−1.66	1.09
Central America	−0.14	−0.18	−0.02
South Africa	−0.99	−2.29	0.76
Southeast Asia	−0.05	0.4	−0.45
Canada	0.09	0.52	−0.16
United States	−0.16	−0.43	0.12
China	−0.27	−1.07	0.72
Middle East	−0.42	−2.43	1.78
Japan and Korea	−0.01	0.83	−0.82
Central Asia	−0.14	−0.21	−0.02

Source: Research findings based on SIMPLE-G.

Table 5. Impact of groundwater sustainability policy on virtual blue water exports by source.

	%		10^6 m^3	
	SW	GW	SW	GW
East Europe	1.41%	−11.90%	59	−152
North Africa	4.87%	−48.39%	103	−249
Sub-Saharan Africa	1.42%	−8.96%	25	−74
South America	2.12%	−14.26%	144	−431
Brazil	2.84%	−17.61%	152	−551
Australia and New Zealand	1.99%	−18.02%	52	−170
Europe	2.29%	−23.14%	238	−928
South Asia	5.26%	−26.74%	779	−2984
Central America	3.23%	−21.42%	239	−647
South Africa	6.30%	−28.41%	88	−141
South East Asia	3.47%	−23.54%	246	−845
Canada	1.25%	−10.30%	10	−25
United States	3.62%	−27.65%	885	−2,935
China	3.54%	−21.20%	65	−228
Middle East	4.39%	−31.38%	138	−299
Japan and Korea	0.76%	−5.40%	2	-6
Central Asia	1.48%	−14.71%	48	−171
World	3.35%	−23.48%	3,271	−10,835

Source: Research findings based on SIMPLE-G.

for combining labour and other non-land inputs. Thus, the demand for labour input follows the changes in demand for non-land composite input due to changes in intensifications. Figure 9 illustrates the spatial pattern of change in employment. The pattern is slightly different from changes in cropped area and production. This is due to the differing contributions of non-land factors of production among countries. The initial numbers for gridded employment were obtained from subregional rates of labour per hectare of cropland. Section A6 in the supplemental data online provides more information on the average number of workers per hectare and per ton of corn-equivalent output in rainfed and irrigated agriculture.

Table 6 provides the model results for the aggregated impacts on employment in crop production activities around the world. The findings suggest that employment in irrigated agriculture will decline by about 1.5 million jobs globally. However, employment is expected to increase in the more extensive rainfed agriculture by 2.5 million jobs, creating a net gain of 1 million new jobs globally. Overall, these results suggest an increase more 0.5 million jobs in China and South Asia, where domestic cropping area increases with the move to more rainfed production. This is due to the expansion of the more extensive rainfed crop production which requires more labour to work the land. Finally, note that the average worker per hectare of land will be different for the marginal land endogenously added or removed from crop production due to substitutions between the non-land and land–water composites. As the land–water composite will be more scarce and costly, an intensification is expected which increases the average number of workers per hectare of land.

Theoretically, irrigated agriculture requires more labour per acre, because irrigating the same crop in the same location would require more labour. However, in some cases the regional average number of workers can be higher for rainfed agriculture compared with irrigated agriculture. There are several different reasons for this. First, irrigation is small in some regions, for example, in Brazil and Canada most of the crop production is

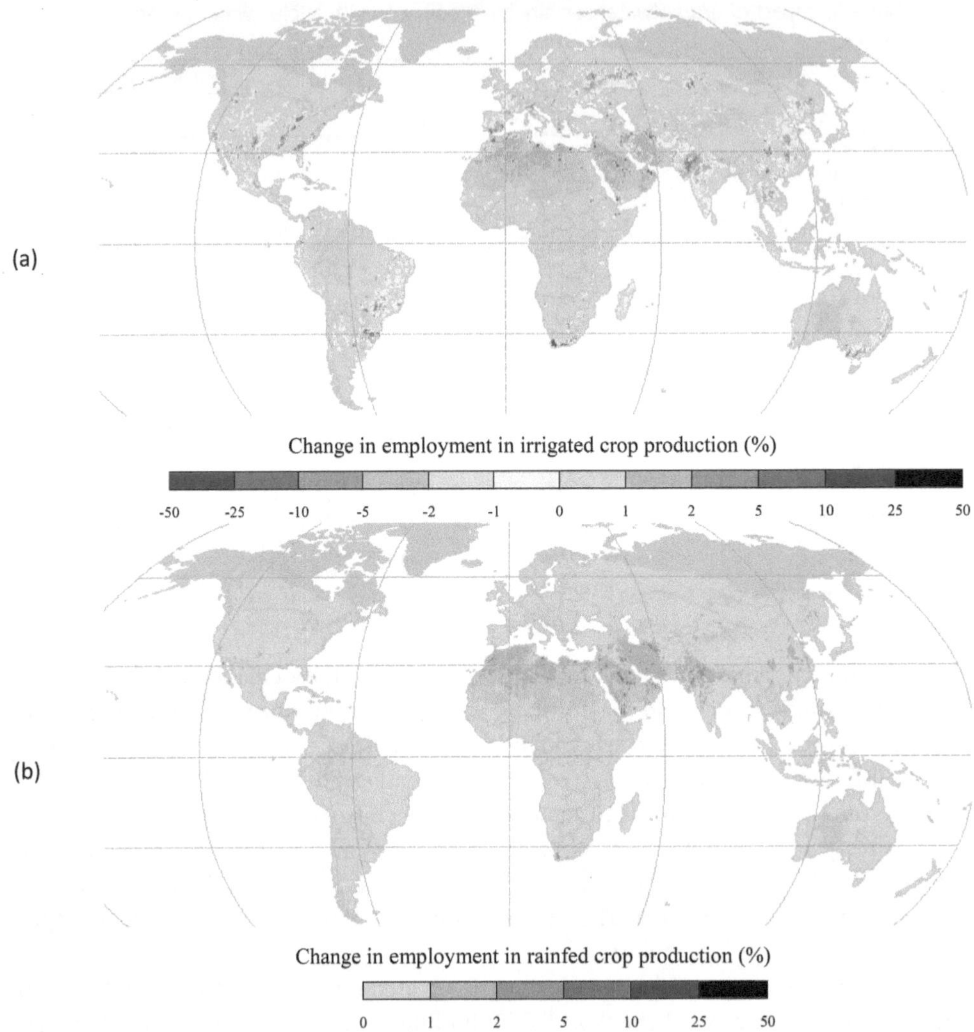

Figure 9. Global changes in agricultural employment in response to global groundwater sustainability restrictions. The map shows the percentage change in (a) employment on irrigated land and (b) employment on rainfed land. Red and yellow colouring means reduction in employment and green colouring represents increases in employment. While the patterns of change are similar to changes in cropped area and production, the reductions in employment are generally smaller while the increases in employment are larger than the percentage change in cropped area.

rainfed. Second, the crop composite can be different between irrigated and rainfed land within a region. For example, many labour-intensive hand-picked crops in Washington State, Florida, Michigan and Oregon are rainfed while corn is irrigated in parts of Nebraska with very low labour intensity. Thus, the outcome at the regional level depends on the significance of each crop and their labour intensity. Finally, more labour might be employed in rainfed production as farmers might decide to have more fertilizer and chemical application events to offset part of irrigation yield gap.

Table 6. Impact of groundwater sustainability policy on employment in crop production, by practice and region.

	% Change			Change in the number of jobs		
	Total	Irrigated	Rainfed	Total	Irrigated	Rainfed
East Europe	0.16%	−0.40%	0.25%	6107	−1979	8086
North Africa	0.28%	−0.26%	1.03%	28,500	−15,631	44,131
Sub-Saharan Africa	0.12%	−0.04%	0.12%	154,544	−1790	156,336
South America	0.18%	−0.15%	0.34%	7979	−2246	10,226
Brazil	0.23%	−0.96%	0.34%	32,697	−11,238	43,935
Australia and New Zealand	0.32%	−0.24%	0.37%	582	−38	620
Europe	0.22%	−0.45%	0.36%	13,417	−4827	18,244
South Asia	0.11%	−0.69%	0.83%	289,680	−858,248	1,147,920
Central America	0.20%	−0.44%	0.39%	14,867	−7281	22,148
South Africa	0.31%	−1.73%	0.55%	2885	−1654	4538
South East Asia	0.09%	−0.40%	0.24%	83,296	−83,868	167,200
Canada	0.37%	0.01%	0.38%	690	1	690
United States	0.23%	−0.83%	0.45%	4233	−2626	6863
China	0.14%	−0.41%	0.69%	296,608	−425,672	722,272
Middle East	0.52%	−0.16%	0.88%	46,601	−4980	51,581
Japan and Korea	0.06%	−0.01%	0.11%	662	−52	714
Central Asia	0.17%	−0.09%	0.36%	10,355	−2271	12,626
World	0.13%	−0.52%	0.50%	993,701	−1,424,400	2,418,129

Source: Research findings based on SIMPLE-G.

Interactions amongst the adaptation margins

Assessing the sensitivity of these findings to key parameter values reveals an interesting finding. The strength of any given individual margins of adaptation is dependent on the other margins. For example, when surface water substitution is limited, there is a stronger price response and we observe larger adaptation responses from trade and relocation. Or when substitution in trade is less possible, or conversion to rainfed is not feasible, global irrigated cropped area tends to respond more strongly and there are bigger responses in surface water use. As a consequence, while the component parts of Figure 5 are quite sensitive to model parameters, the final impact on global production of this bounding analysis is less than 1% and the increase in price for more rigid assumptions is also less than 1%. The Appendix in the supplemental data online provides details about these sensitivity analyses and the interaction amongst the various adaptation margins.

Discussion and conclusions

Stress on groundwater is expected to increase due to income and population growth, along with global warming increasing agronomic water requirements. The rising number of drying wells and land subsidence around the world has increased attention to sustainable groundwater use (Befus et al., 2017; Jasechko & Perrone, 2020, 2021; Klasic et al., 2022). However, little is known about the local and global implications of restricting groundwater use to sustainable levels. This study sheds light on the land use and agricultural production implications of such a scenario.

Recent studies to evaluate the impact of restricting groundwater irrigation on agriculture are either geographically limited (United States) or limited in their treatment of global impacts (Baldos et al., 2020; Graham et al., 2021). This study demonstrates that local impacts are largely ameliorated in the long run at the global level due to local, regional and global adaptations. We have also added to this literature by expanding the analysis to include virtual water trade and employment impacts of groundwater governance. The decline in unsustainable groundwater irrigation is accompanied by a decline in virtual trade in blue water, while employment in agriculture increases in the wake of expanding rainfed production. However, even in the long run, the local impacts can be very significant for locations with a high dependency on unsustainable groundwater. The situation would eventually happen when they run out of water. The production and employment in some locations can decline by up to 90%. In locations without alternative sources for employment and food supply, this would have significant local impacts leading to over 1 million jobs lost and over 1 million families affected. And these local impacts will have adverse impacts on the broader economy – a point which our partial equilibrium analysis ignores by holding income constant. Addressing these economy-wide impacts of sustainability policies requires a general equilibrium model, which would typically be implemented at a national level (Calzadilla et al., 2010; Golub & Hertel, 2012; Liu et al., 2014).

The findings suggest that restricting groundwater irrigation could have a significant impact on local irrigated production in hotspots of non-renewable groundwater irrigation. This directly causes a substantial reduction in agricultural production and employment for these communities. However, the impact of global groundwater sustainability policies is likely to be overestimated if the dynamic responses to restricted groundwater consumption are ignored. We show that these responses at the local and global levels will lessen the negative impact on food prices and production at the regional and global levels. The changes in relative prices at different scales will motivate changes in decisions and market outcomes. This includes compositional effects at the local level, changes in the location of crop production, changes in surface water irrigation, changes in irrigation extent, changes in irrigation intensity and changes in international trade of food commodities. Of course, these adjustments are costly and can cause environmental issues. Overall, the findings suggest that the land use and deforestation implications are small and the long-run impact on food production is less than 1% at the global level. The long-run change in global cropland is also limited to +0.1% corresponding to +1.2 million ha of cropland. The expansion coming from deforestation and cropping of marginal lands could lead to environmental degradation. In addition, as rainfed zones tended to be richer in carbon and biodiversity (Taheripour et al., 2013), other environmental implications of this policy should be studied more carefully with ecological concerns in mind. This is a critical finding when comparing the likely benefits and costs of such a policy.

Our decomposition framework highlights the possible adaptation mechanisms and demonstrates why ignoring market responses may lead to the overestimation of the costs of adopting sustainability policies. In addition, the findings are important for understanding the implications of sustainable groundwater use for virtual trade of blue water. Because of the likely changes in farmers' decisions and economic responses, it is necessary to consider wider market responses in evaluating the impacts of conservation policies. Global environmental and agricultural models that neglect the economic modelling of international trade may overestimate the production losses by not accounting for changes in exports and imports based on economic decisions.

Considering the socio-economic responses in the studies of impact analysis of sustainability policy will face another key challenge. While the sustainability issues are typically very localized, and thus require high-resolution land use and water use models, analysing the social responses and economic decisions at a fine-scale level is difficult due to lack of information. Unfortunately, to the best of our knowledge, there are few global models capable of high-resolution economic modelling. For example, the GLOBIOM model from the International Institute for Applied Systems Analysis (Valin et al., 2013) is a recursive-dynamic optimization model. However, due to its large size and complexity, the model is not resolved at the individual grid cell level, rather it is applied to representative groupings of grid cells – making analysis of local responses to sustainability restrictions more challenging. Similarly, the Global Change Assessment Model provides a framework for modelling groundwater resources and land use, deploying a range of supply- and demand-driven adaptive responses (Turner et al., 2019a, 2019b), but is implemented at the river-basin level. However, high-resolution analysis is important due to heterogeneity of land and water use among grid-cells within aquifers (for example, the sustainability restrictions and responses would be different between Northern Plains and southern regions of Ogallala aquifer in the United States). Here, SIMPLE-G introduces one of the first frameworks capable of modelling land use and water use decisions at the grid-cell level while also considering grid-cell specific responses within local, regional, and global markets. This was possible thanks to close collaboration with hydrologists in the Water Systems Analysis Group of the University of New Hampshire. Further research is required to improve the ability of global models to integrate economic and environmental sciences considering cross-system feedback.

The reliance on virtual water trade carries with it the risk that many countries will not intervene to restore groundwater stability, but rather will avoid the political and administrative challenges of regulation – safe in the knowledge (or more likely, hope) that virtual water will continue to compensate for reduced local production. That route will eventually generate far more severe local (and global) consequences. Intervening to restore stability, which we have endeavoured to model in this paper, reduces the *average* water available from groundwater but crucially preserves the flexibility and emergency buffer resource that renewable groundwater provides. If mining continues, that function will disappear because aquifers will effectively run dry – salinizing or becoming so deep that recharge no longer reaches the saturated zone in a useful timeframe, or the cost of abstraction is prohibitive. The reduction in water availability implied in that scenario – a loss of both volume and flexibility – will have far more profound consequences locally and globally. The implication of this is that while introducing governance 'now' is relatively cheap, the failure to do so will be very expensive.

With increasing risks of megadroughts in the future, groundwater resources play an important role by acting as a buffer in times of extreme drought when there is no surface water available. They provide a temporary source for essential needs (including drinking water). So by incurring negligible production loss today, countries can keep their resource for the time most valuable. This opens a further route for future research.

In closing, this paper has provided some insights into the long-run costs and benefits of groundwater sustainability policy. However, we did not model the transition costs and long-run benefits of sustainable groundwater use. During the transition period, and while the farmers are adjusting their decisions based on market outcomes, there can be a high reduction in local production, employment, and farm income. We do find that aggregate employment will increase in the long run. This assures communities with sustainable water use will avoid the risk of running out of groundwater in the time most needed.

Disclosure statement

No potential conflict of interest was reported by the authors.

Funding

Haqiqi and Hertel acknowledge support from the National Science Foundation [grant numbers CBET-1855937, CBET-1805808, OISE-2020635 and DEB-1924111] and the US Department of Agriculture, National Institute of Food and Agriculture [grant number 2019-67023-29679].

ORCID

Iman Haqiqi http://orcid.org/0000-0001-7331-4615
Chris J. Perry http://orcid.org/0000-0002-9003-7477
Thomas W. Hertel http://orcid.org/0000-0002-7179-7630

References

Allan, J. A. (1997). *'Virtual Water': A long term solution for water short middle eastern economies?* School of Oriental and African Studies, University of London.

Allan, J. A. (2003). Virtual water – The water, food, and trade nexus. Useful concept or misleading metaphor? *Water International*, *28*(1), 106–113. https://doi.org/10.1080/02508060.2003.9724812

Allan, J. A. (2011). *Virtual water: Tackling the threat to our planet's most precious resource.* Bloomsbury Publishing.

Baldos, U. L. C., Haqiqi, I., Hertel, T., Horridge, M., & Liu, J. (2020). SIMPLE-G: A multiscale framework for integration of economic and biophysical determinants of sustainability. *Environmental Modelling & Software, 104805.* https://doi.org/10.1016/j.envsoft.2020.104805

Befus, K. M., Jasechko, S., Luijendijk, E., Gleeson, T., & Bayani Cardenas, M. (2017). The rapid yet uneven turnover of earth's groundwater. *Geophysical Research Letters*, *44*(11), 5511–5520. https://doi.org/10.1002/2017GL073322

Bierkens, M. F. P., & Wada, Y. (2019). Non-renewable groundwater use and groundwater depletion: A review. *Environmental Research Letters*, *14*(6), 063002. https://doi.org/10.1088/1748-9326/ab1a5f

Bredehoeft, J. D. (2002). The water budget myth revisited: Why hydrogeologists model. *Groundwater*, *40*(4), 340–345. https://doi.org/10.1111/j.1745-6584.2002.tb02511.x

Calzadilla, A., Rehdanz, K., & Tol, R. S. J. (2010). The economic impact of more sustainable water use in agriculture: A computable general equilibrium analysis. *Journal of Hydrology*, *384*(3–4), 292–305. https://doi.org/10.1016/j.jhydrol.2009.12.012

Chapagain, A. K., & Ysbert Hoekstra, A. (2011). The blue, green and grey water footprint of rice from production and consumption perspectives. *Ecological Economics*, *70*(4), 749–758. https://doi.org/10.1016/j.ecolecon.2010.11.012

Chatterjee, R., & Ray, R.K. (2016). A proposed new approach for groundwater resources assessment in India. *Journal of the Geological Society of India*, *88*(3), 357–365. https://doi.org/10.1007/s12594-016-0498-2

Chinnasamy, P., & Agoramoorthy, G. (2016). India's groundwater storage trends influenced by tube well intensification. *Groundwater*, *54*(5), 727–732. https://doi.org/10.1111/gwat.12409

Dalin, C., Wada, Y., Kastner, T., & Puma, M. J. (2017). Groundwater depletion embedded in international food trade. *Nature*, *543*(7647), 700–704. https://doi.org/10.1038/nature21403

Dee, D. P., M□ Uppala, S., Simmons, A. J., Berrisford, P., Poli, P., Shinya Kobayashi, U. A., Balmaseda, M. A., Bauer, D. P., Bauer, D. P., Bauer, P., Bechtold, P., Beljaars, A. C. M., van de Berg, L., Bidlot, J., Bormann, N., Delsol, C., Dragani, R., Fuentes, M., Geer, A. J., . . . Vitart, F. (2011). The ERA-interim reanalysis: Configuration and performance of the data assimilation system. *Quarterly Journal of the Royal Meteorological Society*, *137*(656), 553–597. https://doi.org/10.1002/qj.828

Dizard, J. (2022). Ukraine war disrupts global market for grains. *Financial Times*. February 26, 2022. https://www.ft.com/content/b6712657-d6b7-4d56-95f7-849a653d5a66

Famiglietti, J. S. (2014). The global groundwater crisis. *Nature Climate Change*, *4*(11), 945–948. https://doi.org/10.1038/nclimate2425

Galloway, D. L., & Burbey, T. J. (2011). Regional land subsidence accompanying groundwater extraction. *Hydrogeology Journal*, *19*(8), 1459–1486. https://doi.org/10.1007/s10040-011-0775-5

Gleeson, T., Wada, Y., Bierkens, M. F. P., & van Beek, L. P. H. (2012). Water balance of global aquifers revealed by groundwater footprint. *Nature*, *488*(7410), 197–200. https://doi.org/10.1038/nature11295

Golub, A. A., & Hertel, T. W. (2012). Modeling land-use change impacts of biofuels in the GTAP-BIO framework. *Climate Change Economics*, *3*(3), 1250015. https://doi.org/10.1142/S2010007812500157

Graham, N. T., Iyer, G., Hejazi, M. I., Kim, S. H., Patel, P., & Binsted, M. (2021). Agricultural impacts of sustainable water use in the United States. *Scientific Reports*, *11*(1), 1–9. https://doi.org/10.1038/s41598-021-96243-5

Grogan, D. S. 2016. Global and Regional Assessments of Unsustainable Groundwater Use in Irrigated Agriculture. PhD Thesis, University of New Hampshire.

Grogan, D. S., Wisser, D., Prusevich, A., Lammers, R. B., & Frolking, S. (2017). The use and re-use of unsustainable groundwater for irrigation: A global budget. *Environmental Research Letters*, *12*(3), 034017. https://doi.org/10.1088/1748-9326/aa5fb2

Grogan, D. S., Zuidema, S., & Glidden, S. (2022, February). Water Balance Model (WBM) open source release version 1.0.0 ancillary data. *Data Catalog*. https://dx.doi.org/10.34051/d/2022.2

Haqiqi, I. (2019). Irrigation, adaptation, and water scarcity. Doctoral dissertation, Purdue University Graduate School.

Haqiqi, I., Grogan, D. S., Bahalou Horeh, M., Liu, J., Lantz Baldos, U., & Hertel, T. 2020. Environmental stressors can intensify the impacts of pandemics on earth's natural resources and global food systems. In *AGU Fall Meeting Abstracts*, GH023–05.

Haqiqi, I., Grogan, D. S., Hertel, T. W., & Schlenker, W. (2021). Quantifying the impacts of compound extremes on agriculture. *Hydrology and Earth System Sciences*, *25*(2), 551–564. https://doi.org/10.5194/hess-25-551-2021

Jasechko, S., & Perrone, D. (2020). California's central valley groundwater wells run dry during recent drought. *Earth's Future*, *8*(4), e2019EF001339. https://doi.org/10.1029/2019EF001339

Jasechko, S., & Perrone, D. (2021). Global groundwater wells at risk of running dry. *Science*, *372* (6540), 418–421. https://doi.org/10.1126/science.abc2755

Klasic, M., Fencl, A., Ekstrom, J. A., & Ford, A. (2022). Adapting to extreme events: Small drinking water system manager perspectives on the 2012–2016 California drought. *Climatic Change*, *170* (3), 1–25. https://doi.org/10.1007/s10584-021-03305-8

Liu, J., Hertel, T. W., Lammers, R. B., Prusevich, A., Baldos, U. L. C., Grogan, D. S., & Frolking, S. (2017). Achieving sustainable irrigation water withdrawals: Global impacts on food security and land use. *Environmental Research Letters*, *12*(10), 104009. https://doi.org/10.1088/1748-9326/aa88db

Liu, J., Hertel, T. W., Taheripour, F., Zhu, T., & Ringler, C. (2014). International trade buffers the impact of future irrigation shortfalls. *Global Environmental Change*, *29*, 22–31. https://doi.org/10.1016/j.gloenvcha.2014.07.010

Martens, B., Miralles, D. G., Lievens, H., Van Der Schalie, R., De Jeu, R. A. M., Fernández-Prieto, D., Beck, H. E., Dorigo, W. A., & Verhoest, N. E. C. (2017). GLEAM v3: Satellite-based land evaporation and root-zone soil moisture. *Geoscientific Model Development, 10*(5), 1903–1925. https://doi.org/10.5194/gmd-10-1903-2017

Mekonnen, M. M., & Hoekstra, A. Y. (2020). Sustainability of the blue water footprint of crops. *Advances in Water Resources, 143*(September), 103679. https://doi.org/10.1016/j.advwatres.2020.103679

Rienecker, M. M., Suarez, M. J., Gelaro, R., Todling, R., Bacmeister, J., Liu, E., Bosilovich, M. G., Schubert, S. D., Takacs, L., Kim, G.-K., Bloom, S., Chen, J., Collins, D., Conaty, A., da Silva, A., Gu, W., Joiner, J., Koster, R. D., Lucchesi, R., . . . Woollen, J. (2011). MERRA: NASA's modern-era retrospective analysis for research and applications. *Journal of Climate, 24*(14), 3624–3648. https://doi.org/10.1175/JCLI-D-11-00015.1

Rosa, L., Danilo Chiarelli, D., Chengyi, T., Cristina Rulli, M., & D'Odorico, P. (2019). Global unsustainable virtual water flows in agricultural trade. *Environmental Research Letters, 14*(11), 114001. https://doi.org/10.1088/1748-9326/ab4bfc

Saha, S., Moorthi, S., Xingren, W., Wang, J., Nadiga, S., Tripp, P., Behringer, D., Hou, Y.-T., Chuang, H.-Y., Iredell, M., Ek, M., Meng, J., Yang, R., Mendez, M. P., van den Dool, H., Zhang, Q., Wang, W., Chen, M., & Becker, E. (2014). The NCEP climate forecast system version 2. *Journal of Climate, 27*(6), 2185–2208. https://doi.org/10.1175/JCLI-D-12-00823.1

Siebert, S., & Döll, P. (2008). The Global Crop Water Model (GCWM): Documentation and first results for irrigation crops. *Frankfurt Hydrology Paper, 7*, 1–42. Institute of Physical Geography, University of Frankfurt. http://www2.uni-frankfurt.de/45217788/FHP_07_Siebert_and_Doell_2008.pdf

Siebert, S., & Döll, P. (2010). Quantifying blue and green virtual water contents in global crop production as well as potential production losses without irrigation. *Journal of Hydrology, 384*(3–4), 198–217. https://doi.org/10.1016/j.jhydrol.2009.07.031

Taheripour, F., Hertel, T. W., & Liu, J. (2013). The role of irrigation in determining the global land use impacts of biofuels. *Energy, Sustainability and Society, 3*(1), 4. https://doi.org/10.1186/2192-0567-3-4

Turner, S. W. D., Hejazi, M., Calvin, K., Kyle, P., & Kim, S. (2019a). A pathway of global food supply adaptation in a world with increasingly constrained groundwater. *Science of the Total Environment, 673*, 165–176. https://doi.org/10.1016/j.scitotenv.2019.04.070

Turner, S. W. D., Hejazi, M., Yonkofski, C., Kim, S. H., & Kyle, P. (2019b). Influence of groundwater extraction costs and resource depletion limits on simulated global nonrenewable water withdrawals over the 21st century. *Earth's Future, 7*(2), 123–135. https://doi.org/10.1029/2018EF001105

Valin, H., Havlík, P., Forsell, N., Frank, S., Mosnier, A., Peters, D., & van den Berg, M. (2013). Description of the GLOBIOM (IIASA) model and comparison with the MIRAGE-BioF (IFPRI) model. *Crops, 8*(3.1), 10.

Wada, Y., Van Beek, L. P. H., & Marc, F. P. B. (2012). Nonsustainable groundwater sustaining irrigation: A global assessment. *Water Resources Research, 48*, 6. https://doi.org/10.1029/2011WR010562

Wada, Y., Wisser, D., & Bierkens, M. F. P. (2014). Global modeling of withdrawal, allocation and consumptive use of surface water and groundwater resources. *Earth System Dynamics, 5*(1), 15–40. https://doi.org/10.5194/esd-5-15-2014

Willmott, C. J., & Matsuura, K. (2001). Terrestrial air temperature and precipitation: Monthly and annual time series (1950–1999) version 1.02. *Center for Climatic Research, University of Delaware, Newark*, 1155. https://psl.noaa.gov/data/gridded/data.UDel_AirT_Precip.html

Wisser, D., Balazs, M. F., Vörösmarty, C. J., & Schumann, A. H. (2010). Reconstructing 20th century global hydrography: A contribution to the Global Terrestrial Network – Hydrology (GTN-H). *Hydrology and Earth System Sciences, 14*(1), 1–24. https://doi.org/10.5194/hess-14-1-2010

Woo, J., Zhao, L., Grogan, D. S., & Haqiqi, I. (2022). C3F: A collaborative container-based model coupling framework. *Proceedings of PEARC 22, Practice and Experience in Advanced Research Computing*, Article no. 7. Association for Computing Machinery (ACM). https://doi.org/10.1145/3491418.3530298

OPEN ACCESS

The problem with water footprints outside of irrigated drylands

Mark Mulligan

ABSTRACT
Virtual water – embedded in an agricultural commodity by virtue of evapotranspiration during crop growth – is effectively consumed by the purchaser. In water-poor economies, it breaks their dependence on local water-for-food and acts as a political stabilizer. Tony Allan's concept remains highly relevant in the irrigated drylands for which it was developed, but calculating water footprints for agriculture elsewhere should take account of the non-agricultural counterfactual – which may consume more water than the agriculture that replaces it. Otherwise, use of water footprinting, for example, in trade policy, will be misleading and counterproductive.

Tony Allan was always a very humble, kind, selfless but brilliant academic and communicator. I am grateful for the years that I worked alongside him at King's College London, the conversations we had, the annual water meetings that he convened, the introductions that he made, and the PhD students we shared.

Introduction

Tony Allan introduced the term *virtual water* in 1993 and highlighted how the water needed for local production of water-intensive goods, in particular crops, could be replaced by import of these goods from countries with the required water resources (Allan, 1993). In this way, food trade can effectively redistribute water between those areas that have it and those that do not – decoupling national food security from national water security (Tayia et al., 2022). In the same way, countries with low water security but good thermal conditions for crop growth often export water to much wetter regions (Allan, 2011). The virtual water concept underpins the *water footprint* methodology (Hoekstra et al., 2011; Hoekstra & Huynen, 2002) which is widely applied in the academic literature and, increasingly, in corporate and national water accounting to understand the *water cost* of products and support water-use calculations, including in life cycle assessment (LCA) studies. These water footprints may be applied to a process, product, company, sector, individual, community or nation-state, and go well beyond evaporative water use to include direct and indirect water uses and water pollution from the producer all the way to the consumer.

Hydrologically, virtual water is the water consumed in the production of a product by evapotranspiration during the growth of the associated crop and/or the water contaminated during extraction and processing operations. We can separate evaporative water footprints from water pollution footprints and food processing footprints, continuing all the way through the supply chain to the water footprint of cooking. The total water footprint of a product includes all of these. Water consumed or contaminated is no longer available for downstream use. In this way, the production of food for distant markets takes water out of the local terrestrial hydrological cycle (placing it in the atmosphere or contaminating it) for the benefit of sometimes distant users, denying that water to other downstream local uses but often supporting their local and national economy through trade payments.

Water footprinting is now widely applied (Aldaya et al., 2010; Hoekstra & Hung, 2005; Oki & Kanae, 2004; Da Silva et al., 2016), increasingly by agricultural and industrial interests (Godfrey & Chalmers, 2012; Morrison et al., 2010). There has been criticism of its limitations for food and water policy (e.g., Wichelns, 2010, 2015), although it is recognized as a very effective communication tool. However, surprisingly little is written on the physical hydrology underlying water footprinting. Perry (2014, 2019) highlights the key differences between water footprints and carbon footprints. In particular: carbon footprints are globally relevant whereas water footprints have relevance only to the local hydrological basin (since the opportunity costs for water use are only in that basin, downstream of where it is used); water management and mismanagement is case specific and winners and losers are also highly variable between basins. He also points out that water footprints are often not measured relative to the counterfactual (evapotranspiration by natural vegetation); footprints separate flows into *blue* (water stores used in irrigation), *green* (rainfed soil water) and *grey* (water required to assimilate pollutants) components as if these were separate and independent, whereas they are hydrologically connected; and that footprints based on potential rather than actual evapotranspiration (AET) are also inflated because AET is commonly much lower than potential.

There are clearly some *hydrological* challenges in the application of the virtual water concept to calculating water footprints. Some of these are fundamental and hard to resolve: others are a consequence of a naivety of hydrological accounting. The hydrological challenges can be summarized as follows:

- *Counterfactuals matter*: Most of global evapotranspiration flows through natural vegetation which, in many environments transpires more than the crops that replace it. Overall evapotranspiration under forests is greater than croplands (Guo & Mo, 2007; Verstraeten et al., 2005), reflecting the greater leaf area and rooting depth of trees. However, there is significant spatial variability, depending on the climate (Zhang et al., 2017). Of the components of AET, transpiration and interception loss tend to be higher under forest, whilst soil evaporation can be higher under cropland (Verstraeten et al., 2005). Nevertheless, perennial vegetation tends to consume more water compared with annual vegetation, particularly in seasonal environments (Radersma & de Ridder, 1996). Thus, many agricultural water footprints may be negative relative to the counterfactual, because the crops or pasture consume less water than the forest or scrubland that went before them. This means that intensification of agriculture in a basin can lead to *greater* water quantity available for downstream uses, not less as is implicit in the footprint approach.

- *Context matters*: High water footprint land uses in areas with significant nearby downstream demand are more of a concern than high footprints elsewhere. Areas with the same footprint can have very different downstream contexts and these matter for understanding the impact and opportunity cost of the footprint.
- *Scale matters*: At a big enough scale, even evaporated water is returned to land (Berger & Finkbeiner, 2013) so consumptive use is only consumptive in the sense that the spatial distribution of the beneficiaries of the water changes: evapotranspiration means these beneficiaries may no longer be immediately downstream of where the evapotranspiration occurred but could be anywhere downwind or downstream at a larger basin scale. Similarly, at a large enough scale, water quality impacts are diluted and become less significant to downstream water uses (Mulligan, 2021).
- *Water quality is key to water use*: Agriculture affects water quality as well as quantity and this, rather than impacts on quantity, may be the main footprint of agriculture on water in settings such as input-intensive agriculture.
- *Evapotranspiration is not the only flux*: The complex dynamics of infiltration and runoff change with land use and also determine the hydrological impact of a crop relative to the land use that went before. Even in dry environments with low original forest cover, if crop cover or management increases infiltration, the enhanced infiltration can improve the seasonal distribution of water resources downstream which can buffer the effect of increased evapotranspiration under the crop, especially for dry season water availability. Similarly, poor soil management can decrease infiltration and dry season flows in favour of storm flows. Therefore, water footprints should consider the impact on water regulation as well as water quantity and quality.

Here, I will focus on the importance of *counterfactuals* and *context* for water footprinting. Many players are currently grappling with water accounting in the context of environmental and social governance (ESG), including through the CEO [chief executive officer] Water Mandate (Greenwood et al., 2012) and corporate water accounting more generally (Morrison et al., 2010), but focused on accounting for water use in-house and in supply chains.

Methods

In examining the global distribution of AET and its relationship with agriculture globally and by country, the global distribution of *context* is assessed by calculating upstream average agricultural land cover for thousands of water intake points. Finally, the WaterWorld model (v2.92) has been applied globally (at 10 km spatial resolution) to understand the implications for the evaporative water footprint of hypothetical scenarios for cropping under simulated rainfed and irrigated conditions. This in order to better understand the geographical distribution of *counterfactual* impacts on evaporative water footprints – that is, volumes of water that are consumed during the production of the product of interest. It is implicit in the footprint idea that these volumes would be consumed in the absence of the production; thus, footprints are normally positive volumes of water. This paper introduces the concept of a *negative footprint*: the magnitude by which the product consumes less water than the counterfactual natural land cover. A high positive footprint refers to much water consumption by the agricultural production (with negative consequences for those downstream), whereas a big negative footprint refers to a much lower consumption than the counterfactual (with possibly positive consequences for those downstream). A small negative footprint refers to consumption similar to the counterfactual – that is, little change.

Characterizing current evapotranspiration and land cover

To characterize the relationship between current AET and land cover/use we use a 10-year climatology of satellite-derived AET from 2000 to 2010 using the MODIS (MODerate resolution Imaging Spectrometer) MOD16 global evapotranspiration product (Mu et al., 2011). This product applies an algorithm based on the Penman–Monteith (Monteith, 1964) equation, with surface resistance parameterized as an effective resistance to evaporation from the land surface and to transpiration from the plant canopy. The annual satellite swath data were extracted from the hierarchical data format (HDF) files,[1] merged into a global mosaic and averaged over the 10-year period to produce an AET climatology. For compatibility with the other datasets in the global version of the WaterWorld model, data were converted to the Geographical Coordinate System (WGS84 datum) and resampled to 10 km resolution (by pixel averaging).

For land cover/use we use the European Space Agency Sentinel 2020 (Zanaga et al., 2021) land cover classification which maps croplands at 10 m resolution globally. The cropland class was aggregated up to 10 km resolution as a continuous (rather than categorical) dataset representing the percentage of each 10 km pixel covered by cropland. For irrigated agriculture, Siebert et al. (2007) was again aggregated to 10 km resolution as a percentage cover. Global land cover classifications do not include grazing lands (pasture) as a cover class, though grazed lands are probably the globally most significant agricultural footprint. Pasture was defined as a per pixel cover percentage using a two-step process. In the first step, the presence of grazing is determined using four input layers at 1 km spatial resolution: climatically potential forest cover (Hengl, 2019), current tree cover (Zanaga et al., 2021), the headcount of wildland grazers, that is, those not in feedlots (Gilbert et al., 2018) and current herb cover (Zanaga et al., 2021). Grazing is a Boolean and is considered true where a cell has the climatic potential to be forest but has current tree cover $< 40\%$ *or* where current herbaceous cover $> 40\%$) *and* (where wildland livestock headcount > 0). A series of land covers that cannot be grazing land was defined from the Zanaga et al. (2021) dataset: cropland, snow, urban, water and mangroves (*defined covers*). The land cover percentage for grazing in a pixel is thus calculated as:

$$(100 - d) * \mathrm{G} * \min(\mathrm{Wh}/500, 1), \tag{1}$$

where d is defined covers (%); G is grazing present (Boolean); Wh is headcount of wildland grazers; and $500/\mathrm{km}^2$ is considered the headcount at which 100% wildland grazing cover is achieved in a pixel. In this way, grazing land cover varies between 0% and 100% according to the presence of grazing, scaled by the wildland grazer headcount in the pixel.

Non-agricultural land was defined as all land not under cropland or pasture. To understand AET per land cover/use type, we calculate the percentage of global terrestrial AET in each cover class using the Mu et al. (2011) dataset. Cover classes were defined categorically as any pixel with the relevant cover fraction > 0. As an estimate of the local human population whose water resources may be impacted by AET under each cover type, we calculate the population in each cover class using the LandScan™ Global Population Database (2017), p. 1 km resolution population dataset, again aggregated to 10 km resolution. We examine AET in relation to total annual precipitation and mean annual temperature using the WorldClim climatology (Hijmans et al., 2005).

To understand the geographical variation in AET under different land uses we use the WaterWorld *zone of interest* (ZOI) and *view-by* tools to calculate the AET under different land uses by masking the AET grid by each Boolean land use (ZOI), and applying zonal statistics to calculate the per cent of total national AET coinciding with each different land cover, by country. The country boundary dataset used by WaterWorld is derived from FAO (2014).

Characterizing downstream context

Water intakes are considered to be sites at which upstream water use has the potential for significant impacts on downstream users. To characterize the downstream context of croplands, pastures and irrigated areas we define 4776 global water intakes by bringing together 1438 intakes from the City Water Map (McDonald, 2016) with 9298 points tagged *water_works* in the Open Street Map database (downloaded December 2020). Each WaterWorld 10 km pixel with one or more water intakes is assigned an intake identifier and the intakes within that pixel are merged to produce a total of 4776 water intake points across 161 countries. To understand the potential impact of land use on downstream water intakes, we use WaterWorld to calculate the upstream average cover of cropland, pasture, irrigation and non-agricultural (natural) land for each water intake. In this case, non-agricultural (natural) land is calculated continuously as the percentage of each pixel that is not cropland or pasture, for pixels that are not classified as water or built-up.

Upstream averaging involves working from watersheds down flowlines (in this case based on HydroSHEDS; Lehner et al., 2008) to flow sinks and calculating the total number of land pixels and the sum of the variable in question. The upstream average of a variable at a point is the sum of the variable for all upstream cells divided by the number of upstream cells, and represents a simple indicator of the likely influence of upstream land uses on flows into the cell. The upstream average cropland, pasture, irrigated agriculture and natural land is calculated for each of the 4776 water intakes to understand the upstream context of these areas at sites of significance for water resources.

Scenario-modelling with WaterWorld

WaterWorld is a sophisticated, spatially explicit and physically based model for hydrological ecosystem services (HES) under current (baseline) and scenario conditions. WaterWorld uses remotely sensed and globally available datasets for hydrological analysis and decision-making, and is especially useful in ungauged and/or data-poor environments. The model can be applied globally at 10 km resolution, applied nationally or for major basins at 1 km resolution or applied locally (100 × 100 km tiles) at 1 ha resolution, and uses higher resolution datasets for finer scale analyses. The model (v2.92) simulates the hydrological baseline as an average for the period 1950–2000 and can be used to calculate the hydrological impacts of scenarios of climate change, land-use change, and land and water management options relative to the baseline. It includes routines for wind-driven rainfall, fog interception, snowmelt, evapotranspiration, water balance and runoff. The model is self-parameterizing in the sense that all spatial data required for model application anywhere in the world are provided with the model. The model's equations and processes are described by Mulligan and Burke (2005) and Mulligan (2013a, 2013b). The model is not routinely calibrated to observed flows as it is designed for hydrological scenario analysis and for use in ungauged basins: calibration is not usually possible under these circumstances (Sivapalan et al., 2003). Furthermore, it is considered that if a physically based model has undergone verification and is able to replicate the current situation using exclusively physical relationships, it will probably do so under scenario conditions since the physical relationships would remain the same (Mulligan, 2013b).

The one-dimensional surface water balance in WaterWorld v2.92 (for detail, see Mulligan, 2013a) is:

$$WB = WDR + Fo + Sn - AET \tag{2}$$

where *WB* is water balance (mm); *WDR* is wind-driven rainfall (mm); *Fo* is fog inputs (mm); *Sn* is snowmelt (mm); and *AET* is actual evapotranspiration (mm).

The latter is calculated on the basis of the energy available (the net radiation received) and the surface area available for transpiration and wet canopy evaporation. Because of the monthly time step and large spatial scales (100 m to 10 km pixels) used in the model, full surface, soil and wet canopy water balances are not possible. Available surface area (LAI) is considered a good surrogate for the availability of water through transpiring stomata or wet canopy evaporation. Thus, AET is calculated based on available energy and atmospheric demand to give potential evaporation and this is then combined with the non-self-shaded surface area available for the interception of radiation/evaporation of water to give an approximation of AET, which is responsive to vegetation type and cover as well as climate conditions. AET is thus calculated as:

$$Ea = (611 * exp((17.27 * NT)/(273.15 + NT)))/1000 \tag{3}$$

$$SSCK = (4098 * Ea)/sqr(273.15 + NT)$$

$$PotE = (SSCK/(SSCK + 0.066)) * NWM$$

$$PotE = PotE * (60 * 60/1000000)$$

$$PotE > 0: \ PotE = \ (PotE/2.45); \ PotE \leq 0: \ PotE = 0;$$

$$PotE > 0: \ ActE = \ PotE * EtFrac; \ PotE \leq 0: ActE = 0$$

where *NT* is air temperature (°C); *Ea* is vapour pressure (KPa); *SSCK* is slope of the saturation vapour pressure curve (Kpa/C); *NWM* is net radiation receipt (W/m^2); 2.45 is the latent heat of vaporization of water (MJ/kg); and *ETFrac* is the non-self-shaded area of tree and herbaceous leaves, which combines the cover of trees and herbaceous in the pixel with the leaf area index for trees/shrubs and herbaceous vegetation (set to 3.33 and 2.0, respectively).

Each pixel in WaterWorld is made up of continuous cover of three land covers: tree, herbaceous and bare, which must sum to 100%. Land use for each pixel may be one of cropland, pasture, natural, built-up, road, mining, oil and gas, and water.

WaterWorld has a range of scenario tools including for developing land cover and use scenarios. The WaterWorld scenarios outlined here are designed to help map the areas in which water footprints of agriculture (changes in AET relative to the baseline) would be positive relative to current land cover (i.e., the crop would consume more water than the antecedent vegetation cover) and where they would be negative (i.e., the crop would consume less water than the antecedent vegetation cover). Two scenarios are generated: one that simulates irrigated cropping (with an irrigation-supported herbaceous cover set to 100%) and one that simulates rainfed cropping with tree cover removed and replaced with a herbaceous cover close to the baseline cover for the area cropped. In both cases, the baseline cover of the tree, herbaceous and bare ground in each pixel is replaced with herbaceous cover, representing the crop. At the same time, land use is changed from the baseline land use to cropping. In WaterWorld, land cover affects water balance whereas land use affects water quality. Here, we focus on water quantity.

The area available for agricultural development was defined by creating a ZOI in WaterWorld using three rules: agricultural suitability (FAO, 2012) > 0%, built-up area < 50% (Zanaga et al., 2021), and water < 50% (Zanaga et al., 2021). These three rules ensure that cropland only replaces existing cover in agriculturally suitable areas that are not dominantly built up or water bodies. The resulting area assessed as suitable for cropping is shown in Figure 1. This ZOI is then used to constrain the scenarios applied.

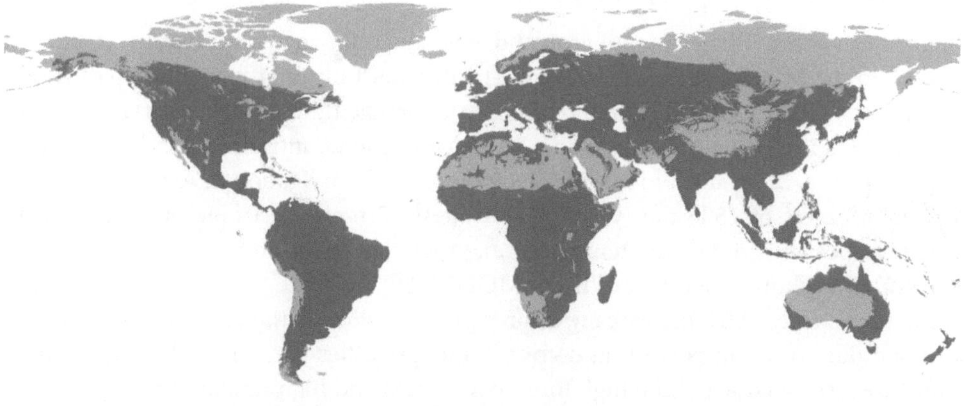

Figure 1. Areas considered suitable for cropping in the WaterWorld scenarios.

The world under irrigated agriculture

The irrigation scenario is a for large-scale irrigated cropping. It increases herbaceous cover beyond the currently measured values to 100% cover supported by irrigation. The irrigation is assumed to be drip-type irrigation that maximizes productive water use and minimizes wastage to interception loss (common for overhead irrigation) or free-water evaporation (common for flood irrigation). The scenario applied is a WaterWorld *land use and cover change > define your own rule* scenario in which herb cover is set to 100% in all areas for which the suitability mask (ZOI) is true, and land use in these areas is changed from its baseline use to *cropping*. The model thus reduces tree cover and bare cover in these pixels to zero. This scenario results in global loss of tree cover of approximately 21% (from baseline: 35.7% to 14.9%), a gain in herbaceous cover of about 28% (from baseline: 35.8% to 60.5%) with cropland increasing by 38% from baseline: 7.4% to 45%). Figure 2(a–c) shows the spatial distribution of these changes. The scenario represents high-efficiency irrigation in which AET results largely from productive transpiration through vegetation cover rather than interception loss from water sprayed on leaves or free water evaporation from ponded irrigation water.

The world under rainfed agriculture

The rainfed scenario is for large-scale rainfed cropping using herbaceous and bare covers in line with their current, rainfall-supported, ratios. The scenario applied is a WaterWorld *land use and cover change > define your own rule* scenario in which tree cover is reduced by 100% in all areas in which the suitability mask (ZOI) is true. Herbaceous and bare covers are set to fill the space formerly occupied by trees, according to their ratios in the baseline land cover. Land use in these areas is changed from its baseline use to *cropping*. This scenario results in global loss of tree cover of approximately 21% (from baseline: 35.7% to 14.9%), a gain in herbaceous cover of about 19% (from baseline: 35.81% to 54.57%) with cropland increasing by approximately 32% from baseline: 7.4% to 38.9%). Figure 3(a–c) shows the spatial distribution of these changes.

Results

Characterizing current evapotranspiration and land cover

Figure 4 shows the distribution of cropland, pasture and irrigated agriculture globally in relation to the MOD16 AET climatology. The distributions of cropland and pasture are similar, with pastures much more common in drylands, though often overlapping with croplands. Irrigation is most common in high-income countries, particularly Europe, North America, India and China with much more patchy irrigation in African countries. AET is in excess of 1500 mm/year in the wettest parts of the tropics and varies with climate (especially rainfall and temperature), vegetation cover and irrigation.

Figure 5(a) shows the relationship of MOD16 AET to total annual precipitation and tree cover. Clearly, AET increases with precipitation, though the relationship is heteroscedastic due to the impact of land cover, amongst other variables. The highest AET values are associated with both high forest cover (red) and high rainfall. The relationship of AET with mean annual temperature (Figure 5b) indicates a sharp increase in AET at

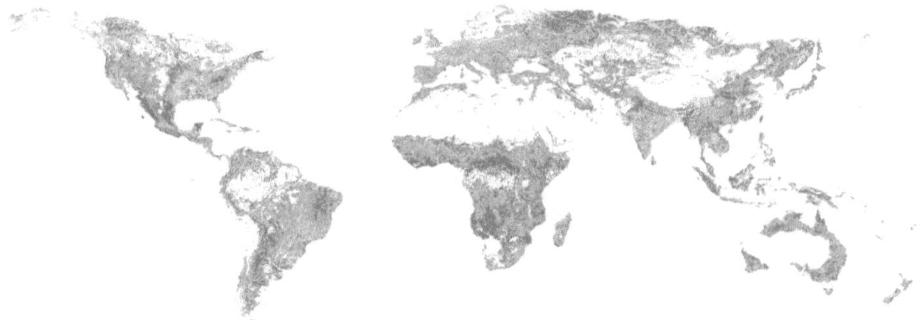

(a) Change in tree cover (-20.9), red=0%, blue=-100%

(b) Change in herb cover (+27.7%),red=+100%,blue=0%

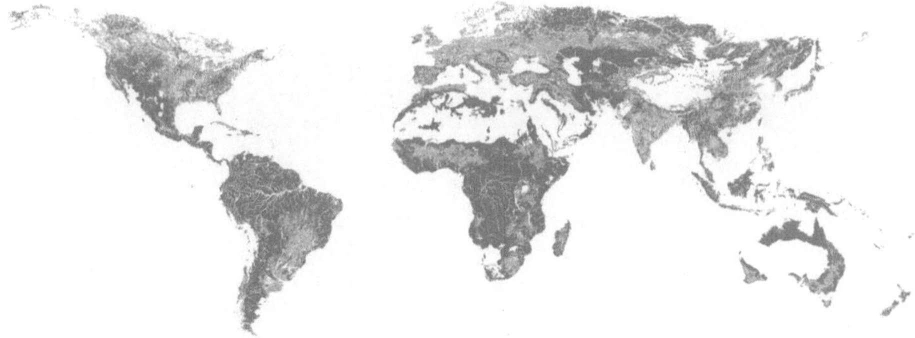

(c) Change in cropland (+38.), red=+100%, blue=0%

Figure 2. Spatial distribution of changes in tree, herb and cropland cover under the irrigated agriculture scenario: (a) change in tree cover (−20.9), red = 0%, blue = −100%; (b) change in herb cover (+27.7%), red = +100%, blue = 0% ;and (c) change in cropland (+38.0), red = +100%, blue = 0%.

temperatures > 20°C, again with much higher AET for tree covered areas (red) compared with those without tree cover (blue) at the same temperature.

The relationship between MOD16 potential and AET (Figure 5c) indicates that where tree cover is low and thus water is limiting (blue points), AET is low at all levels of

(a) Change in tree cover (-20.8%), red=0%, white=-100%

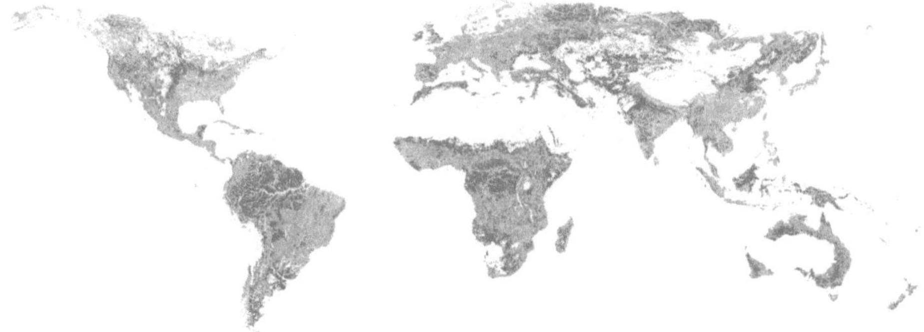

(b) Change in herb cover (+19.2%), red=+100%, white=0%

(c) Change in cropland (+31.8), white=0%, red=+100%

Figure 3. Spatial distribution of changes in tree, herb and cropland cover under the rainfed agriculture scenario: (a) change in tree cover (−20.8%), red = 0%, white = −100%; (b) change in herb cover (+19.2%), red = +100%, white = 0%; and (c) change in cropland (+31.8), white = 0%, red = +100%.

potential evapotranspiration (PET). However, where tree cover is high (red) and water not limiting, there is a direct linear increase in AET with PET.

Figure 6 shows the proportion of national AET occurring under different land uses, by country. European countries dominate in the proportion of their AET through croplands

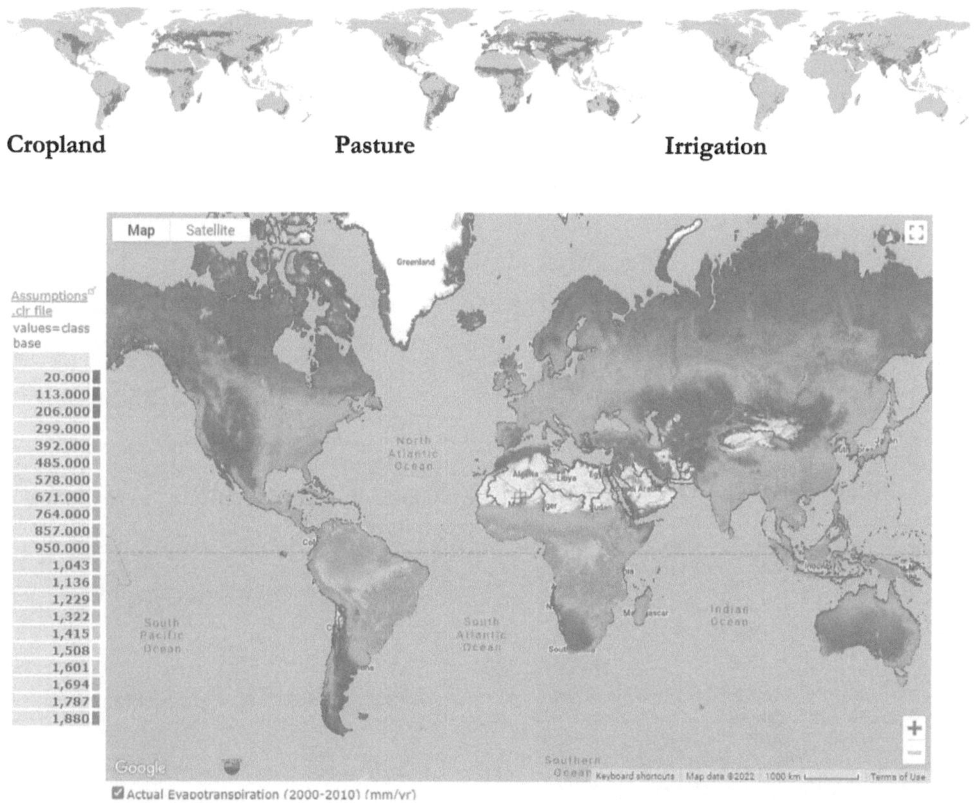

Figure 4. Distribution of cropland, pasture, irrigation and actual evapotranspiration (AET) for the baseline analysis.
Source: Map data ©2022 Google.

(Figure 6a), with up to 97% of some country's AET occurring through croplands. Central Asian and South American countries join Europe as having a high proportion of their national AET under pastures (Figure 6b). For irrigated agriculture, (Figure 6c), Mediterranean Europe, Central and South Asia dominate with 50–90% of their AET through this land cover. Canada, Russia and parts of western Africa and Southeast Asia have the highest proportion of their AET through non-agricultural (natural) land.

Table 1 shows the distribution of land, of global terrestrial and freshwater MOD16 AET and of human population between croplands, pastures, irrigated areas and uncultivated (natural) land. Some 23% of global land (at 10 km resolution) contains *some* cropland and approximately 31% contains *some* pasture land. A total of 14% of land contains some irrigated agriculture and about 45% contains no agriculture. When the proportion of cover in each pixel is counted (figures in brackets), agricultural covers are much lower and natural cover much higher. The difference between categorical and continuous calculation of land areas is significant and is a key reason why WaterWorld uses continuous representations of land covers. AET per unit land area is greatest under irrigated land (1.04% of global AET per per cent of global land cover). Nature and pasture are similar in their proportion of global AET per per cent of global land cover, with

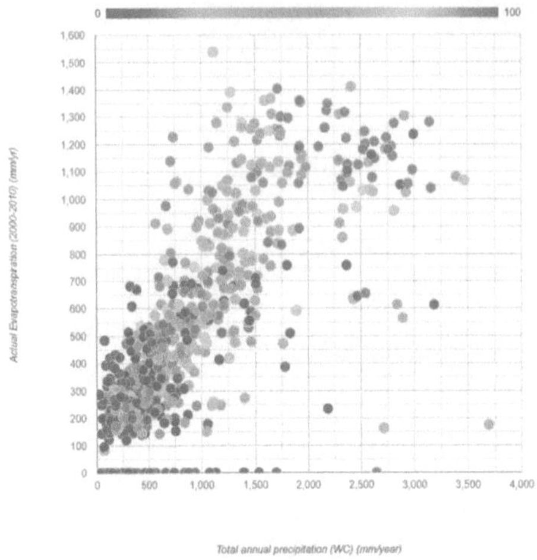

(a) Global terrestrial AET vs total annual precipitation. Marker colour: Cover of tree-covered ground (Sentinel. 2020) (%) Mean

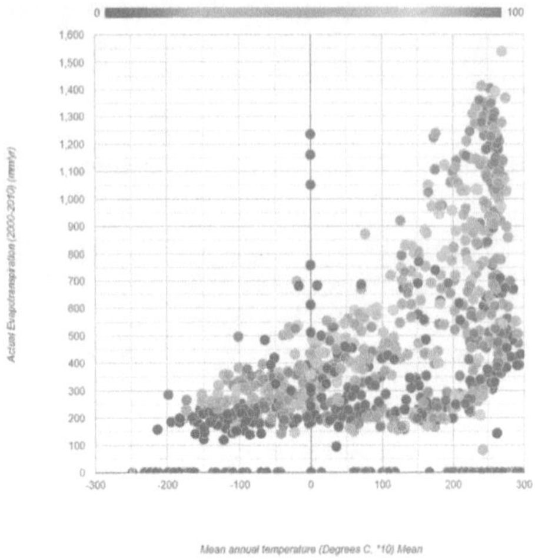

(b) Global terrestrial AET vs mean annual temperature. Marker colour: Cover of tree-covered ground (Sentinel. 2020) (%) Mean

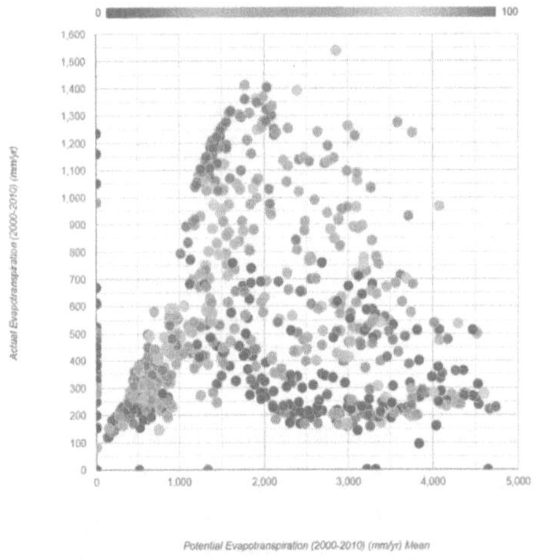

(c) Global potential ET vs actual ET. Marker colour: Cover of tree-covered ground (Sentinel. 2020) (%) Mean

Figure 5. Plots of climatic and land cover variables in relation to baseline actual evapotranspiration (AET): (a) global terrestrial AET versus total annual precipitation; marker colour: cover of tree-covered ground (Sentinel 2020) (%) mean; (b) global terrestrial AET versus mean annual temperature; marker colour: cover of tree-covered ground (Sentinel 2020) (%) mean; and (c) global potential evapotranspiration versus actual evapotranspiration; marker colour: cover of tree-covered ground (Sentinel 2020) (%) mean.

cropland slightly higher. Note that nature-dominated areas in this analysis include tropical forests but also deserts and cold and ice-covered areas.

Around 50% of the global population occurs in pixels with some cropland or some pasture, 61% in areas with some irrigated agriculture and 98% in areas with some nature.

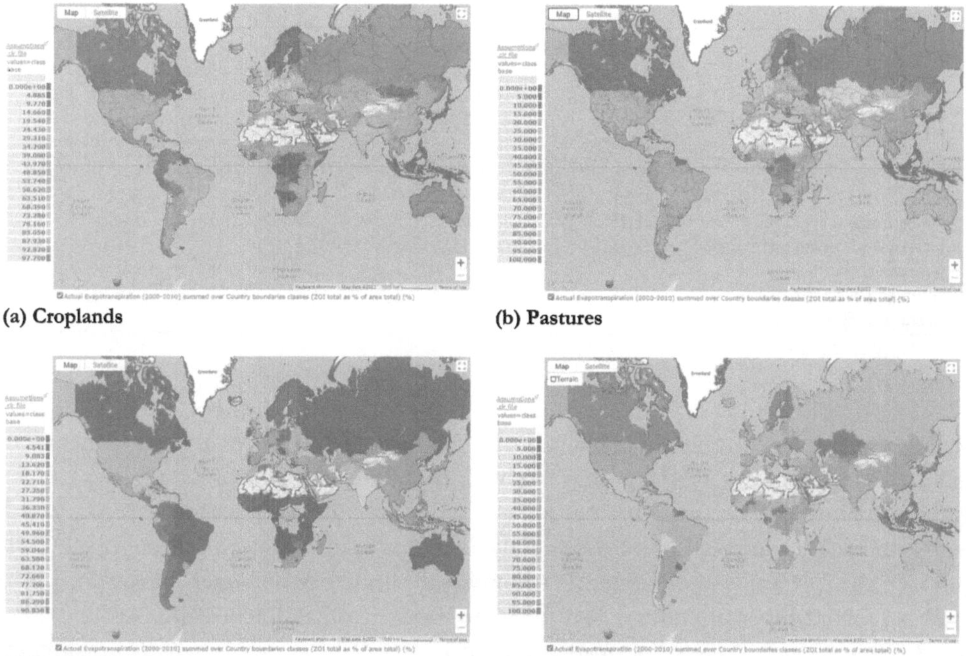

(a) **Croplands** (b) **Pastures** (c) **Irrigated agriculture** (d) **Non-agricultural land**

Figure 6. Proportion of total national actual evapotranspiration (AET) coincident with the distribution of cropland, pasture, irrigated agriculture and natural vegetation, by country: (a) croplands; (b) pastures; (c) irrigated agriculture; and (d) non-agricultural land. Source: Map data ©2022 Google.

Table 1. Agricultural land, actual evapotranspiration (AET) and population.

Land use	% total land area	% total global actual evapotranspiration (2000–10)	% AET per % land area	% global population (persons)
Pixels with some cropland (rainfed and irrigated)	23.01 (7.35)	23.34	1.01	52.36
Pixels with some pasture (rainfed and irrigated)	31.25 (1.53)	30.97	0.99	52.10
Pixels with some nature	45.75 (83.16)	45.68	0.99	98.95
Pixels with some irrigated agriculture	14.32 (7.38)	14.97	1.04	61.43

Note: Figures in parentheses for land cover indicate global averages calculated using continuous fields rather than a binary classification.

These zones overlap and are not mutually exclusive. Agricultural land covers have the potential to impact water resources for large numbers of people locally.

Characterizing downstream context

Table 2 shows that water intakes have a median upstream cropland cover of 3.92% with upper quartile of 28.26%. For pasture, on the other hand, intakes have a median upstream

Table 2. Distribution statistics for upstream average land cover for 4775 water intakes globally.

	Upstream average of cropland (Sentinel 2020) (unique id), mean	Upstream average of pastures (ESA/GLW, 2020) (unique id), mean	Upstream average of FAO irrigation percentage (unique id), mean	Upstream average of % of land remaining under non-agricultural cover/use) (unique id), mean
Median	3.92	0.99	1.00	77.00
Lower quartile	0.00	0.00	0.00	54.00
Upper quartile	28.26	3.26	6.00	94.00
% intakes with upstream average of zero	36.9	35.13	38.7	0.04
% of intakes with upstream average > 50%	10.7	0.4	3.3	78.2

Note: ESA/GLW, European Space Agency/Global Livestock of the World.

pasture cover of 0.99% and upper quartile of 3.26%. These differences are due to the more intensive nature of cropland (more pixels with high intensity of cropland than pasture) and the greater spatial association of cropland with human population centres compared with extensive pastures. Intakes have a median upstream irrigated agriculture of 1% (upper quartile 6%), whilst the dominant upstream influence on water intakes for most intakes remains non-agricultural (natural) (median 74% of land upstream). Around 36% of intakes have no influence of upstream cropland, 35% have no influence of upstream pasture, 39% have no influence of upstream irrigated agriculture and only 0.04% have no influence of upstream natural land. About 11% have > 50% influence of cropland compared with 0.4% for pasture.

The world under irrigated agriculture

This scenario represents a conversion of suitable land to high-efficiency irrigation in which AET results largely from productive transpiration through vegetation cover, rather than wasteful interception loss from water sprayed on leaves or free water evaporation from ponded irrigation water. Conversion of global crop-suitable areas to irrigated agriculture leads to decrease in evapotranspiration relative to the baseline land cover for most areas (Figure 7a) representing a global average decrease in AET of (−31.3 mm/year, −3.8% of baseline). Some areas show an increase in AET, though they are few. The changes vary from −400 to +400 mm/year (Figure 7a) or −30% to +50% of baseline AET (Figure 7b). The spatial footprint of areas in which irrigated agriculture leads to increases in evapotranspiration compared with the baseline are shown in Figure 7(c) and are restricted to deserts, particularly North Africa, the Middle East and West Asia (representing 9.9% of the Earth's surface). All other agriculturally suitable areas (34.5% of the Earth's surface) show a decrease in AET (a negative crop water footprint) with this scenario due to the replacement of tree cover with a less-evaporative herbaceous cover (Figure 7d).

Examining the changes in AET relative to the driving changes in tree cover (Figure 8a), where herb cover replaces existing tree cover (tree cover declines), there is a significant decrease in AET (a negative water footprint) and where a herb cover does not replace tree cover (i.e., tree cover does not change because there are no trees to start with) there is a significant increase in AET (positive water footprint). In Figure 8(b) we see that, under the irrigated cropping scenario, when precipitation is < 500 mm/year, irrigated cropping always leads to a positive water footprint (in this situation there is no tree cover to be replaced). Above 500 mm/year the preceding vegetation cover usually has greater AET than the crop that replaces it so the evaporative water footprint becomes negative. The greater the tree cover loss (blue points) the more-negative the water footprint. The relationship with temperature (Figure 8c) shows that footprints get more positive and more negative as temperature increases. Positive footprints occur only where there is no tree cover replacement, negative footprints increase as the magnitude of tree cover replacement increases. Areas in which the footprint becomes more positive with temperature are hot, dry areas (deserts) whereas those in which the footprint becomes more negative with temperature are hot, humid areas (tropical forests).

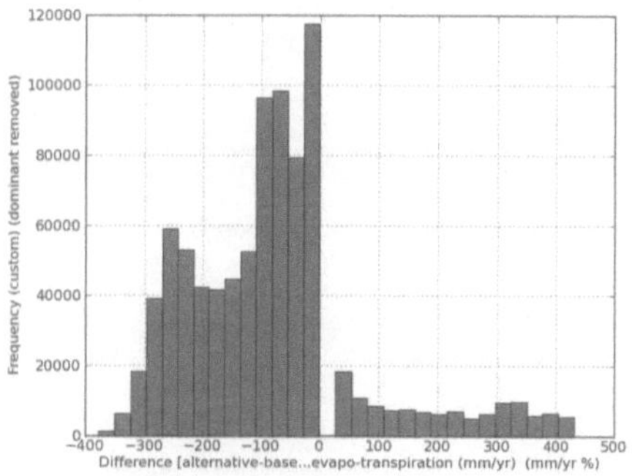

(a) Change in AET

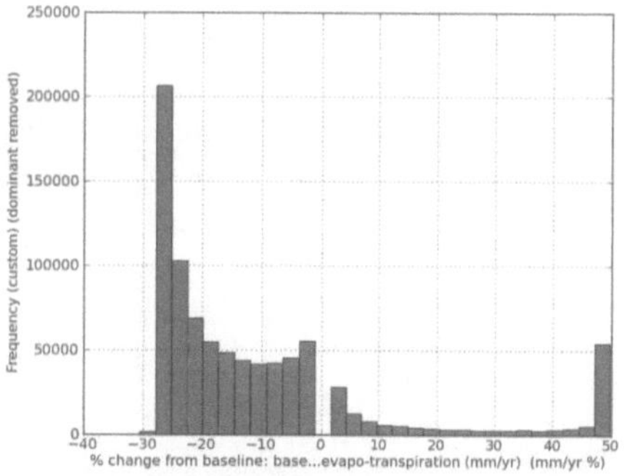

(b) Per-cent change in AET

(c) Increases in AET (areas of positive water footprint of irrigated crops) . **Map data ©2022 Google.**

Figure 7. Actual evapotranspiration (AET) outcomes of irrigated agriculture scenario: (a) change in AET; (b) per cent change in AET; (c) increases in AET (areas of positive water footprint of irrigated crops); and (d) decreases in AET (areas of negative water footprint of irrigated crops). Source: Map data ©2022 Google.

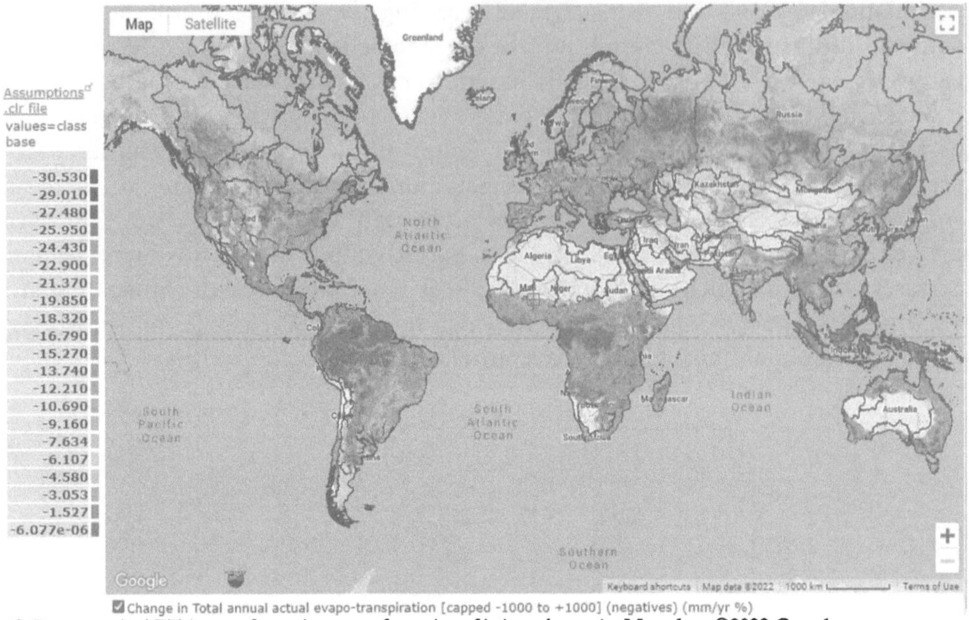

(d) Decreases in AET (areas of negative water footprint of irrigated crops) . **Map data ©2022 Google.**

Figure 7. (continued).

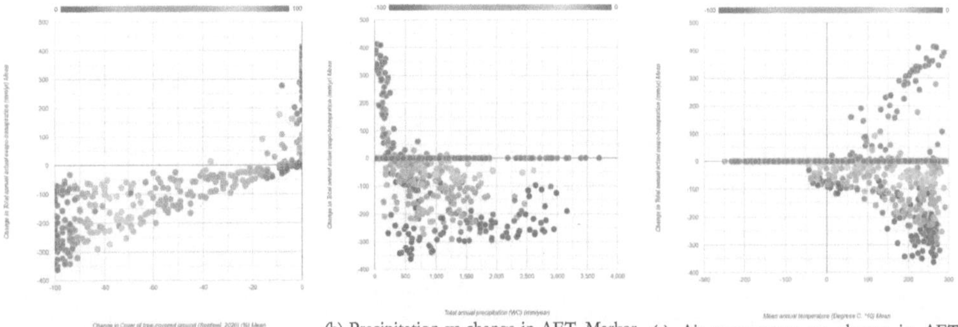
(a) Change in tree cover vs change in AET. Marker colour: Change in Cover of herb-covered ground (Sentinel, 2020) (%) Mean

(b) Precipitation vs change in AET. Marker colour: Change in Cover of tree-covered ground (Sentinel, 2020) (%) Mean

(c) Air temperature vs change in AET. Marker colour: Change in Cover of tree-covered ground (Sentinel. 2020) (%) Mean

Figure 8. Change in actual evapotranspiration (AET) with land cover and climate for the irrigation scenario: (a) change in tree cover versus change in AET; marker colour: change in cover of herb-covered ground (Sentinel 2020) (%) mean; (b) precipitation versus change in AET; marker colour: change in cover of tree-covered ground (Sentinel 2020) (%) mean; and (c) air temperature versus change in AET; marker colour: change in cover of tree-covered ground (Sentinel 2020) (%) mean.

The world under rainfed agriculture

Conversion of global crop-suitable areas to rainfed agriculture (Figure 9) leads exclusively to decreases in evapotranspiration relative to the baseline land cover (negative water footprints), representing a global average decrease in AET of 47.4 mm/year,

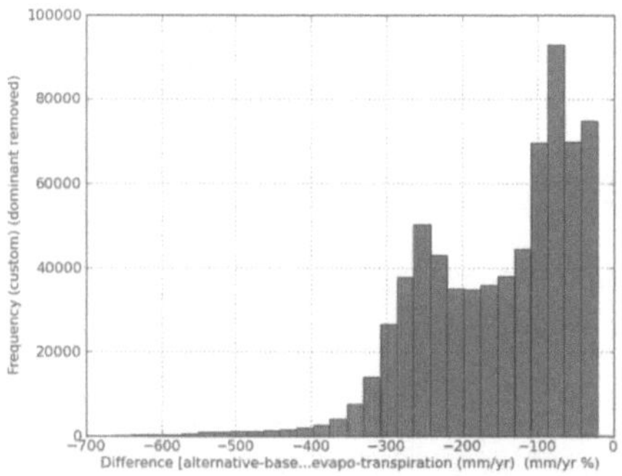

(a) Change in AET

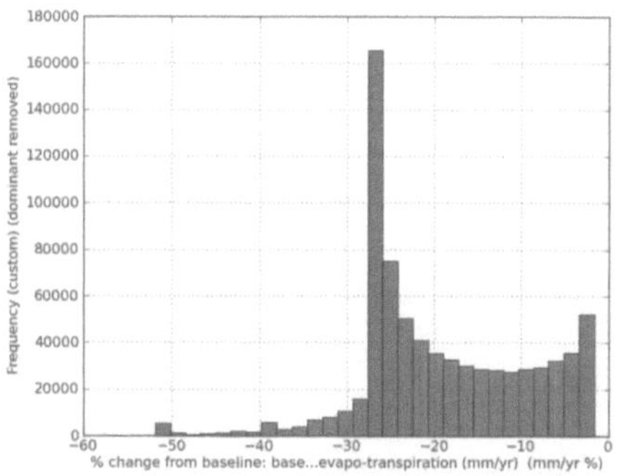

(b) Per cent change in AET

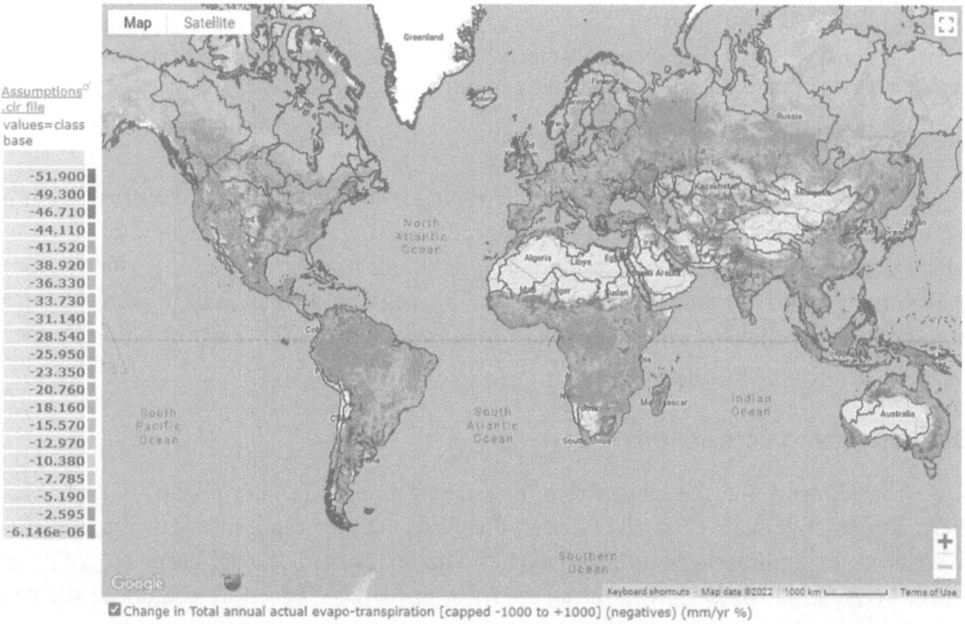

(c) Decreases in AET (areas of negative water footprint of irrigated crops) . **Map data ©2022 Google.**

Figure 9. Change in actual evapotranspiration (AET) for the rainfed agriculture scenario*: (a) change in AET; (b) per cent change in AET; and (c) decreases in AET (areas of negative water footprint of irrigated crops). Source: Map data ©2022 Google.

6.45% of baseline. The changes vary from −550 to 0 mm/year or −50% to −0% of baseline AET (Figure 9b). The spatial footprint of areas in which rainfed agriculture leads to decreases in consumptive water use over the baseline are shown in Figure 9(c) and represent all areas in which the scenario is applied. Decreases in AET tend to be very small in drylands or treeless areas and much greater where rainfed agriculture has replaced forests.

We can see from Figure 10(a) that, under the rainfed cropping scenario, the greater the decrease in tree cover the greater the decrease in AET (and the more negative the evaporative water footprint) because tree cover is replaced by herbaceous cover.

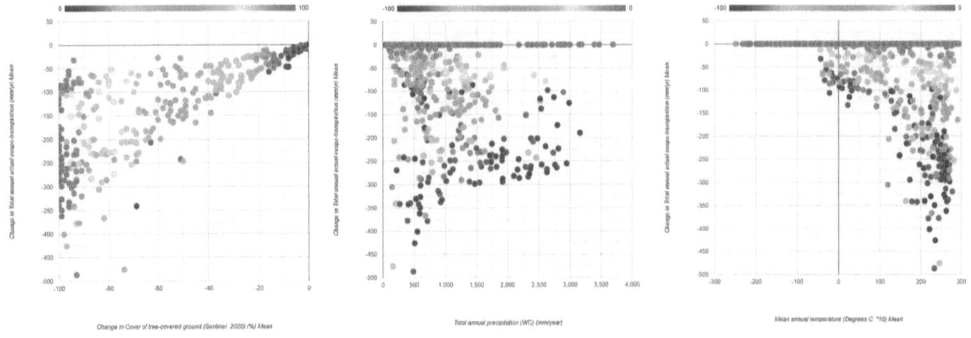

Figure 10. Change in actual evapotranspiration (AET) with land cover and climate for the rainfed scenario: (a) change in tree cover versus change in AET; marker colour: change in cover of herb-covered ground (Sentinel 2020) (%) mean; (b) precipitation versus change in AET; marker colour: change in cover of tree-covered ground (Sentinel 2020) (%) mean; and (c) temperature versus change in AET; marker colour: change in cover of tree-covered ground (Sentinel 2020) (%) mean.

Figure 10(b) shows that, under the rainfed cropping scenario, there is no clear relationship between precipitation and change in AET but rather, change in AET is greatest for pixels in which the most forest has been replaced. The relationship with temperature (Figure 10c) shows that evaporative water footprints get more negative as temperature increases and become most negative in hot, humid areas where tree cover loss on cropping is the greatest.

Discussion and conclusions

It is widely known that the total AET of forests is greater than of croplands (Guo & Mo, 2007; Verstraeten et al., 2005). The cover, diversity, rooting depth and leaf area index of most types of natural vegetation is greater than for rainfed annual crops so AET for natural vegetation is often greater than AET for croplands. Indeed, if techniques such as drip irrigation are used to ensure applied water is productively used in transpiration rather than wasted in loss to interceptions or evaporation from ponded water, even those croplands that are irrigated can have lower AET than an established natural vegetation. However, there is little work at the global scale within the context of water footprinting to understand the implications of conversion to agriculture of a natural land cover for evaporative losses or gains and, thus, crop water footprints.

Calculating water footprints with PET is misleading in a complex way

In characterizing current global AET using a climatology derived from the MOD16 product, we find that AET increases with mean annual precipitation and temperature. In both cases, higher AET values occur under forest cover. Differences in AET between forested and non-forested areas at the same mean annual precipitation can be as high as 600 mm/year (i.e., forested areas some 50% higher AET). Where forest cover is high, AET can be three times higher at the same temperature. AET is significantly less than PET: the global median AET (323 mm/year) is 27% of PET (1181 mm/year) with the lower quartile of AET (213 mm/year) some 53% of PET (400 mm/year) and the higher quartile AET (555 mm/year) some 22% of AET (2483 mm/year). Thus, water footprints based on PET rather than AET will significantly inflate evapotranspiration estimates, as highlighted by Perry (2014). AET is greatest where PET is approximately 2000 mm/year and declines with increasing PET thereafter (Figure 5c). Below 2000 mm/year PET, AET is highest where tree cover is greatest, > 2000 mm/year PET, tree cover is generally low to absent so AET declines significantly. The ratio AET/PET is highest at low PET (> 0.75 at < 500 mm/year PET), decreasing to < 0.25 at 2000 mm/year PET in some areas and 3000 mm/year in all areas. Thus using PET for water footprinting instead of AET not only overestimates water use, but does so in an inconsistent way across environments.

Context is critical

The evaporative footprint of agriculture is large, but highly variable between countries (Figure 6): some 54% of global terrestrial AET occurs in areas with some agriculture. It is difficult to attribute the exact amounts passing through agriculture since our 10 km pixels contain a mix of land covers, each with potentially different AET. Nevertheless, the global footprint of agriculture (about 9% measured as continuous cover fractions, much higher in categorical accounting), combined with its geographical distribution generate a high evaporative footprint. The evaporative footprint of agriculture varies significantly between countries, with close to all AET coincident with cropland or pasture land uses for some countries and close to all coincident with natural land for other countries. Thus, the context of evaporative loss varies significantly between countries and basins, depending on the extent of agriculture. The local and downstream impact of the evaporative footprint for an individual crop is thus highly dependent on whether the country or basin is dominated by agriculture: the national and basin context is thus key to purposeful footprinting (Ridoutt & Huang, 2012) and this context is to be understood upstream–downstream rather than in administrative units (Feng et al., 2010). The risk of evaporative water footprints to downstream water intakes varies with the distribution of agriculture upstream of those intakes: we show that the median upstream cropland cover for global water intakes (3.92%) is greater than that for pastures (0.99%) reflecting the greater proximity of croplands to population centres and the more intensive nature of cropland. Water intakes are still dominated by upstream non-agricultural land uses (77%), including built-up land.

Negative evaporative water footprints are the norm outside drylands

Our agriculturalization scenarios indicate that *decreases* in AET on conversion of (largely perennial) natural vegetation to (largely annual) croplands are the norm outside of drylands. Conversion to an irrigated crop (supporting 100% herbaceous cover irrespective of local climate) using high efficiency techniques that minimise free water and interception losses, leads to a global decrease in AET of 3.8%. Under this scenario, 34.5% of the global land surface shows decreases in AET and the other 9.9% shows increases (the total land area of new cropping was 35%). Areas in which irrigated agriculturalization leads to positive water footprints consistent with the logic of water footprints, are confined to deserts (Figure 7c), coincident with the environments for which the virtual water concept was developed. Negative water footprints are greatest where a full tree cover is replaced by a full herb cover. Only areas without pre-agricultural tree cover show a positive water footprint on conversion (Figure 8a). Positive water footprints on conversion to irrigated agriculture are confined to low rainfall, warm and cold deserts (Figure 8b,c). Conversion to rainfed cropping (all tree cover replaced by herbaceous and bare cover at their original ratios) produces exclusively negative evaporative water footprints, even in deserts, and a global decrease in AET of 6.45%. Positive footprints are not possible in this context since rainfed annual herbaceous crops never reach the AET values of deeper rooted, higher leaf area perennial trees and shrubs. Water footprints are most negative in areas where a full tree cover is replaced with a full herbaceous cover (Figure 10a) across the full range of rainfall values, but particularly at temperatures > 20°C (Figure 10b,c).

Where overhead or flood irrigation is used there is the potential for agricultural AET to be higher than AET for natural vegetation (except for natural wetlands) because of water wasted to AET from interception losses (spray) and free water evaporation (flood). Neither of these is a productive use of water. Indeed, water footprints for crops grown with flood irrigation may be closer to the potential evapotranspiration rates sometimes calculated rather than the AET rates used here. This does not mean that footprinting should focus on this worst-case water use scenario, but rather it should recognize the significant differences that exist between different irrigation methods and calculate accordingly.

Clearly, agriculture has other water footprints than the evaporative footprint, for example, the water pollution load of agricultural inputs (fertilizers, herbicides, pesticides, manures) and the water cost of food processing operations. Whilst these do not suffer the counterfactual problem (agricultural water contamination will always be greater than what went before), they do suffer the context problem since the impact of water pollution depends on the available water, the downstream demand for that water and the decay in the hydrological influence of pollution away from source (Mulligan, 2021).

In conclusion, agriculture and agricultural AET is complex and highly variable. It is a non-constant fraction of PET, varying with land cover and climate characteristics. Under rainfed conditions, there is almost nowhere in the world where a conversion from native vegetation to a crop at the same rainfed-supported density increases AET (water footprint) compared with the natural vegetation. Irrigation, which supports increases in cropland vegetation cover (to 100% herbaceous) relative to the preceding natural vegetation, can generate positive evaporative water footprints but only for drylands where the preceding vegetation was less evaporative (low cover, low to zero tree cover). Flood or overhead irrigation can waste a lot of water, leading to positive footprints.

Tony Allan developed the concept of virtual water to better understand how water-poor economies can be supported by evapotranspiration in wetter environments through international food trade. It was not conceived as a means of understanding water security in wet environments or measuring the water consumption of a product or service. The per product water footprints in common use today are a very effective environmental communication tool but are a simplification of hydrological process, context and counterfactual. It is possible to provide greater realism to these metrics but to do so requires that they are measured at a fine spatial scale, taking into account the local, upstream and downstream context and counterfactuals, rather than as single global, regional or national per product values – which can be highly misleading and, as a policy tool, potentially counterproductive. Greater realism will require spatial hydrologists to engage with the concept of water footprinting and apply greater hydrological precision to its calculation.

Note

1. See MOD16A3.105_MERRAGMAO; http://files.ntsg.umt.edu/data/NTSG_Products/MOD16/MOD16A3.105_MERRAGMAO/.

Acknowledgments

In addition to Tony Allan, I am grateful to colleagues, in particular Martin Keulertz and David Dent, who continue the thought-provoking meetings and discussions that Tony started. I also acknowledge all those colleagues listed in the references who provide global datasets for others to use, and also the taxpayers and students who support their, and our, work.

Disclosure statement

No potential conflict of interest was reported by the author.

ORCID

Mark Mulligan http://orcid.org/0000-0002-0132-0888

References

Aldaya, M. M., Novo, P., Rodríguez-Casado, R., Llamas, M. R., Varela-Ortega, C., & Garrido, A. (2010). *Water footprint and virtual water trade in Spain: Policy implications*. Springer.

Allan, J. A. (1993). *Fortunately there are substitutes for water: Otherwise our hydropolitical futures would be impossible*. Proceedings of the Conference on Priorities for Water Resources Allocation and Management, Overseas Development Administration

Allan, J. A. (2011). *Virtual water: Tackling the threat to our planet's most precious resource*. Bloomsbury.

Berger, M., & Finkbeiner, M. (2013). Methodological challenges in volumetric and impact-oriented water footprints. *Journal of Industrial Ecology*, *17*(1), 79–89. https://doi.org/10.1111/j.1530-9290.2012.00495.x

Da Silva, V. D. P. R., De Oliveira, S. D., Hoekstra, A. Y., Dantas Neto, J., Campos, J., Braga, C., de Araújo, L., Aleixo, D., de Brito, J., de Souza, M., & de Holanda, R. (2016). Water footprint and virtual water trade of Brazil. *Water*, *8*(11), 517. https://doi.org/10.3390/w8110517

FAO. (2012). *Agro-ecological suitability and productivity. Crop suitability index (class). Global agro-ecological zoning version 4 (GAEZ v4)*. https://gaez.fao.org/

FAO. (2014). *Global Administrative Unit Layers (GAUL)*. https://data.apps.fao.org/map/catalog/srv/eng/catalog.search?id=12691#/metadata/9c35ba10-5649-41c8-bdfc-eb78e9e65654

Feng, K., Hubacek, K., Minx, J., Siu, Y. L., Chapagain, A., Yu, Y., Guan, D., & Barrett, J. (2010). Spatially explicit analysis of water footprints in the UK. *Water*, *3*(1), 47–63. https://doi.org/10.3390/w3010047

Gilbert, M., Nicolas, G., Cinardi, G., Van Boeckel, T. P., Vanwambeke, S. O., Wint, G. R. W., & Robinson, T. P. (2018). Global distribution data for cattle, buffaloes, horses, etc., in 2010. *Nature Scientific Data, 5*(1), 180227. https://doi.org/10.1038/sdata.2018.227

Godfrey, J. M., & Chalmers, K. (eds). (2012). *Water accounting: International approaches to policy and decision-making.* Edward Elgar.

Greenwood, R., Willis, R., Hoenig, M., Pegram, G., Baleta, H., Morrison, J., & Farrington, R. (2012). *The CEO water mandate guide to water-related collective action.* CEO Water Mandate.

Guo, R. P., & Mo, X. G. (2007). Differences of evapotranspiration on forest, grassland and farmland. *Journal of Applied Ecology, 18*(8), 1751–1757.

Hengl, T. (2019). *Potential distribution of biomes (Potential Natural Vegetation) at 250 m spatial resolution.* https://doi.org/10.5281/zenodo.3526620

Hijmans, R. J., Cameron, S. E., Parra, J. L., Jones, P. G., & Jarvis, A. (2005). Very high resolution interpolated climate surfaces for global land areas. *International Journal of Climatology, 25*(15), 1965–1978. https://doi.org/10.1002/joc.1276

Hoekstra, A. Y., Chapagain, A. K., Aldaya, M. M., & Mekonnen, M. M. (2011). *The water footprint assessment manual: Setting the global standard.* Earthscan.

Hoekstra, A. Y., & Hung, P. Q. (2005). Globalisation of water resources: International virtual water flows in relation to crop trade. *Global Environmental Change, 15*(1), 45–56. https://doi.org/10.1016/j.gloenvcha.2004.06.004

Hoekstra, A. Y., Huynen, M. (2002). Balancing the world water demand and supply. In P. Martens & J. Rotmans (Eds.), *Transitions in a globalising world* (pp. 17–35). Swets & Zeitlinger.

LandScan^{TM} Global Population Database. (2017). Oak Ridge National Laboratory. http://www.ornl.gov/landscan/

Lehner, B., Verdin, K., & Jarvis, A. (2008). New global hydrography derived from spaceborne elevation data. *Eos, Transactions, AGU, 89*(10), 93–94. https://doi.org/10.1029/2008EO100001

McDonald, R. (2016). City water map (version 2.2). *KNB Data Repository.* Resource Watch. www.resourcewatch.org

Monteith, J. L. (1964). *Evaporation and environment. The state and movement of water in living organisms 205–234.* Symposium of the society of experimental biology, Vol. 19, Cambridge, Cambridge University Press

Morrison, J., Schulte, P., & Schenck, R. (2010). *Corporate water accounting. An analysis of methods and tools for measuring water use and its impacts.* UN Environment Programme, UN Global Compact & Pacific Institute.

Mulligan, M. (2013a). *WaterWorld v2.x documentation.* http://www.policysupport.org/waterworld

Mulligan, M. (2013b). WaterWorld: A self-parameterising, physically-based model for application in data-poor but problem-rich environments globally. *Hydrology Research, 44*(5), 748–769. https://doi.org/10.2166/nh.2012.217

Mulligan, M., & Burke, S. M. (2005, December). *Global cloud forests and environmental change in a hydrological context.* http://tiny.cc/2005CloudForestRep

Mulligan, M. (2021). Mapping hydrological ecosystem services and impacts of scenarios for deforestation and conservation of lowland, montane and cloud-affected forests, and R. W. Myster (Ed.), *The Andean cloud forest* (pp. 189–218). Springer.

Mu, Q., Zhao, M., & Running, S. W. (2011). Improvements to a MODIS global terrestrial evapotranspiration algorithm. *Remote Sensing of Environment, 115*(8), 1781–1800. https://doi.org/10.1016/j.rse.2011.02.019

Oki, T., & Kanae, S. (2004). Virtual water trade and world water resources. *Water Science and Technology, 49*(7), 203–209. https://doi.org/10.2166/wst.2004.0456

Perry, C. (2014). Water footprints: Path to enlightenment, or false trail? *Agricultural Water Management, 134*, 119–125. https://doi.org/10.1016/j.agwat.2013.12.004

Perry, C. (2019). Learning from water footprints: Who loses, who wins, and who cares? *Policy Quarterly, 15*(3), 72. https://doi.org/10.26686/pq.v15i3.5690

Radersma, S., & de Ridder, N. (1996). Computed evapotranspiration of annual and perennial crops at different temporal and spatial scales using published parameter values. *Agricultural Water Management*, *31*(1–2), 17–34. https://doi.org/10.1016/0378-3774(95)01235-4

Ridoutt, B. G., & Huang, J. (2012). Environmental relevance – The key to understanding water footprints. *Proceedings of the National Academy of Sciences*, *109*(22), E1424–E1424. https://doi.org/10.1073/pnas.1203809109

Siebert, S., Döll, P., Feick, S, Hoogeveen, J., & Frenken, K. (2007). *Global map of irrigation areas version 4.0.1*. University of Frankfurt (Main) and FAO.

Sivapalan, M., Blöschl, G., Zhang, L., & Vertessy, R. (2003). Downward approach to hydrological prediction. *Hydrological Processes*, *17*(11), 2101–2211. https://doi.org/10.1002/hyp.1425

Tayia, A. Collins, A., & Gilmont, M. (2022). The role of virtual-water decoupling in achieving food-water security: lessons from Egypt 1962–2013. *Water International*, *47*(7), 1118–1139. https://doi.org/10.1080/02508060.2022.2133835

Verstraeten, W. W., Muys, B., Feyen, J., Veroustraete, F., Minnaert, M., Meiresonne, L., & De Schrijver, A. (2005). Comparative analysis of the actual evapotranspiration of Flemish forest and cropland, using the soil water balance model WAVE. *Hydrology and Earth System Sciences*, *9*(3), 225–241. https://doi.org/10.5194/hess-9-225-2005

Wichelns, D. (2010). Virtual water and water footprints offer limited insight regarding important policy questions. *International Journal of Water Resources Development*, *26*(4), 639–651. https://doi.org/10.1080/07900627.2010.519494

Wichelns, D. (2015). Virtual water and water footprints do not provide helpful insight regarding international trade or water scarcity. *Ecological Indicators*, *52*, 277–283. https://doi.org/10.1016/j.ecolind.2014.12.013

Zanaga, D., Van De Kerchove, R., De Keersmaecker W., Souverijns, N., Brockmann, C., Quast, R., Wevers, J., Grosu, A., Paccini, A., Vergnaud, S., Cartus, O., Santoro, M., Fritz, S., Georgieva, I., Lesiv, M., Carter, S., Herold, M., Li, L., Tsendbazar, N-E., Ramoino, F., & Arino, O. (2021). *ESA WorldCover 10 m 2020 v100*. https://doi.org/10.5281/zenodo.5571936 © ESA WorldCover project 2020/Contains modified Copernicus Sentinel data (2020) processed by ESA WorldCover consortium]

Zhang, Y., Chiew, F. H., Peña-Arancibia, J., Sun, F., Li, H., & Leuning, R. (2017). Global variation of transpiration and soil evaporation and the role of their major climate drivers. *Journal of Geophysical Research: Atmospheres*, *122*(13), 6868–6881. https://doi.org/10.1002/2017JD027025

Virtual water, international relations and the new geopolitics of food

Eckart Woertz

ABSTRACT

Food security and virtual water considerations interact in a corporatized third food regime, particularly in the Middle East and North Africa (MENA), the world's largest grain importing region. Globalization has created asymmetric trade interdependence that can be weaponized by states that control global economic hubs, yet countries in the Global South have actively exploited the rivalries of big powers, temporarily capturing chokepoints and entering new alignments. The virtual water paradigm that Tony Allan first coined can be applied creatively to international relations that are increasingly characterized by 'weaponized interdependence'.

Virtual water and MENA food imports

The virtual water paradigm that Tony Allan coined was first rehearsed using the Middle East and North Africa (MENA) as a case study (Allan, 2001). Its scope and policy implications are global, but its relevance to this part of the world continues to be profound. Together with oil and gas exports, the region's water scarcity and resulting dependence on food imports defines its geoeconomic position.

Tony Allan famously observed that the import of virtual water via food trade added a 'second River Nile' to the region's water balance. He offered a new way of thinking about its water dilemmas by pondering the implications of policies pursued by rulers unaware of their practical significance. From the 1950s onwards, the MENA increasingly absorbed structural grain surpluses dumped on markets in the developing world by the United States and, later, the European Economic Community. Export-promotion schemes such as PL 480 – renamed the *Food for Peace* programme under the John F. Kennedy administration – provided food imports at subsidized rates (Wallerstein, 1980). For recipient countries such as Egypt, they constituted cheap inputs (Burns, 1985): they fed the growing urban workforces created by the import-substituting industrialization drives that were prevalent in the developing world at that time. Nobody thought of them as mitigating water scarcity – which in fact they did – but the virtual water paradigm switched attention from conflicting claims over blue water resources to the unexpected remedy of food imports. It critically questioned MENA countries' aspirations of self-sufficiency that flew in the face of resource endowments. From the 1970s onwards, renewable water resources were no longer sufficient to grow the required food; extremely water-scarce parts of the region – notably Israel, Palestine and desert Libya – had already reached that point in the 1950s.

With its focus on food trade, the virtual water paradigm was compatible with trade-based notions of food security that called for specialization according to comparative advantage to earn the hard currency needed for food imports. Yet, Tony Allan was keenly aware that markets do not just exist, they are made – and their creation is an eminently political process. In his later work, he explored how the global food system affects the water–energy–food nexus (Allan et al., 2015) and how societies cope with these challenges (Allan et al., 2018).

Agriculture accounts for some 92% of global water consumption (as opposed to water withdrawal, some of which can be used more than once in any one hydrological cycle; Allan et al., 2015). Farmers are the principal stewards of water resources, yet they have little voice in a global food system orientated towards procurement and distribution, and they communicate with agribusiness via contracts that ignore or misprice water and its economic externalities. It is this global agribusiness sector that has provided most of the food imports for the MENA from relatively few exporter nations. And with the global food crisis of 2007/08, these imports suddenly became tenuous.

The global food crisis of 2007/08 and the new geopolitics of food

The surge in global food prices in 2007/08 prompted some exporters (e.g., Argentina, India, Russia and Vietnam) to announce export restrictions out of concern for their own food security. This eroded trust in global food markets and supply chains, and food-importing countries in Asia and the Middle East reacted by investing in farmland abroad to secure privileged bilateral access to commodities. In parallel to this foreign-investment drive, domestic agriculture was downsized in many countries, for example, Saudi Arabia, started to phase out subsidized wheat production, which it had begun to ramp up in the 1980s (Elhadj, 2022 in the first part of this special issue). The connection between food imports and the domestic water balance was explicitly acknowledged in such policies: the virtual water paradigm had moved from academe to the corridors of power. Egypt in fact now incorporates virtual water in its national water balance accounting (Tayia et al., 2022, in this issue).

In comparison with the world food crisis of 1972–75 that prompted the spurious Saudi wheat bonanza and ill-fated plans to develop Sudan as an Arab breadbasket, the global food system had transformed by the time the events of 2007/08 came around. Whereas the production increases of the Green Revolution relied heavily on intensification of inputs in the 1960s–80s, the role of higher total factor productivity via agricultural technology and improved inputs became more important for agricultural production growth since the 1990s, accounting for some three-quarters of it globally in the 2000s, before slightly declining in the 2010s (Fuglie, 2018). The crucial role of productivity growth was reflected in the importance of globally traded input factors such as seeds and fertilizers, two-thirds of which came from foreign suppliers on average – a trade that interacted closely with food trade itself and technology adoption (Farrokhi & Pellegrina, 2022; Hertel et al., 2020). Africa where the Green Revolution was not implemented to any large degree compared with Asia and Latin America became a focus of attention for future global production growth (Thurow, 2010). However, a much-touted boost from genetically modified organisms was hotly debated in view of false promises and the detrimental ecological effects of industrial agriculture, motivating calls for a Green Revolution 2.0 with better inclusion of smallholders and conservationist concerns (Conway & Barbier, 2009; Pingali, 2012). Most importantly, the value chains of the global food system had become increasingly corporatized, which prompted some to speak of the emergence of a *third food regime* since the 1970s (Burch & Lawrence, 2009). It entailed a supermarket revolution that included developing countries as well, where it was driven by rising incomes of the middle classes and increased refrigerator ownership on the demand side and growing foreign direct investment (FDI) and improved logistics and inventory management on the supply side (Reardon et al., 2003).

In contrast to the second food regime of the post-war years that was characterized by the disposal of surplus production from North America and Europe and their national institutions, new actors had now emerged. International corporations presided over changing diets and the rise of packaged foods; they dominated trade, processing, distribution and the provision of inputs to the food system. Thus, they became crucial actors in the allocation of virtual water. An increasing share of exports now emanated from tropical countries, such as palm oil from Indonesia and Malaysia and corn and soybeans from Brazil (Woertz & Keulertz, 2015). Russia regained its historic position as a net grain exporter, seeking to leverage it by establishing a national grain-trading house. With the breakdown of international commodity agreements in the 1970s, stockpiles of staple crops for price stabilization had diminished; agriculture was included in the Uruguay Round of the General Agreement on Tariffs and Trade (GATT) in 1986; and the US and the European Union decoupled farm subsidies from price-support schemes in 1996 and 2003, respectively (Winders, 2011). Food aid was now mainly administered through the World Food Programme (WFP), not by national bodies (Paarlberg, 2010).

In the 1960s and 1970s, the US allotted or withheld food aid depending on political compliance. It contemplated food embargos in retaliation for the Arab oil boycott and the Iranian hostage crisis, and actually implemented one against the Soviet Union in the wake of the latter's invasion of Afghanistan. But the food weapon backfired: it mainly hurt the export interests of American farmers and gave way to a depoliticization of food trade under the Ronald Reagan administration (Woertz, 2013b). The 2007–08 crisis reignited importers' memories of their strategic vulnerability. Many foreign agri-investments in its wake focused on Africa and Latin America, in food-insecure countries with insecure land rights and problematic governance that account for most unused global land banks and have substantial freshwater reserves (Deininger et al., 2011). Even countries with physical water shortages in the MENA, Central Asia and South Asia ranked among target destinations and potential exporters of virtual water.

A political backlash ensued. Non-governmental organizations (NGOs), farmers and holders of customary land rights organized bottom-up resistance. In exporter countries with developed agri-markets such as Australia, Brazil and Thailand, there was a different kind of resistance, driven by governments and parts of the domestic business class: ceilings on foreign land ownership were implemented or discussed to keep a strategic industry in national hands.

There was a considerable implementation gap in Gulf – and also Chinese – agri-investments in developing countries (Brautigam, 2015; Woertz, 2013b). Especially in Africa and for staples, commercial crops such as palm oil, rubber and sugar fared better.

Here the share of large-scale land acquisitions in global production was noticeable (Lay et al., 2021). Concerns about infrastructure, commercial viability, political uncertainties, corruption and the reliability of supplies loomed large. More than a decade after foreign agri-investments in many developing countries had been announced, they still did not show up in trade statistics; food imports continued to come from established agri-exporters. When Gulf states did genuinely put money on the table, it was in such countries and further up the value chain in food processing and distribution, rather than in politically controversial farmland acquisitions. The Saudi Agricultural and Livestock Investment Company (SALIC, owned by the Saudi government) teamed up with international grain-trader Bunge and acquired a majority stake in the privatized Canadian wheat board. In Asia, food-traders such as Chinese Noble or Singapore-based Olam and Wilmar became major global players, challenging the market share of established rivals in the West such as ADM, Bunge, Cargill and Dreyfus (Keulertz & Woertz, 2015).

The crisis of 2007/08 created mistrust in the global food system. Importers sought to circumvent the multilateral trading environment by gaining direct bilateral access via farmland investments yet, when push came to shove, they often settled for muscling their way into the increasingly corporatized global value chains of the third food regime. This raises the question of how food security and virtual water considerations play out in the new geopolitics of food and, thus, interact with international relations.

Virtual water and international relations

The importance of hard-power politics decreased with the end of the Cold War. That of economic statecraft to further geopolitical objectives grew meanwhile (Luttwak, 1990). Besides economic and financial sanctions, this spans trade, investment and monetary policies, energy and commodities, cyber regulation and aid (Blackwill & Harris, 2016) – what might be termed *geoeconomics*: the strategic use of economic power to control spheres of influence, over and above the broader concept of international political economy (Wigell et al., 2019). Geoeconomics can enrich the three prevalent strands of International Relations (IR) theory. It transcends realism by stressing the importance of economic means in power competition and the impact of geography on policy, liberal institutionalism by insisting that economic interdependence and conflict can co-occur and constructivism by noting that beyond ideological discourses material factors matter. This has implications for the conceptualization of food trade as well, whose weaponization has a long history, ranging from the First World War to the recent Russo-Ukrainian war.

Globalization led to the mushrooming of financial markets and trade interdependence. However, such interdependence has been asymmetric and states have sought to exploit such imbalances more than the hegemonic neoliberal narratives of the 1990s and early years of the new century ever imagined. Farrell and Newman (2019) have argued that *weaponized interdependence* enables states that control global economic hubs to exert power over others. This pertains not only to material flows such as energy but also to financial and information markets; the international payment system SWIFT or the infrastructure of the Internet such as undersea cables are cases in point. Panopticon effects (i.e., they can see what we are doing but we cannot see them) enable states to glean critical knowledge to be used in negotiations, to counter adversaries and to create political frames; *chokepoint effects* occur when states are in a position to limit or penalize access to the hubs that they control.

Weaponized interdependence effectively grants agency only to those who preside over global economic networks – meaning China, the EU and the US. Yet countries in the Global South have actively sought to take their destinies into their own hands by exploiting the rivalries of big powers, temporarily capturing chokepoints, entering new alignments and creating revised narratives in an attempt to shape prevailing norms and institutions. Brazil managed to capture a larger share of the Chinese market for soybeans in the wake of trade wars between the US and China, Turkey used its geographical position by weaponizing migration flows in its adversarial dealings with the EU. Serbia entered closer health collaboration with China when the EU restricted supplies of hospital goods to non-member states at the beginning of the COVID-19 pandemic. India leveraged its 70% share of global hydroxychloroquine production during the same pandemic to grant preferential access to cooperating countries when a debate arose whether the drug might be effective against COVID-19 (Narlikar, 2021). More recently, Saudi Arabia and the United Arab Emirates were anxious to preserve the OPEC + deal with Russia, maintain a neutral position in the Ukraine conflict and resisted Western calls to raise oil production to calm global markets (Aboudouh, 2022).

Apart from outright military blockade, as in the First and Second World Wars (Collingham, 2011; Schatkowski-Schilcher, 1989, 1992), there are other means to restrict food imports. Secondary sanctions leverage the global financial system to target trading partners of countries that are under direct sanctions. They have had considerable impact on the food imports of countries such as Iran and Syria, even if food and medicine were formally excluded from sanctions regimes. Barred from payment channels and their assets frozen, they struggled to find takers for their grain tenders. The crisis of 2007/08 showed food net importers the perils of asymmetric and potentially weaponized interdependence when they faced food-trade restrictions vis-à-vis exporting countries. In response, they tried to bypass the prevalent liberal trade order by gaining privileged bilateral access to commodities, but they also attempted to use the system to their advantage by buying into the increasingly corporatized value chains of the global food system. They built up national food-trading houses and formed new trading relationships as Russia and tropical countries came to play an increasingly important role in global food exports. When seeking to buy into farmland assets or crucial inputs such as fertilizer they could face strong resistance, as shown by the aborted takeover of the Canadian Potash Corp by Chinese state-owned Sinochem in 2010 (Massot, 2011).

And in seeking to forge new narratives of collaboration, the outsiders had to grapple with counternarratives that accused them of land-grabbing. Qatar tried to instigate a *Global Dryland Alliance* between similarly affected countries and Gulf States backed proposals by Japan and Switzerland to sanction export restrictions at the World Trade Organization (WTO), an institution that had traditionally been preoccupied with reduction of import rather than export barriers (Woertz, 2013a). They might pursue more such initiatives at international organizations in the future, such as the *Net Food Importing Developing Countries* that formed an interest group within the GATT to lobby for affordable food imports during the Uruguay Round (Narlikar, 2003).

With global food trade growing faster than global food production, cross-national virtual water linkages more than doubled between 1986 and 2007 and important shifts occurred towards Asia, where China imported increasing quantities of soy from Brazil (Dalin et al., 2012). Quantifications stemming from the virtual water paradigm, such as the *Water Footprint Network* launched in 2008 by Arjen Hoekstra, have argued for ecologically optimized trade flows based on water endowments and virtual water considerations (Water Footprint Network, 2021). This view has been criticized by economists who have pointed out that the virtual water approach does not factor in local opportunity costs. Comparative advantage is not based solely on one production factor such as water, but several, and water efficiency rather than water abundance is empirically a more decisive factor of water allocation as is the available arable land per capita. Hence decisions on water allocation and their impact are 'fundamentally local decisions to be determined in the context of local opportunity costs, property rights, political and market power' (Baylis et al., 2021, p. 16.11; Wichelns, 2015). Virtual water flows are often indeed not correlated with the actual water-resource endowments. Water-scarce countries such as Kazakhstan can be net exporters and water-rich countries such as Finland net importers of virtual water. Whether and to what extent virtual water considerations should inform multilateral trade governance is hence a contentious issue between economic and ecological sets of literature. There is, however, a certain consensus that trade can help adaptation to local water scarcity and mitigate related conflicts.

This is also exemplified in the water conscious food trade initiatives of food net importers in the MENA described above. We see that the virtual water paradigm has entered the political calculations of food net importers, who now explicitly conceptualize food trade as a means of mitigating water scarcity at home, and we may conclude that past concerns that such scarcity might lead to 'water wars' were overblown. Water is a unique commodity. It flows and yet is bulky, ill-suited for long-distance transportation. Unlike an oil well or a diamond mine, one cannot take lasting possession of it or it does not make commercial sense to do so; hence, there is no *blood water* that would finance rogue militias.

Water is an unwieldy object of warfare, even though it can still become a target or tool of war. However, the absence of war does not mean the absence of conflict; competing claims of different communities to scarce water resources can lead to local conflict. At a higher level, water can encourage cooperation between states, although such collaboration might become brittle in the future as it is often asymmetric and lacks a firm legal footing in international agreements (Chellaney, 2013, pp. 41–43). Even so, competitions over water hegemony pursued by diplomatic and political means are a more likely scenario than either outright military conflict or peaceful cooperation (Zeitoun & Warner, 2006).

There are significant hurdles for interstate water wars. In comparison international grain trade has played a prominent role in interstate conflict ever since its proliferation in the wake of steam power and steeply declining transportation costs in the 1870s. The naval blockades of the two World Wars or the politicization of food trade in the 1970s are cases in point when the US threatened a grain embargo in retaliation to the Arab oil boycott of 1973 and the Tehran hostage crisis of 1979 and actually implemented one in response to the Soviet Union's invasion of Afghanistan (Woertz, 2018). The costs for American farmers were substantial as competing suppliers such as Argentina and Australia picked up the slack. So, the US changed course in the 1980s and sought to re-establish its reputation as a reliable supplier of grains, abjuring the use of food trade in sanctions within the 'Reagan Doctrine of Agricultural Trade' (Woertz, 2013b, ch. 4). Yet, Iraq suffered tremendously from 1990 to 2003 when a multilateral UN embargo curtailed its oil exports and access to foreign assets and finance, severely hampering the import of food and medicine items that were formally excluded from sanctions (Gordon, 2010; Woertz, 2019).

The current Russian war on Ukraine threatens to add another chapter to the MENA's rich experience with the food weapon. If the former Soviet Union was a grain net importer for its livestock programme, some of its successor states have regained their position as major grain exporters to world markets over the past two decades, a position their territory already occupied in the late 19th and early 20th centuries. About half of the Arab world's wheat imports come from Ukraine and Russia. Some countries such as Egypt, Lebanon and Sudan have exposures well above 80% (Woertz, 2022). Saudi Arabia is the world's second largest importer of barley, mostly from the two countries, to use it as feedstock for sheep and other animals. The World Food Programme (WFP) struggles to source supplies to its assistance programmes to most food-insecure countries in the region such as Yemen and Syria (WFP, 2022).

The war has already prompted much feared export restrictions and will limit technology transfer to Russia in the field of agriculture as well. It will affect the harvest of Ukraine whose main export outlet the Black Sea port of Odessa is blocked. Russian exports are affected as well by soaring transaction costs. Transportation costs and insurance premia have increased, and sanctions curtail the access to trade finance. The impact goes beyond the two countries as both are major producers of fertilizers. Soaring costs and limited availability might affect the production elsewhere, limit margins of farmers and put a question mark behind a supply side response to rising food prices (Benton et al., 2022).

There is a debate whether the detrimental impact is a mere derivative of the war or if Russia is intentionally weaponizing food supplies to bring Ukraine to its heels. The G7 even speaks of a 'grain war' of Russia with the intention to create destabilizing food and refugee crises globally as a means of hybrid warfare (Euronews, 2022). A war of narratives emerges with the West blaming Russia's war and the latter Western sanctions and Ukrainian defensive mining of maritime access points for the looming food crisis (Lederer, 2022). For MENA countries this renewed episode of food trade politicization shows that food trade and its implicit virtual water carries promises but also risks as it can be more easily weaponized and interrupted than blue water flows.

Conclusions

Today the MENA is the world's largest grain importer and mitigates its water scarcity via virtual water imports. The large majority of this virtual water trade is based on green water. Agriculture accounts for over 90% of consumptive water use globally and some 70% of global crops are produced with rainfed agriculture. The paradigm that Tony Allan established sheds doubts on the popular talk about interstate 'water wars' and cautions against excessive focus on blue water resources and related allocation conflicts. On the surface it is compatible with a trade-based approach to food security, yet Tony Allan stressed the political nature of such markets.

The food regime in which the virtual water exchange occurs has been transformed since the 1970s with greater corporatization of value chains, growing role of the WFP in the allocation of food aid compared with national bodies and emergence of new export power houses, most notably Brazil for soybeans and corn, Malaysia and Indonesia for palm oil, and Russia and Ukraine for grains and sunflower oil. MENA countries that are dependent on food imports engage with this international food regime and try to influence it to their advantage as the global food crisis 2007–08 has shown. Their perception that the food and virtual water trade entails vulnerabilities beside its benefits is informed by the historic precedents of its politicization. This calls for a better conceptualization of it in IR scholarship.

The weaponization of blue water faces significant hurdles yet has attracted a lot of attention. The food trade that alleviates the MENA's water balance, by the gains of virtual water, at the same time offers opportunities for asymmetric and weaponized interdependence. The implications of virtual water are worth more consideration by national planners, trade negotiators and foreign policy personnel. Almost 30 years after Tony Allan first introduced it, the virtual water paradigm can be applied in new contexts worldwide and can enrich the analysis of international relations and related theories. It will continue to surprise us.

Disclosure statement

No potential conflict of interest was reported by the author.

Funding

This work was supported by the Bundesministerium für Bildung und Forschung, World Order Narratives of the Global South (WONAGO) [01UC2106B].

ORCID

Eckart Woertz http://orcid.org/0000-0001-9526-2668

References

Aboudouh, A. (2022, 26 April). *Russia's war in Ukraine is making Saudi Arabia and the UAE rethink how they deal with US pressure over China.* Atlantic Council. https://www.atlanticcouncil.org/blogs/menasource/the-ukraine-war-is-making-saudi-arabia-and-the-uae-rethink-how-they-deal-with-us-pressure-over-china/

Allan, T. (2001). *The Middle East water question: Hydropolitics and the global economy.* I.B. Tauris.

Allan, T., Bromwich, B., Keulertz, M., & Colman, A. (Eds.). (2018). *The Oxford Handbook of food, water and society.* Oxford University Press.

Allan, T., Keulertz, M., & Woertz, E. (2015). The water–food–energy nexus: An introduction to nexus concepts and some conceptual and operational problems. *International Journal of Water Resources Development, 31*(3), 301–311. https://doi.org/10.1080/07900627.2015.1029118

Baylis, K., Heckelei, T., & Hertel, T. W. (2021). Agricultural trade and environmental sustainability. *Annual Review of Resource Economics, 13*(1), 379–401. https://doi.org/10.1146/annurev-resource-101420-090453

Benton, T., Froggatt, A., Wellesley, L., Graham, O., King, R., Morisetti, N., Nixey, J., & Schröder, P. (2022). *The Ukraine war and threats to food and energy security: Cascading risks from rising prices and supply disruptions*. Chatham House. https://www.chathamhouse.org/2022/04/ukraine-war-and-threats-food-and-energy-security

Blackwill, R. D., & Harris, J. M. (2016). *War by other means: Geoeconomics and statecraft*. The Belknap Press of Harvard University Press.

Brautigam, D. (2015). *Will Africa feed China?* Oxford University Press.

Burch, D., & Lawrence, G. (2009). Towards a third food regime: Behind the transformation. *Agriculture and Human Values, 26*(4), 267–279. https://doi.org/10.1007/s10460-009-9219-4

Burns, W. J. (1985). *Economic aid and American policy toward Egypt, 1955–1981*. State University of New York Press.

Chellaney, B. (2013). *Water, peace, and war: Confronting the global water crisis*. Rowman & Littlefield.

Collingham, E. M. (2011). *The taste of War: World War Two and the battle for food*. Allen Lane.

Conway, G., & Barbier, E. (2009). *After the green revolution: Sustainable agriculture for development*. Earthscan.

Dalin, C., Konar, M., Hanasaki, N., Rinaldo, A., & Rodriguez-Iturbe, I. (2012). Evolution of the global virtual water trade network. *Proceedings of the National Academy of Sciences, 109*(16), 5989–5994. https://doi.org/10.1073/pnas.1203176109

Deininger, K., Byerlee, D., Lindsay, J., Norton, A., Selod, H., & Stickler, M. (2011). *Rising global interest in farmland. Can It yield sustainable and equitable benefits?* World Bank.

Elhadj, E. (2022). Ozymandias in the desert: irrigation in Saudi Arabia. *Water International, 47*(6), 969–978. https://doi.org/10.1080/02508060.2022.2123612

Euronews. (2022, 14 May). *G7: Russia extending Ukraine military war to 'grain war', says German FM Baerbock, Euronews*. https://www.euronews.com/2022/05/14/ukraine-war-grain-exports-blocked-by-russia-threaten-to-bring-hunger-and-famine-g7-warns

Farrell, H., & Newman, A. L. (2019). Weaponized interdependence: How global economic networks shape state coercion. *International Security, 44*(1), 42–79. https://doi.org/10.1162/isec_a_00351

Farrokhi, F., & Pellegrina, H. S. (2022). Trade, technology, and agricultural productivity. *STEG Working Paper. Structural Transformation and Economic Growth (STEG) Programme*. https://steg.cepr.org/publications/trade-technology-and-agricultural-productivity

Fuglie, K. (2018). R&D capital, R&D spillovers, and productivity growth in world agriculture. *Applied Economic Perspectives and Policy, 40*(3), 421–444. https://doi.org/10.1093/aepp/ppx045

Gordon, J. (2010). *Invisible war: The United States and the Iraq sanctions*. Harvard University Press.

Hertel, B. T. W., Baldos, U. L. C., & Fuglie, K. O. (2020). Trade in technology: A potential solution to the food security challenges of the 21st century. *European Economic Review, 127*, 103479. https://doi.org/10.1016/j.euroecorev.2020.103479

Keulertz, M., & Woertz, E. (2015). States as actors in international agro-investments. *International Development Policy, 6*(1). https://doi.org/10.4000/poldev.2274

Lay, J., Anseeuw, W., Eckert, S., Flachsbarth, I., Kubitza, C., Nolte, K., & Giger, M. (2021). *Taking stock of the global land rush: Few development benefits, many human and environmental risks. Analytical Report III. Land matrix analytical reports*. Bern Open Publishing. https://doi.org/10.48350/156861

Lederer, E. M. (2022, May 20). US accuses Russia of weaponizing food in Ukraine war. *AP News*. https://apnews.com/article/russia-ukraine-politics-united-nations-antony-blinken-5296aae2dec1db42f17bd98efb627b91

Luttwak, E. N. (1990). From geopolitics to geo-economics: Logic of conflict, grammar of commerce. *The National Interest, 20*, 17–23. https://www.jstor.org/stable/42894676

Massot, P. (2011). Chinese state investments in Canada: Lessons from the potash saga. *Canada-Asia Agenda, 16*. https://www.scribd.com/document/83349468/Chinese-State-Investments-in-Canada-Lessons-from-the-Potash-Saga-January-2011-Pascale-Massot

Narlikar, A. (2003). *International trade and developing countries: Bargaining coalitions in the GATT and WTO*. Routledge.

Narlikar, A. (2021). Must the weak suffer what they must? The global south in a world of weaponized interdependence. In D. W. Drezner, H. Farrell, & A. L. Newman (Eds.), *The uses and abuses of weaponized interdependence* (pp. 289–304). Brookings.

Paarlberg, R. L. (2010). *Food politics: What everyone needs to know*. Oxford University Press.

Pingali, P. L. (2012). Green Revolution: Impacts, limits, and the path ahead. *Proceedings of the National Academy of Sciences*, *109*(31), 12302–12308. https://doi.org/10.1073/pnas.0912953109

Reardon, T., Timmer, C. P., Barrett, C. B., & Berdegué, J. (2003). The rise of supermarkets in Africa, Asia, and Latin America. *American Journal of Agricultural Economics*, *85*(5), 1140–1146. https://doi.org/10.1111/j.0092-5853.2003.00520.x

Schatkowski-Schilcher, L. (1989). Die Weizenwirtschaft des Nahen Ostens in der Zwischenkriegszeit: Der Einfluß der Ökonomie auf die Politik am Beispiel Syriens. In L. Schatkowski-Schilcher & C. Scharf (Eds.), *Der Nahe Osten in der Zwischenkriegszeit, 1919–1939. Die Interdependenz von Politik, Wirtschaft und Ideologie* (pp. 241–259). Franz Steiner.

Schatkowski-Schilcher, L. (1992). The famine of 1915–1918 in Greater Syria. In J. Spagnolo (Ed.), *Problems of the Modern Middle East in historical perspective: Essays in honor of Albert Hourani* (pp. 229–258). Ithaca Press (for Garnet Publishing Ltd.).

Tayia, A., Collins, A. M., & Gilmont, M. (2022). The role of virtual-water decoupling in achieving food–water security: lessons from Egypt, 1962–2013. *Water International*, *47*(7), 1118–1139. https://doi.org/10.1080/02508060.2022.2133835

Thurow, R. (2010). The fertile continent: Africa, agriculture's final frontier. *Foreign Affairs*, *89*(6), 102–111. https://www.foreignaffairs.com/articles/africa/2010-10-21/fertile-continent

Wallerstein, M. B. (1980). *Food for war–food for peace: United States food aid in a global context*. MIT Press.

Water Footprint Network. (2021). *Homepage*. https://waterfootprint.org/en/water-footprint/

WFP. (2022). *Food security implications of the Ukraine conflict*. UN World Food Programme. https://www.wfp.org/publications/food-security-implications-ukraine-conflict

Wichelns, D. (2015). Virtual water and water footprints do not provide helpful insight regarding international trade or water scarcity. *Ecological Indicators*, *52*, 277–283. https://doi.org/10.1016/j.ecolind.2014.12.013

Wigell, M., Scholvin, S., & Aaltola, M. (Eds.). (2019). *Geo-economics and power politics in the 21st century: The revival of economic statecraft*. Routledge.

Winders, B. (2011). The food crisis and the deregulation of agriculture. *The Brown Journal of World Affairs*, *XVIII*(1), 83–95. https://www.jstor.org/stable/24590778

Woertz, E. (2013a). The governance of Gulf agro-investments. *Globalizations*, *10*(1), 87–104. https://doi.org/10.1080/14747731.2013.760932

Woertz, E. (2013b). *Oil for food. The global food crisis and the Middle East*. Oxford University Press.

Woertz, E. (2019). Iraq under UN embargo, 1990–2003: Food security, agriculture, and regime survival. *Middle East Journal*, *73*(1), 92–111. https://doi.org/10.3751/73.1.15

Woertz, E. (2022). *The Russian War against Ukraine: Middle East Food Security at Risk. GIGA Focus*(2). German Institute for Global and Area Studies. https://www.giga-hamburg.de/en/publications/giga-focus/the-russian-war-against-ukraine-middle-east-food-security-at-risk

Woertz, E., & Keulertz, M. (2015). Food trade relations of the Middle East and North Africa with tropical countries. *Food Security*, *7*(6), 1101–1111. https://doi.org/10.1007/s12571-015-0502-5

Woertz, E. (2018). Geopolitics, food and agriculture. In R. Zurayk, E. Woertz, & R. Bahn (Eds.), *Crisis and conflict in agriculture* (pp. 28–39). CABI.

Zeitoun, M., & Warner, J. (2006). Hydro-hegemony: A framework for analysis of transboundary water conflicts. *Water Policy*, *8*(5), 435–460. https://doi.org/10.2166/wp.2006.054

The role of virtual-water decoupling in achieving food–water security: lessons from Egypt, 1962–2013

Ahmed Tayia, Alexandra M. Collins and Michael Gilmont

ABSTRACT

Since the 1970s, many economies increasingly rely on 'importing' water embedded in food imports, a process referred to here as virtual-water 'imports'. In water-scarce countries, virtual-water 'imports' are used to protect the economy's own water which would otherwise be consumed in food production to meet the growing population and economy food needs, or to support the population's food needs beyond that sustainable by internal water resources. This process is referred to here as virtual-water decoupling. This paper examines the role of virtual-water decoupling to achieve a version of food–water security for water-scarce societies, with a particular focus on Egyptian virtual-water decoupling policy during the period 1962–2013.

Introduction[1]

Global awareness of the critical state of freshwater resources has generated much research over recent decades. This concern is a consequence of the challenge of securing supplies of water sufficient to meet the growing water needs of humanity (Falkenmark, 1989; Jägerskog & Phillips, 2006; Tayia, 2019; UNDP, 2006; UNESCO, 2006; UN-Water, 2008). Since the 1970s, many water-scarce economies ran out of the water needed to secure their food and fibre needs, referred to here as *food–water* (Allan, 2013). They have increasingly relied on water embedded in food and feed imports, identified by Allan as *virtual-water 'imports'*, that protected local physical water that would have been consumed in food production in a process referred to here as virtual-water decoupling. The UNEP (2011, 2014, 2015) promoted the concept of decoupling and this Egyptian case study has enabled the utility of the idea to be exemplified. Given the growing role of virtual-water decoupling in food–water security policymaking worldwide, this study focuses on the role of virtual-water decoupling in achieving food–water security.

Egypt was chosen as a case study economy that has a history of virtual-water imports to achieve food–water security. It provides an example of a country that has had to introduce virtual-water decoupling to maintain a version of food–water security in very challenging demographic and hydro-political circumstances. It has succeeded in maintaining a relatively efficient water policy despite facing a seriously worsening status of internal water self-sufficiency. Egypt depends on water sourced from 10 other upstream Nile riparian countries, many themselves increasing their abstraction, within an already variable river system (Tayia et al., 2021). The predicament of the Egyptian economy has become even more complicated as it has had to adapt to economic development imperatives against a background of a growing population as well as to the predicted consequences of climate change. Therefore, investigation of such a case study enables the identification of factors that could influence the ability of a country to adopt virtual-water decoupling as a policy instrument to maintain water security, providing potentially generalizable insights to enable the analysis of other case studies and the possibility of informing policy in these and other countries.

The remainder of the paper is structured as follows. The next section outlines the foundations of the virtual-water decoupling conceptual framework used in this study. The third section briefly outlines the methods used to examine virtual-water decoupling policies in Egypt during the study period 1962–2013. The fourth section examines the overall quantitative metrics of the volume of virtual-water 'trade' in Egypt during the study period. It then links this trade to the overall volume of water embedded in food and feed consumption to trace the annual volumes of virtual-water decoupling in Egypt during the study period. The fifth section then dives deeper into the qualitative trends of decoupling to identify the key crops involved in virtual-water decoupling in Egypt during the study period, and the social and trade policies that can also be associated with the decoupling trends. It then identifies its key determinants of virtual-water decoupling during the study period. The sixth section discusses the main findings of the research before drawing some generalizable lessons from the Egyptian experience for water scarce countries that adopt virtual-water decoupling policies. The seventh section concludes.

Conceptual framework

This study uses the concept of virtual-water decoupling to analyse the role of water embedded in food and feed imports to meet growing food and fibre needs. The virtual-water decoupling concept has evolved gradually from the virtual-water concept identified by Allan in the early 1990s.

The concept of *virtual water* has been framed by Allan (1993, 1994) to describe the water embedded in products traded in the global food system. The term 'virtual' is used to highlight the fact that water is embedded within the production of food but does not involve the physical transportation of water from production to consumption location. The concept of virtual water identifies specifically the water consumed in the production of a product, not the water that is physically embodied in that product (Hoekstra, 2003). Allan (1997) coined the term *virtual-water 'imports' to* refer to the water embedded in food and feed imports. Conversely, the term *virtual-water 'exports'* was coined to refer to the water consumed in the production of a product that is associated with the process of 'exporting' food and fibre (Allan, 2003).

This articulation of the virtual-water 'imports' and 'exports' has given rise to the concept of virtual-water 'trade' to trace the water embedded in the global trading (Allan, 2003; Hoekstra et al., 2008). Subsequently, virtual-water 'trade' has been integrated into a water-resources decoupling policy planning model introduced by UNEP (2015). This model aimed to assist governments of water-scarce economies to reduce their reliance on their own freshwater resources by reducing national freshwater resources consumed per unit of an economic activity (UNEP, 2015). In this model, virtual-water 'trade' was identified as the water resource decoupling mechanism that reduce an economy's reliance on their freshwater resources (UNEP, 2015).

This use of virtual-water 'trade' as a water resource decoupling mechanism has been reframed by Gilmont (2014, 2015) from a water resource decoupling policy mechanism to a policy analysis framework. In his water resource decoupling policy analysis framework, he indicated that the reliance on virtual-water 'trade' does not mean necessarily that the country experiencing water-resource decoupling (Gilmont, 2014). A country can achieve water-resource decoupling using virtual-water 'trade' only when this country increases its reliance on virtual-water 'imports' to meet its domestic food and feed consumption (Gilmont, 2015). He used this water resource decoupling framework to analyse to relation between virtual-water 'trade' to the total water embedded in domestic food and feed consumption in the Middle East and North Africa (MENA) countries (Gilmont, 2015; Gilmont et al., 2017).

This water resource decoupling policy analysis framework has then been further developed by Tayia et al. (2019). They identified a *virtual-water decoupling model* that analyses the relationship between virtual-water 'trade' and the physical water consumed in producing food in an economy. This model comprises two mechanisms that can link virtual-water 'trade' to water resource decoupling, namely *virtual-water decoupling* and *recoupling*. They have used the term *virtual-water decoupling* to describe the process by which a country attempts to achieve physical water security by increasing the virtual-water 'import' share of the total water embedded in domestically consumed food and fibre (Figure 1a).

Conversely, the term *virtual-water recoupling* was used to describe the opposite process by which a country retreats from physical water resource decoupling by increasing its reliance on local physical water to produce national food and feed needs while reducing its reliance on virtual-water 'imports' in meeting its domestic food and feed consumption (Figure 1b).

This paper used this *virtual-water decoupling model* to analyse the role of virtual-water 'trade' in the relationship between the volume of 'imported' virtual water and the total water embedded in food consumption in Egypt. It complements the analysis carried out by Allan (1997, 2003) and Gilmont (2015) in analysing the role of virtual-water 'trade' in circumventing water scarcity in Egypt. The novelty of this study is that its scope is not limited to the identification the overall trend of virtual-water decoupling/recoupling in Egypt. Rather it enables a detailed quantitative and qualitative analysis of the Egyptian virtual-water inflows and outflows in order to map their role in achieving food–water security. It also identified the promoting factors of and the constraints on virtual-water decoupling in Egypt.

Methods

The study has examined both the quantitative and qualitative patterns of virtual-water decoupling in achieving food–water security in Egypt during the period 1962–2013. The study uses food trade (import and export) and domestic production available in the United Nations Food and Agriculture Organization (FAO). Metrics were all sourced

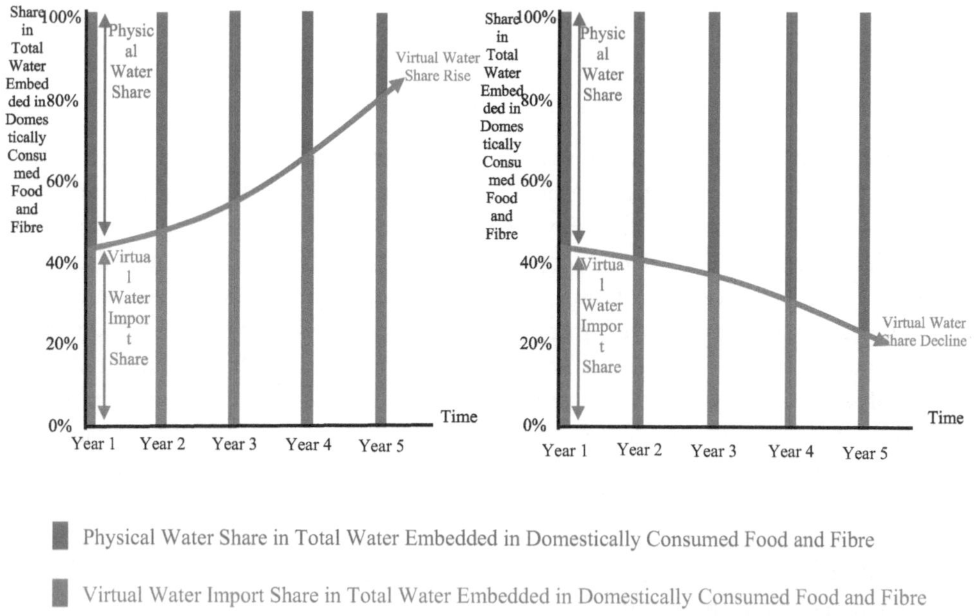

(a)
Virtual-Water Decoupling

(b)
Virtual-Water Recoupling

Figure 1. Virtual-water decoupling/recoupling conceptual framework – self-elaborated: (a) virtual-water decoupling; and (b) virtual-water recoupling.

from the FAOSTAT trade database of the FAO (2019). A brief overview of the methods used is provided here. A more detailed explanation of the methods used in each section of the analysis will be provided in each section.

The first stage of the analysis comprises a quantitative role of Egypt's virtual-water decoupling in achieving food–water security. The volumes of virtual-water 'imports' and 'exports' were traced to determine the pattern of virtual-water trade in Egypt. The resultant virtual-water 'trade' pattern to the pattern is then related to the total water embedded in food and feed domestically consumed to quantify the role of Egypt's virtual-water decoupling in achieving food–water security. The analysis then identifies and analyses the factors that determined the quantitative pattern of virtual-water decoupling including internal water resources (namely Nile flows), economic diversity and trade balance, and general national economic performance.

The next element of the methodology relates to the qualitative pattern of virtual-water decoupling in Egypt is examined. This involves an analysis of the crop structure of virtual-water decoupling/recoupling in Egypt in order to identify the crops involved in virtual-water decoupling, and their changes through time. Using these metrics, the main determinants of the qualitative pattern of virtual-water decoupling in Egypt have been identified.

A review of the quantitative trends, their causal factors and the emergent qualitative nature of the trends is synthesized to elicit policy recommendations not only for Egypt but for other water scarce economies.

Pattern of virtual-water decoupling in Egypt: metrics, trends and determinants

Overall trends in the volumes of virtual-water 'trade' in Egypt

The study shows that successive Egyptian governments relied on virtual-water 'trade' over the study period to meet the growing food needs of its growing population. This paper shows for the first time that Egypt's reliance on virtual-water 'trade' varied over time, including periods of increasing dependence on imports (decoupling) and also times of increased dependence on domestic production (recoupling). These variations were caused by major changes in the environment, such as a period of high annual Nile flows, or in the political economy, such as the period of economic strength and stability. Three phases of virtual-water 'trade' patterns, based on the occurrence of decoupling or recoupling trends have been identified based on the degree of reliance on virtual-water 'trade' and the decoupling/recoupling occurrence: 1962–73, 1973–84 and 1984–2013 (Figures 2 and 3).

During the 1962–1973 phase, Egypt was predominantly a net virtual-water 'exporter'. It was able to operate its historic food self-sufficiency policy predominantly due to a period of relatively high Nile annual floods and the manageable water demands associated with moderate population increases (El Gamal & Zaki, 2017; World Bank, 2020). During the first half of this phase, both virtual-water 'imports' and 'exports' witnessed a similar limited growth from about 2.5 bcm in 1962 to more than 3 bcm in 1967 (Figure 2) (FAO, 2019). Egypt fluctuated between being a net virtual-water

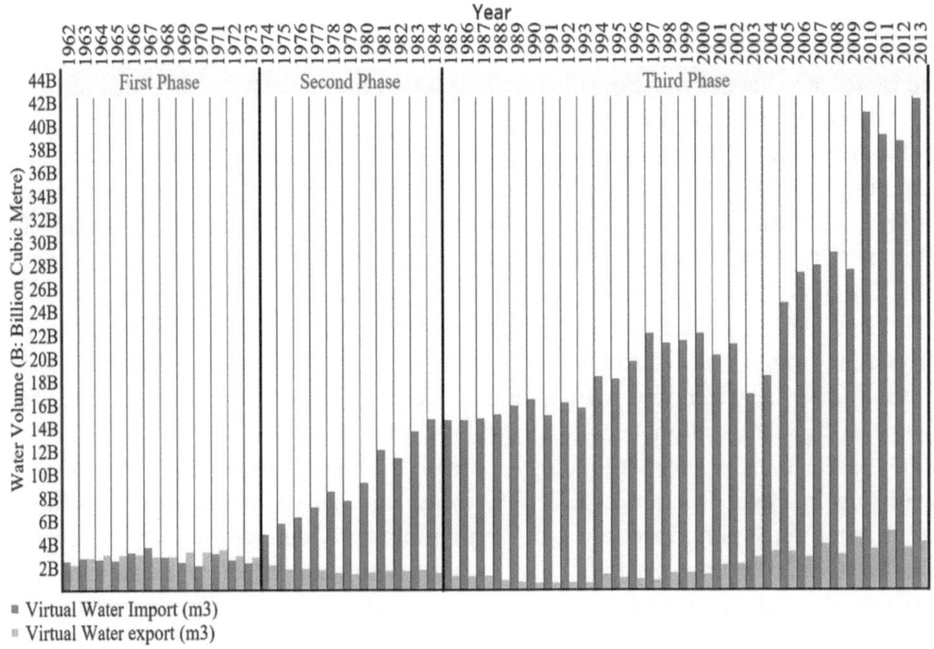

Figure 2. Comparison between virtual-water 'imports' and virtual-water 'exports' during the period 1962–2013, showing the divergence between the level of virtual water 'imports' and of virtual water 'exports' since 1974 – self-elaborated. Source: FAO (2019).

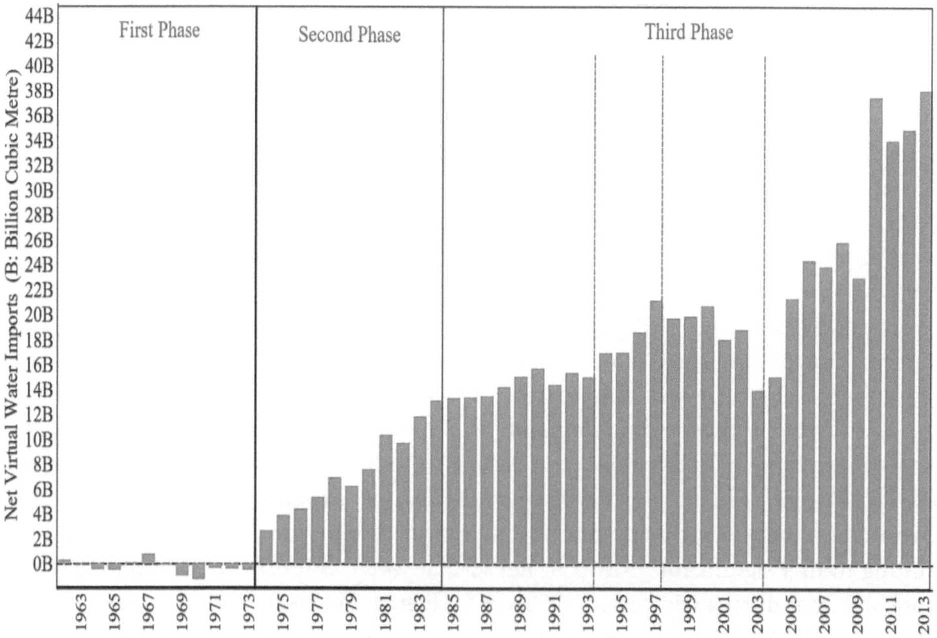

Figure 3. Net virtual-water 'imports', 1963–2013, showing that Egypt became a significant net virtual-water 'importer' from 1973 onwards – self-elaborated. Source: FAO (2019).

'importer' in 1962, 1966 and 1967 and being a net virtual-water 'exporter' in 1962, 1963 and 1964 (Figure 3). From 1968, increasing economic pressures forced Egypt to reduce its virtual-water 'imports' and correspondingly increase reliance on its domestic food production to meet population food needs. Consequently, Egypt's virtual-water 'imports' declined to about 2.5 bcm while its virtual-water 'exports' maintained their level of about 3 bcm (Figure 2). Egypt therefore became a net virtual-water 'exporter' after 1967 until the end of this phase in 1973 (Figure 3).

During the 1973–1984 phase, Egypt increased its virtual-water 'imports' to close its food–water gap. The economic sustainability of the Egypt's food system was challenged by the increasing food demand resulting from population growth and the international food price spike of 1974. Egypt gradually increased its virtual-water 'imports' relying on US food aid and the selective liberalization of its international trade. Virtual-water 'imports' grew from less than 2.5 bcm in 1973 to almost 14.6 bcm in 1984, with an increase of almost 500% (Figure 2) (FAO, 2019). In parallel, Egyptian virtual-water 'exports' witnessed a sharp decline of almost 50%, from about 3 bcm to about 1.5 bcm (Figure 2) (FAO, 2019). This simultaneous rise in virtual-water 'imports' and decline in virtual-water 'exports' transformed Egypt from a net virtual-water 'exporter' in 1973 with a 0.5 bcm surplus into a significant virtual-water 'importer' from 1974 with a 13.1 bcm deficit by 1984) (Figure 3) (FAO, 2018b).

During the 1984–2013 period, Egypt experienced a number of epochs of virtual-water 'trade'. The first epoch, in the 10 years from 1984–1993, has been related to the economic crisis in the Egyptian economy, when its annual growth declined, its foreign debts increased, and its negative trade balance deficit became significant (Abou-Mandour &

Abdel Hakim, 1995). Motivated by its desire to redress the budgetary and foreign exchange pressures, Egypt wanted to curb its increasing reliance on food imports. Due to this policy, Egypt witnessed only a slight increase in virtual-water 'imports' from around 14.6 bcm to about 15.6 bcm during the period 1984–1993 (Figure 2). In parallel, the Egyptian virtual-water 'exports' witnessed a sharp decline of more than 60%, falling from about 1.5 bcm to about 0.6 bcm in the same period (Figure 2) (FAO, 2019). Combining the effect of both imports and exports, there is a slight increase in net virtual-water 'imports' from about 13.1 bcm to about 15 bcm seen by 1993 (Figure 3) (FAO, 2019).

Between 1993 and 1997 Egypt experienced a short epoch of increased virtual-water 'imports'. Egypt emerged from the stabilization period of its Economic Reform Programme in 1993. It was able to relax its constraints on virtual-water 'imports' to meet its growing domestic food demand. The annual volume of virtual-water 'imports' rose by more than 40%, from about 15.6 bcm to more than 22 bcm, over the four-year epoch 1993–1997 (Figure 2). Egyptian virtual-water 'exports' experienced a sluggish growth from about 0.6 bcm to about 0.8 bcm in this period (Figure 2) (FAO, 2019), resulting in in a significant growth in net virtual-water 'imports' from about 15 bcm to about 21.2 bcm during the period 1993–1997 (Figure 3) (FAO, 2019).

In the 1997–2000 period, there was a series of high Nile floods (Ministry of Water Resources and Irrigation, 2018). The availability of Nile water enabled the previous short epoch of increasing virtual-water 'imports' to end. Domestic food production increased with a decline in virtual-water 'imports' by about 25%, from more than 22 bcm to less than 17 bcm over the period 1997–2003 (Figure 2). Over the same period, Egyptian virtual-water 'exports' increased from the very low level 0.8 bcm to more than 2.8 bcm, a 250% growth (Figure 2) (FAO, 2019). This concurrent decline in virtual-water 'imports' and the growth in virtual-water 'exports' caused net virtual-water 'imports' to decline from more than 21 bcm to about 14 bcm during this epoch (Figure 3) (FAO, 2019).

The *final epoch*, from 2003 to the end of the study period in 2013, witnessed the return of Egypt to virtual-water decoupling. Egypt carried out a second stage of economic reform, encompassing institutional and organizational reforms facilitating strong economic growth (Tayia, 2015). Egypt increased its reliance on virtual-water 'imports' to meet the increasing food and feed consumption associated with strong economic growth. Virtual-water 'imports' grew from less than 17 bcm in 2003 to more than 42 bcm in 2013, recording an increase of more than 150%. Egyptian virtual-water 'exports' also continued to grow, increasing from about 2.8 bcm to about 4 bcm, recording a growth of more than 40% (Figure 2) (FAO, 2019). The strong growth in virtual-water 'imports' relative to 'export' growth resulted in net virtual-water 'imports' tripling from about 14 bcm to about 38 bcm during the epoch 2003–2013 (Figure 3) (FAO, 2019).

Quantitative trends of virtual-water decoupling/recoupling in Egypt

This section traces the evolution of virtual-water decoupling/recoupling in Egypt by relating Egyptian virtual-water 'imports' to the total water embedded in food and fibre consumed in Egypt during the study period. As shown in Figure 1, virtual-water decoupling occurs when the share of virtual-water 'imports' in the water embedded in domestically consumed food exhibit a rising trend. Virtual-water recoupling occurs when the virtual-water 'imports' share of total water embedded domestically consumed food show a declining trend.

During the 1962–73 phase, it is evident that the share of virtual-water imports in domestically consumed food was relatively small, with physical water (sourced within Egypt) comprising over 80% of food water needs during this phase. As Figure 4 shows, Egypt embraced some elements of virtual-water decoupling during the first half of this phase, with the share of virtual-water 'imports' increasing from 14% in 1962 to 18% in 1967. Subsequently, Egypt experienced a virtual-water recoupling trend, with the share of virtual-water 'imports' declining to about 11% in 1973.

During the 1973–1984 phase, Egypt consistently experienced virtual-water decoupling. While food consumption and production increased steadily in this phase, consumption outpaced production growth by 1974. The increasing food gap necessitated the Egyptian government to resort increasingly to virtual-water decoupling; the share of virtual-water 'imports' embedded in domestically consumed food rocketed from about 11% in 1973 to about 40% by the end of the phase in 1984. Egyptian physical water contribution to producing food and fibre within Egypt fell from about 89% at the beginning of this phase to about 60% by the end of this phase.

Although the 1984–2013 phase began with a reversal of the previously adopted virtual-water decoupling, it ended with a 10-year period of increasing virtual-water decoupling (Figure 3). As a consequence of its balance-of-trade problems and enabled by non-conventional-water sources, Egypt increased its domestic food production, resulting in a virtual-water recoupling during the 1984–1993 epoch. During this epoch, physical water embedded in domestically consumed food and fibre increased from 60% in 1984 to

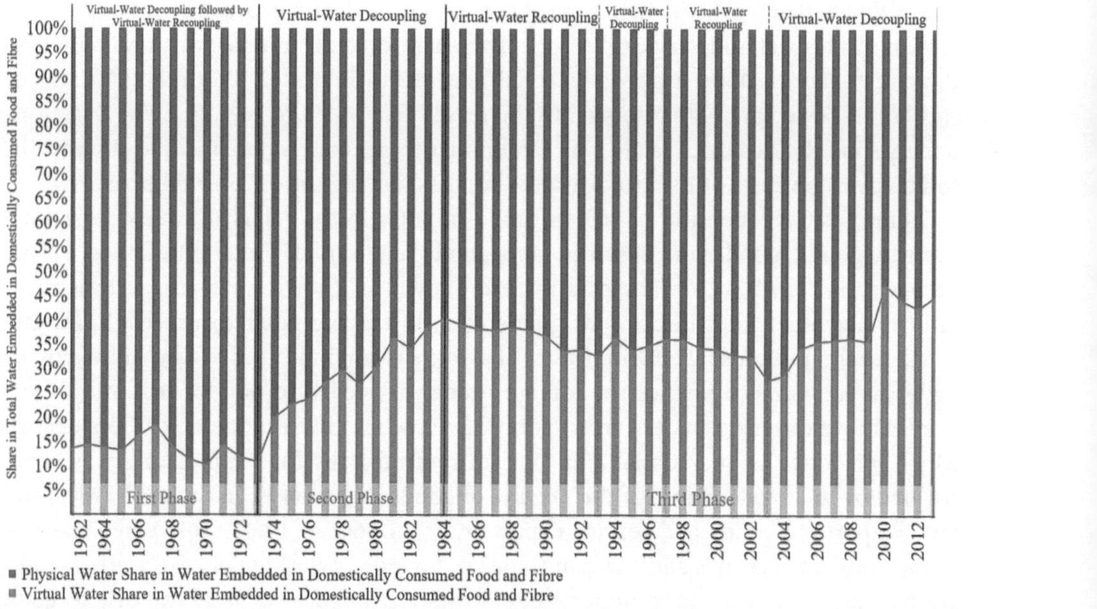

Figure 4. Relative shares of imported virtual water and physical food–water in domestic food–water consumption, showing that Egypt experienced different epochs of virtual-water decoupling/recoupling during the period 1962–2013 – self-elaborated. Source: FAO (2019).

more than 67% in 1993 (Figure 4). Subsequently, Egypt experienced a short epoch of virtual-water decoupling, enabled by the economic reforms carried out in the 1990s. The share of physical water embedded in domestically consumed food fell to about 64%. There followed a short epoch of virtual-water recoupling, which increased the share of physical water embedded in domestically consumed food to more than 72% in 2003. Finally, supported by its economic growth and challenged by physical water scarcity, Egypt experienced a significant period of virtual-water decoupling until the end of the study period in 2013. During this last epoch of decoupling, the share of food–water 'imports' increased, albeit with fluctuations, to its highest level of 47% in 2010 and ended the period at a high level of 45% of water embedded in domestically consumed food and fibre and conversely therefore just 55% of food–water needs met through physical water.

Determinants of virtual-water decoupling/recoupling quantitative trends in Egypt

This section explores the relationship between the trends in decoupling and recoupling and key environmental and political economic factors within Egypt. It demonstrates that the quantitative virtual-water decoupling/recoupling trends in Egypt were considerably influenced, by the Egyptian national economic context and by international economic contexts, that is, trends and shocks in international food trade. The analysis reveals that the level of diversification of the Egyptian economy in general and its ability to generate stable revenues to cover its virtual-water 'imports' considerably influenced the path of virtual-water decoupling in Egypt during the period 1962–2013.

Examining the state of the Egyptian economy from 1973 shows that the weakness of the Egyptian economy constrained the first phase of its virtual-water decoupling policy that began in 1974 and put an end to it by the mid-1980s. The analysis of the 1973–1984 phase shows that Egypt at that time did not have a diversified economy with slowly growing industrial and services sectors. The share of industrial sector grew only from about 25% of gross domestic product (GDP) in 1973 to 30% of GDP in 1984. The services and domestic sector share also slightly increased from 45% to 50% of GDP during the same period. Its export capacity was too limited to generate stable income streams that could cover the costs of its virtual-water 'imports' during the 1973–1984 phase.

Figure 5 shows the ratio of value of food imports to the value of three categories of exports, goods, goods and services, and goods and services and remittance exports between 1977 and 2013.[2] These ratios are known as known as coverage ratios. The ratio of Egyptian food imports to goods exports has escalated to exceed 100% after 1984, reaching almost 225% in 1988. The ratio is an indicator of the ability of the Egyptian economy to finance the foreign exchange needed to pay for its food imports from its revenues from its export of goods. Even when the revenue of services exports, which included the Suez Canal and tourism revenues, the food import bill was hardly covered by the total revenue of the exports of goods and services in the mid-1980s, with food imports to goods and service exports ratio of 86% in 1988. Only when the Suez Canal, tourism revenues and the remittance of Egyptians working abroad were included are the costs of Egypt's food imports comfortably covered, with food imports to goods and service exports, Suez Canal income and the remittance ratio of 56% in 1988. Thus, when the Suez Canal income and the remittance of Egyptians working abroad sharply declined as a consequence of the international post-oil crisis conditions in the 1980s, the

Figure 5. Coverage ratios of food imports in Egypt during the period 1976–2013 – self-elaborated. Source: World Bank (2020).

first virtual-water decoupling policy was reversed in what has been described in this study as virtual-water recoupling.

The relatively strong economic performance of Egypt in the early 1990s enabled the short 1994–1997 epoch of virtual-water decoupling and the prolonged period of very strong virtual-water decoupling from 2004 until the end of the study period (Figure 5). The economic reforms initiated in Egypt in 1991 and accelerated after 2002 had a very significant impact on Egypt's food import policy. since food–water 'imports' are very closely linked with food commodity imports, Egypt's capacity to meet food import costs has enabled Egypt to benefit from under-priced food commodities on international markets. This period of economic growth was based on the achievement of higher levels of economic diversification that led to an increase in the Egypt's exports of both of goods and services. Figure 5 shows that the coverage ratio of the export of goods to food imports was less than 40% by 2006. The coverage ratio of goods and services exports in relation to food imports was less than 25% in the same year. Reliance on the volatile personal remittances decreased. There was a shift to reliance on internal stable sources of income.

The second factor that affected virtual-water decoupling in Egypt was the international economic context. As Figure 4 shows, the international food crisis shocks associated with the 2008 global financial crisis led to decline in virtual-water imports. However, Egypt was able to restore its virtual-water decoupling after the crisis.

Qualitative pattern of virtual-water decoupling: structure & determinants

Structural pattern of virtual-water 'trade' in Egypt

The crops structuring the virtual-water 'imports' and 'exports' are concentrated in a few crops over the study period. Virtual-water 'imports' maintained its concentrated structure over the study period while virtual-water 'exports' have witnessed a degree of diversification towards the end of the study period.

Virtual-water 'imports' were used mainly to meet shortages in food and feed. Six major crops accounted for more than 65% of the annual virtual-water 'imports' during the study period (Figure 6b): wheat, maize, soybean, sugar, coffee and tea. Over the study period, the total volume of virtual-water 'imports' of these six commodities increased from about 2.2 bcm to more than 28 bcm per year, recording an increase of more than 1300% (Figure 6a). In parallel, the total volume of virtual-water 'imports' increased from about 2.5 bcm to 42 bcm in the same period, recording an increase of more than 1700% (Figure 6a). Thus, the total share of virtual-water 'imports' of six-basic commodities dropped from 86% to about 68% by the end of the of the study period (Figure 6b).

Three of these six crops, namely wheat, maize and soybean, accounted for more than 60% of the annual virtual-water 'imports'. This heavy concentration of virtual-water 'imports' in three crops reflected the scale of the consumption concentrations in Egypt. While wheat is the main nutritional ingredient in Egypt, maize and soybean are key animal feed ingredients. The relative shares of the three crops in annual virtual-water 'imports' has changed over the study period. Over the study period, both shares of maize and soybeans in virtual-water imports have grown from about 11% and 7% to about 15% and 22% respectively, while the share of virtual water of wheat has declined from more than 50% to less than 25%.

Egyptian virtual-water 'exports' moved from being highly concentrated in cotton and rice in the 1960s and became more diversified to include fruit and vegetables, from the 1990s onwards (Figure 7). Virtual-water 'exports' grew steadily from 2 bcm to more than 4 bcm by 2013. This growth was a consequence of increases in rice exports after 1984 until Egypt banned these exports in 2008. The expansion of rice cultivation in the 1990s and 2000s led to the contraction in the acreage of cotton as they compete for growing area (Cassing et al., 2009). Cotton production declined sharply, causing the virtual-water 'exports' of cotton to fall from 1.7 bcm to less than 0.3 bcm over the study period (Figure 7a). The share of cotton of virtual-water 'exports' fell from about 80% in 1962 to less than 10% in 2013. As rice exports declined after the export ban in 2008, its share of virtual-water import returned almost to its initial level of 10% by the end of the study period (Figure 7b). Meanwhile, the increase in virtual-water 'exports' continued as a result of the growth in horticultural and vegetables exports, such as oranges, potatoes, olives and grapes (Figure 7a).

Structure of virtual-water decoupling in Egypt

This section presents the finding of the analysis of the structure of virtual-water decoupling/recoupling in Egypt. The analysis focused on the three key components of Egyptian virtual-water 'trade' decoupling, namely wheat, maize and soybeans.

Wheat had a similar virtual-water decoupling pattern to the overall virtual-water decoupling in Egypt (Figure 8a). As the largest component of virtual-water 'imports', it would be expected that wheat would appear as a main driver of overall decoupling trends in Egypt. Wheat experienced three main phases of virtual-water decoupling/recoupling over the study period. The share of virtual-water 'imports' at the beginning of the study period was about 50% of total water embedded in its domestic consumption of wheat.

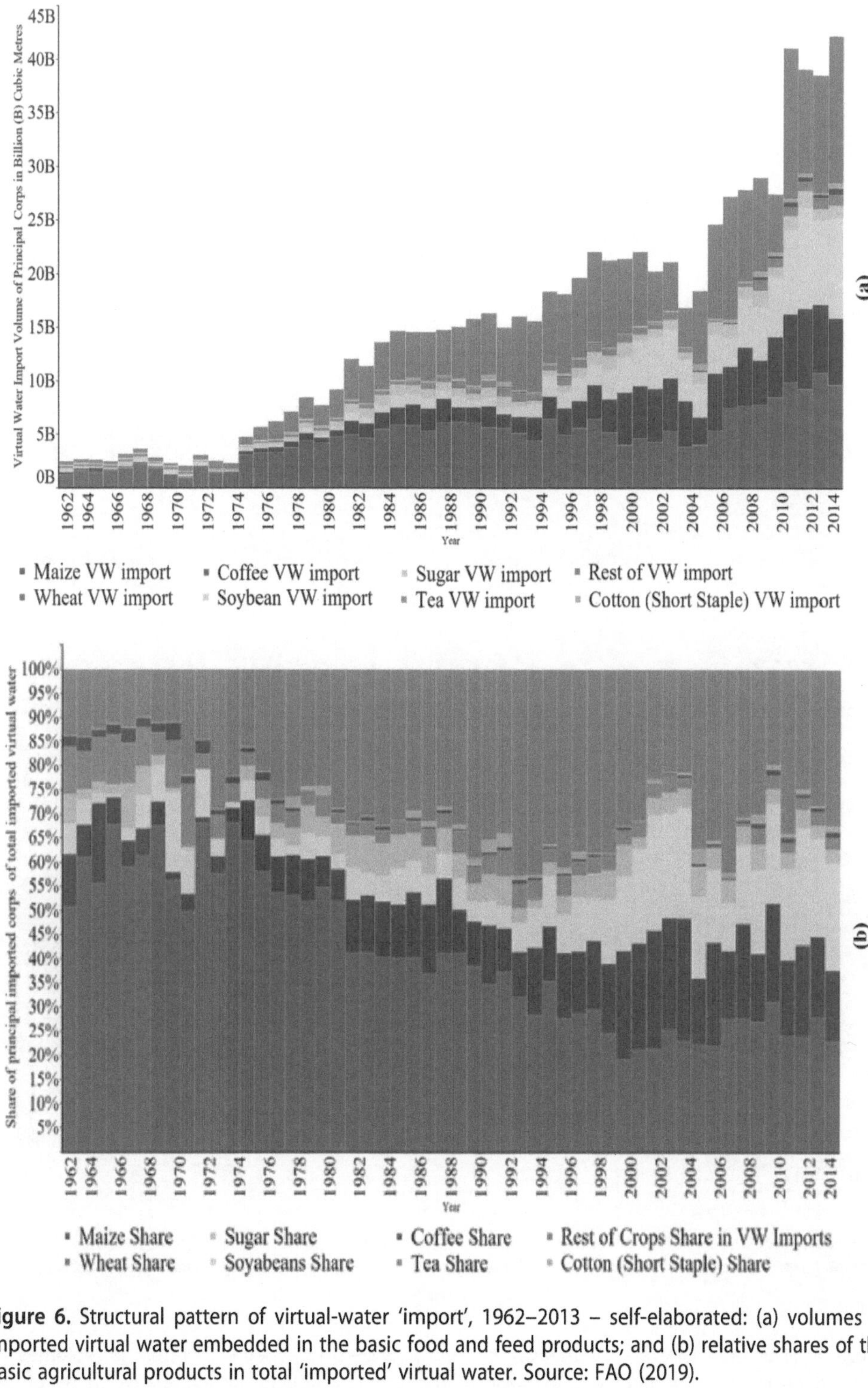

Figure 6. Structural pattern of virtual-water 'import', 1962–2013 – self-elaborated: (a) volumes of imported virtual water embedded in the basic food and feed products; and (b) relative shares of the basic agricultural products in total 'imported' virtual water. Source: FAO (2019).

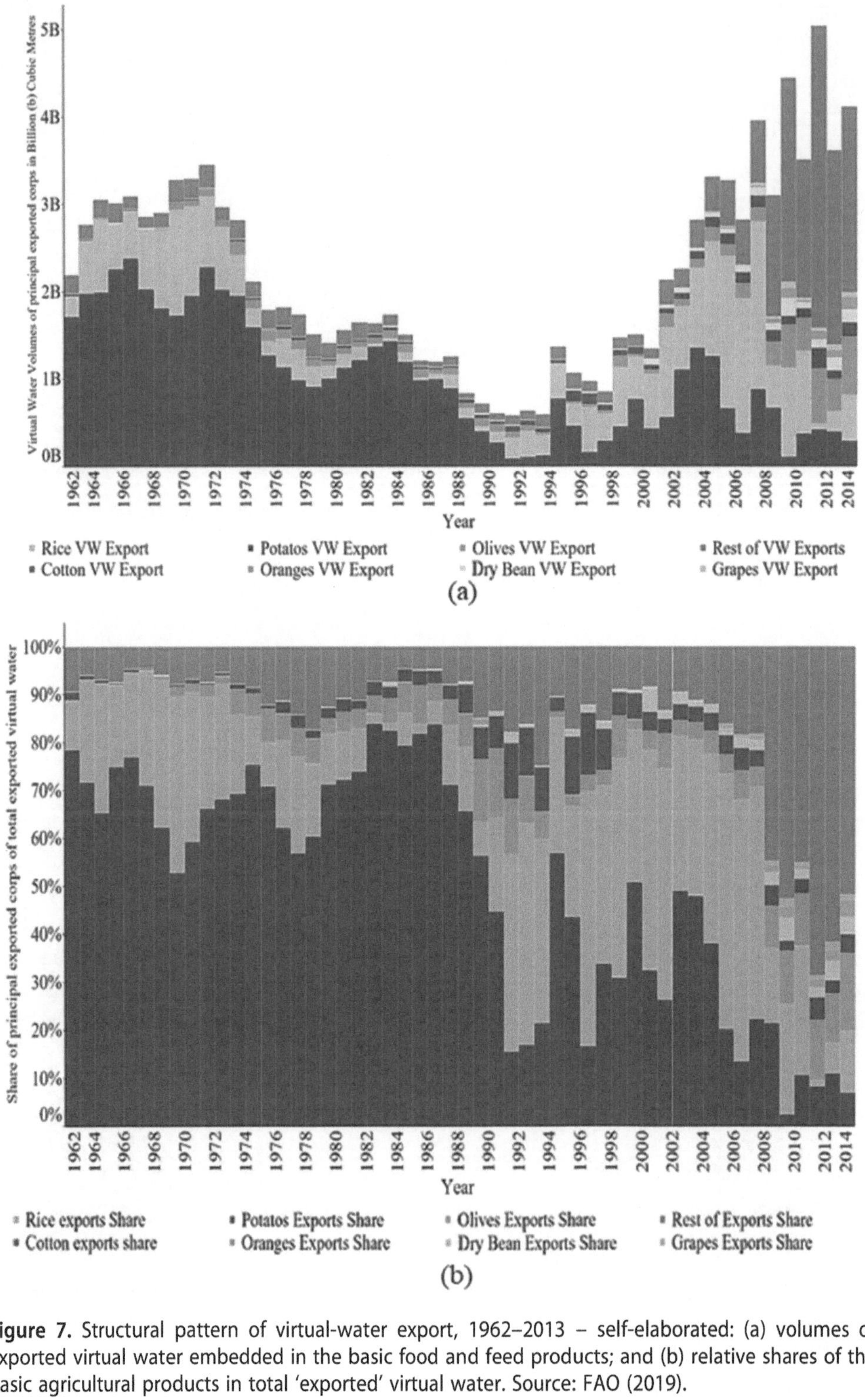

Figure 7. Structural pattern of virtual-water export, 1962–2013 – self-elaborated: (a) volumes of exported virtual water embedded in the basic food and feed products; and (b) relative shares of the basic agricultural products in total 'exported' virtual water. Source: FAO (2019).

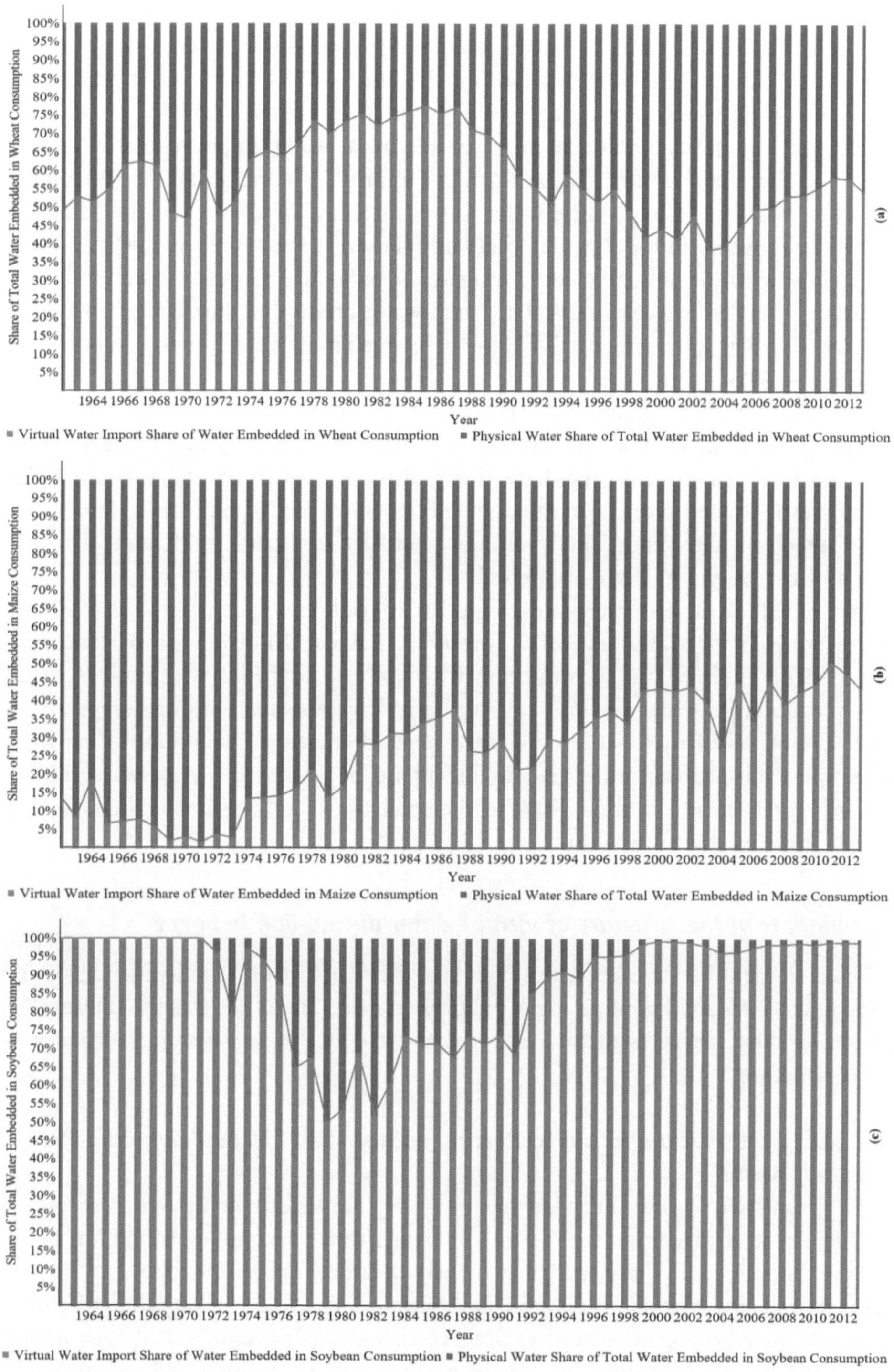

Figure 8. Structure of domestic consumption of food–water embedded in wheat, maize and soybean. Source: FAO (2019).

Wheat experienced strong virtual-water decoupling during the 1973–1984 phase. The share of virtual-water 'imports' of the total water embedded in wheat consumption increased to almost 78% by 1984. During the 1984–2003 phase, wheat experienced a level of virtual-water recoupling where Egypt increased its reliance on domestic physical water after 1984, reducing the share of virtual-water 'imports' to about 55% by the end of the study period. These Egyptian efforts to increase the domestic wheat production with Egyptian physical water was induced by the deeply embedded cultural significance of wheat in the Egyptian diet and the need to reassure the Egyptian people that Egypt was wheat self-sufficient (Megahid, 2016; Siam & Croppenstedt, 2007; Tayia et al., 2019). Finally, wheat witnessed a virtual-water decoupling phase after 2004 where the share of virtual-water 'imports' of water embedded in its total domestic consumption of wheat was 50%.

From 1971 onwards, maize has experienced a consistent and strong virtual-water decoupling trend, albeit with interannual variations (Figure 8b). Beginning the study period at a level of about 15%, the share of virtual-water imports of total water embedded in its domestic consumption of maize declined to 2% by 1969. It maintained this level for a short epoch of four years. The level then gradually rose from 1971, until it reached about 45% by the end of the study period.

Egypt relied mainly on virtual-water 'imports' in its domestic consumption of soybean (Figure 8c). Virtual-water imports represented almost 100% of the of total water embedded in its domestic consumption of soybeans. Soybeans witnessed a virtual-water recoupling during the 1973–1979 phase. Virtual-water 'imports' regressed to about 50% of the of total water embedded in its domestic consumption of soybeans. There was then a virtual-water decoupling phase in which its virtual-water imports constantly grew to reach about 99% of the total the of total water embedded in its domestic consumption of soybeans.

Determinants of the structure of virtual-water decoupling in Egypt

The research identified two key determinants of the structure of virtual-water decoupling in Egypt. One was the Egyptian socio-political context. The other was the international agricultural crop market structure, referred to here as the *international virtual-water market* structure.

Impact of Egypt's socio-political context on virtual-water decoupling in the 1962–2013 study period

The Egyptian social-political context has indirectly influenced virtual-water decoupling policy in Egypt. A very important feature of virtual-water decoupling is that it is hidden from the general public. These invisible virtual-water decoupling processes have the very important capacity to make it possible for legislators to avoid the very difficult politics that would have been generated if the scale of virtual-water 'imports' had been publicly discussed. Political difficulties that could be encountered include political and social perceptions of food insecurity and vulnerability, as well as fears over food price rises (Allan, 1997; Tayia & Madani, 2017). Two key Egyptian socio-political contexts have been very important indeed and have indirectly driven the evolution of the virtual-water decoupling policy in Egypt.

The first factor that promoted the adoption of virtual-water 'trade' decoupling in Egypt was the socio-political context of food consumption in Egypt. The research encountered the strongly held belief in Egypt that the state was the main agent in securing affordable food supplies for society (Tayia et al., 2019). This inherited collective belief has been reinforced by the modern Egyptian state which has assumed the responsibility of delivering a version of food security since the 1950s. As a consequence, any attempt by successive Egyptian governments to relinquish governmental responsibility for ensuring food security have been met with strong popular resistance, including street demonstrations in 1977 (Loewe et al., 2020). Such resistance generated conditions that increased levels of food consumption and in turn promoted, indirectly, the early expansion of virtual-water decoupling in the 1970s before the utility of virtual-water decoupling has been progressively established.

The second link with virtual-water 'trade' that influenced virtual-water decoupling but also prevented it reaching its full potential in Egypt is related to from Egypt's culturally determined very concentrated diet. There is a very strong preference for bread as the main component of daily nutrition.

This bread-based diet was reinforced by US food aid in the 1950s that provided wheat at very low prices, reinforcing the concentrated Egyptian wheat-bread diet. This wheat concentration proved to be a persistent informal element of Egypt's food consumption institutional environment, which successive Egyptian governments failed to reform throughout the study period 1962–2013.

The concentrated wheat-based diet combined with the popular belief of state responsibility for food security constrained virtual-water decoupling processes. As a consequence, water decoupling policy in Egypt has focused on mainly addressed domestic wheat supply deficits, and to a lesser extent those of feed crops, at the expense of a more strategic adoption of decoupling across a wider range of crops.

Impact of the international virtual-water market structure on virtual-water decoupling in the 1962–2013 study period

The study has shown that the structure of the international virtual-water market has influenced the Egyptian virtual-water decoupling policy in Egypt. The concentration of the suppliers of virtual-water exports has made Egypt vulnerable to international trade shock. This risk was reflected in the Egyptian food–water security policy that attempted to balance between relying on the domestic physical water and 'imported' virtual water in its key food crops, especially wheat.

Virtual-water exports tend to be concentrated in a few countries and regions, while imports are dispersed over a very large number of economies (FAO, 2015). In most of commodities, the five largest exporting countries typically account for 70% or more of international export volumes (OECD-FAO, 2016). This unbalanced pattern of virtual-water 'trade' has given exporters much more bargaining power on prices than importers. Fortunately, virtual-water exporters have usually not used their superior bargaining power to raise international food prices. This approach was the result of policies adopted mainly by OECD countries, and especially the United States, as a consequence of their preference for subsidizing the livelihoods of their farmers. The manufacturing-biased economic development policies adopted by developing countries reinforced the process. OECD countries resorted to a combination of protectionist measures and subsidies reflected in under-priced food exports (OECD, 2015). However, this outcome raised concern about possible export interruption caused by natural disasters or by sudden and unexpected policy shifts in the main exporting countries (FAO, 2015).

This concentration of virtual-water suppliers has been very relevant in the Egyptian case study. Although Egypt attempted to diversify its suppliers of virtual-water 'imports' of its key imported commodities, the outcome was limited by the concentration of international suppliers (Figure 9).[3]

As Figure 9a shows, Egypt virtual-water 'imports' associated with wheat have remained concentrated over the study period. The only significant change that Egypt has been able to introduce was gradually shifting its wheat supplies from three key 'Western' suppliers, mainly the United States, Australia and France, to suppliers in Eastern European/Black Sea countries, mainly Russia, Ukraine and Romania after 1990. A similar diversification pattern could be identified for maize from the data shown in Figure 9b. Egypt replaced its two key western suppliers of maize, the United States and France, with three South American and Eastern European suppliers, Argentina, Brazil and Ukraine. Similarly, the United States replaced by Argentina as the top supplier of soybeans to Egypt from 1992 to 2017; the United States restored its top position in 2015 (Figure 9c).

The concentration of wheat, maize and soybeans suppliers in combination with the strategic significance of these three crops in the Egyptian food and fibre virtual-water imports puts Egypt at high risk. Future international market shocks could occur if the food commodity exporters were to change their subsidizing and under-pricing policies. This might explain why Egypt maintained a self-sufficiency policy, aiming at domestically producing 50% of wheat. In the heavily concentrated international virtual-water market, knowing global reliable capacities to produce affordable food will be key to complement the Egyptian domestic capacity of food production. It will also be necessary to know the degree to which international sources of food–water will be reliable in future because they are being produced sustainably in the river basins in which water is being mobilized for food production.

Finding and discussion and lessons beyond Egypt

The study has revealed the dynamics and complexity associated with the adoption of virtual-water decoupling policies to reduce the pressures on physical water resources in water-scarce countries. The analysis of the Egyptian case shows that Egypt has moved – under the pressure of its increasingly growing population and economic development needs – from being a historically significant net-virtual-water exporter to being one of the top net virtual-water 'importers' in the world. Egypt has experienced a growth in annual virtual-water imports, albeit with fluctuations, from about 15% in 1974 to 45% of the total food–water consumed in 2013.

The research has identified a number of determining factors that either promote or constrain the virtual-water decoupling policies. These factors are, first, the national socio-political context; second, the national economic context; third, the international virtual-water market structure; and finally, the international economic context such as international commodity prices.

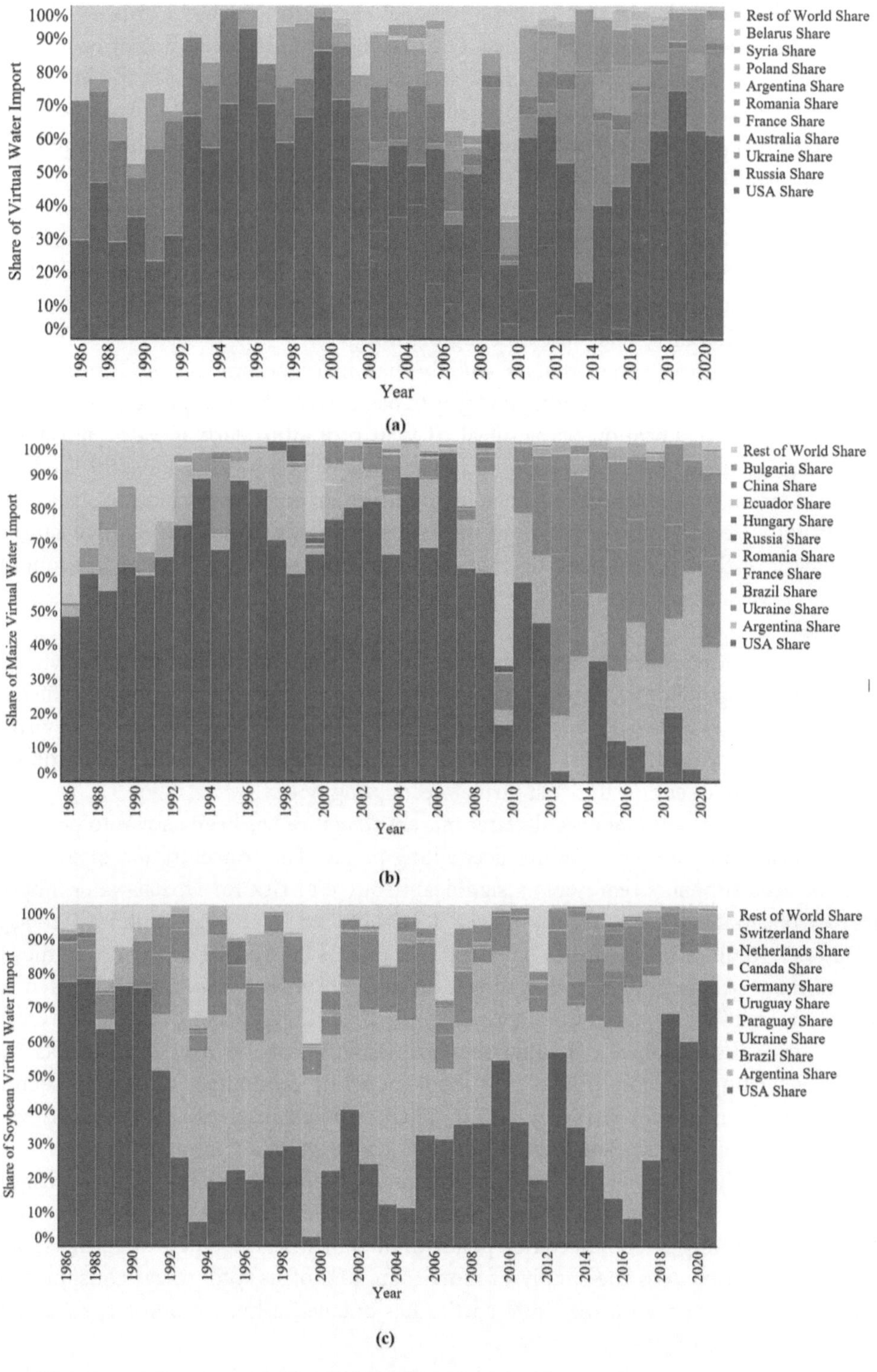

Figure 9. Share of key suppliers in the key imports (wheat, maize and soybean) to Egypt during the period 1962–2013 – self-elaborated. Source: FAO (2019).

The analysis has shown that that the national social context had a mixed effect on Egypt's virtual-water decoupling policy. On the one side, its social pressures for food subsidy promoted the growth of virtual-water decoupling in volume. On the other side, this concentrated dietary pattern constrained the government ability to diversify the crops comprising its virtual-water trade. This constraint stemmed from the desire of the Egyptian government to reduce its vulnerability to the international market risks by having a target for self-sufficiency of 60%. This induced the Egyptian government to promote wheat cultivation over other higher water productivity crops, such as cotton, vegetables and fruit, whose cultivation could have higher returns to water than wheat. The highly wheat-concentrated diet pattern in Egypt has relatively limited the allocative efficiency potential of virtual-water decoupling in Egypt.

Similarly, it has been shown that a favourable national economic context is necessary if a sustained virtual-water decoupling policy is to be achieved. Sustained economic growth and economic diversification play crucial roles in promoting virtual-water decoupling policy. The Egyptian case study provided a clear example of how the state of the economy can affect virtual-water decoupling. The fragile Egyptian economy could not sustain the first virtual-water decoupling policy in the 1980s. However, the relatively high rates of economic growth and economic diversification experienced by the Egyptian economy since the 1990s has enhanced its ability to finance its virtual-water decoupling activities after 2004.

Another key determining factor that has been shown to significantly affect virtual-water decoupling policy are international political and economic conditions. In the case of Egypt, very good US–Egyptian relations enabled the first major phase of virtual decoupling policy in the 1970s. Conversely, the international post-oil crisis conditions of the 1970s put an end to this first virtual decoupling policy.

Finally, the international virtual-water market structure has been shown to be strategically constraining on virtual-water decoupling policy. The concentration of some key food and feed suppliers represents a significant source of risk for virtual-water 'importers'. Virtual-water trade currently makes it possible for importing economies to save money as a result of the subsidies paid by governments of food exporting economies to their farmers (Allan, 1998), be that direct, indirect or 'virtual (e.g., accepting environmental externalities of agricultural production). If food-exporting economies were to reduce their subsidy policies, or suffer significant environmental or political shocks, this could trigger unprecedented shocks for food-importing economies. Countries, such as Egypt, are dependent on virtual-water 'suppliers' who themselves are potentially very vulnerable to such shocks. The current Russia–Ukraine war is a clear manifestation of the impact of such political shocks. With Russia and Ukraine supplying Egypt with more than 85% of its wheat virtual-water 'imports', Egypt has fallen under huge pressure. Despite maintaining a self-sufficiency policy of 50% of wheat, Egypt is suffering supply instability that threatens the supply of more than 40% of its total wheat consumption. This heavy dependency on the crisis parties has entailed a huge economic, social and political burden on Egypt.

In sum, this research has recognized the complexity of food–water security policy-making and the multiplicity of water and non-water factors. It has revealed that while water scarcity pressure and economic growth and diversification promote virtual-water decoupling, national social contexts and international economic contexts can constrain the adoption of decoupling policies. Thus, the potential for virtual-water decoupling to be adopted in an economy depends on its ability to capitalize on the enabling factors while accounting for these constraints. These findings feed into some policy recommendations for water-scarce countries adopting virtual-water decoupling policy. First, a virtual-water water decoupling policy should be a part of a comprehensive economic development policy that accelerates the growth of the economy and deepen its diversification. As the study shows, economic growth and diversification enhance the allocative efficiency of water in the economy and enhance its financial capacity to carry our effective and sustained virtual-water decoupling. Second, the diversification of the diet is another dimension that should be sought in parallel with virtual-water decoupling dimension. However, since the food pattern is socially enshrined, this should be a gradual and long-term process. Third, to mitigate international virtual-water market volatility, virtual-water decoupling should aim at diversifying its international supply of virtual-water 'imports' among suppliers that can provide an affordable and stable food supply. Fourth, the study showed the importance of the adoption of selective virtual-water decoupling policy where there is international market volatility. In a highly concentrated volatile international market, virtual-water imports should be combined with domestic production, while heavier virtual-water dependence should be concentrated on commodities with less shock exposure.

Conclusions

This article has presented a quantitative and qualitative analysis of virtual-water decoupling/recoupling in Egypt. The quantitative analysis examined the overall trend in virtual-water decoupling in Egypt during the 1962–2013 period and its main determinants. The qualitative analysis examined the crops structure of virtual-water decoupling in Egypt during the study period and its main determinants. Using the Egyptian case study, this article has shown that virtual-water decoupling can be become an efficient solution to close the gap between the available physical water resources and the water needed for food and fibre production. The study identified factors that can influence the pattern and the outcome of virtual-water decoupling policy. These factors include the national economic and socio-political contexts, as well as the virtual-water trade pattern and international virtual-water market and the international economic context. This highlights the importance of economic development and diversification policies, along with diversification of diet and the international supply of virtual-water 'imports' to encourage greater virtual-water decoupling whilst increasing resiliency to international market and economic risks. Thus, the understanding and identification of virtual-water decoupling influencing factors identified by this study is essential for the formulation of more effective future virtual-water decoupling policy not only in Egypt but in other water-scarce countries.

Notes

1. This paper was written with significant guidance and input from Professor Tony Allan. The authors would like to dedicate this work to Tony's long-term interest in the water economy of Egypt, the contribution of virtual water to enhancing our understanding and

management of the global food–water economy, and to Tony's incredible encouragement, development and mentorship of his colleagues and students. We hope our continued research will contribute to his enduring legacy to the world.
2. Data are available only since this date.
3. Data are available since 1986 only.

Disclosure statement

No potential conflict of interest was reported by the authors.

ORCID

Ahmed Tayia http://orcid.org/0000-0001-7267-7609
Alexandra M. Collins http://orcid.org/0000-0003-4703-4921
Michael Gilmont http://orcid.org/0000-0001-6822-2206

References

Abou-Mandour, M., & Abdel Hakim, T. (1995). Egyptian agriculture: Its characteristics, historical development and current challenges. *Egyptian Agriculture Profile*, *13*, 3–13.

Allan, T. (1993). Fortunately there are substitutes for water: Otherwise our hydropolitical futures would be impossible. In *Priorities for water resources allocation and management* (July, pp. 13–26). Overseas Development Administration.

Allan, T. (1997). *'Virtual water': A long term solution for water short Middle Eastern economies?* (Vol. 5145, Issue 14). School of Oriental and African Studies.

Allan, T. (1998). Virtual water: A strategic resource global solutions to regional deficits. *Ground Water*, *36*(4), 545–546. https://doi.org/10.1111/j.1745-6584.1998.tb02825.x

Allan, T. (2003). Virtual water – The water, food, and trade nexus useful concept or misleading metaphor? *Water International*, *28*(1), 4–11. https://doi.org/10.1080/02508060.2003.9724812

Allan, T. (2013). Food–water security: Beyond water resources and the water sector. In B. Lankford, K. Bakker, M. Zeitoun, & C. Declan (Eds.), *Water Security: Principles, perspectives and practices* (pp. 321–335). Routledge/Earthscan.

Allan, T. (1994). Overall perspectives on countries and regions. In P. Rogers & P. Lydon (Eds.), *Water in the Arab World: Perspectives and prognoses* (pp. 65–100). Harvard University Press.

Cassing, J., Nassar, S., Siam, G., Moussa, H. (2009). Arab Republic of Egypt. In K. Anderson & W. A. Masters (Eds.), *Distortions to Agricultural Incentives in Africa* (pp. 71–98). World Bank. https://doi.org/10.1596/978-0-8213-7662-1

El Gamal, T., & Zaki, N. (2017). Egyptian irrigation after the Aswan High Dam. In M. Satoh & S. Aboulroos (Eds.), *Irrigated Agriculture in Egypt: Past, Present and Future* (pp. 47–80). Springer Nature.

Falkenmark, M. (1989). The massive water scarcity now threatening Africa: Why isn't it being addressed? *Ambio*, *18*(2), 112–118.

FAO. (2015). *The State of agricultural commodity markets 2015–2016: Nutrition in the trade and food security nexus*. Food and Agriculture Organization of the United Nations.

FAO. (2019). *FAOSTAT*. http://www.fao.org/faostat/en/#home

Gilmont, M. (2014). Decoupling dependence on natural water: Reflexivity in the regulation and allocation of water in Israel. *Water Policy*, *16*(1), 79–101. https://doi.org/10.2166/wp.2013.171

Gilmont, M. (2015). Water resource decoupling in the MENA through food trade as a mechanism for circumventing national water scarcity. *Food Security*, *7*(6), 1113–1131. https://doi.org/10.1007/s12571-015-0513-2

Gilmont, M., Rayner, S., Harper, E., Nassar, L., Tal, N., Simpson, M., & Harper, E. (2017). *Decoupling national water needs for national water supplies: Insights and potential for countries in the Jordan basin.* WANA Institute, Amman, Jordan.

Hoekstra, A. Y. (2003). *Virtual water trade.* Virtual Water Trade: Proceedings of the international expert meeting on virtual water. Research report series no. 12. http://www.waterfootprint.org/Reports/Report12.pdf

Hoekstra, Arjen, & Chapagain, A. (2008). *Globalization of water: Sharing the planet ' s freshwater resources.* Blackwell Publishing Ltd.

Jägerskog, A., & Phillips, D. (2006). *Managing trans-boundary waters for human development.* United Nation Development Programme (UNDP).

Loewe, M., Zintl, T., & Houdert, A. (2020). The social contract as a tool of analysis: Introduction to the special issue on 'Framing the evolution of new social contracts in Middle Eastern and North African countries'. *World Development, xxxx*, 104982. https://doi.org/10.1016/j.worlddev.2020.104982

Megahid, A. (2016 , June 5). Egypt's chasing impossible dream of wheat sufficiency. *The Arab Weekly*, 18. https://www.thearabweekly.com/egypts-chasing-impossible-dream-wheat-sufficiency

Ministry of Water Resources and Irrigation. (2018). *Water resources and uses database.*

OECD. (2015). Trade Dimensions of Food Security. In *OECD Food, Agriculture and Fisheries Papers*, 77, 1–45. https://doi.org/10.1787/5js65xn790nv-en

OECD-FAO. (2016). *OECD-FAO agricultural outlook 2016–2025.* https://doi.org/10.1787/agr_outlook-2016-en

Siam, G., & Croppenstedt, A. (2007). An assessment of the impact of wheat market liberalization in Egypt: A multi-market model approach (*ESA Working Paper*). FAO. (Issues 0715).0715).

Tayia, A. (2015). *The role of institutional change in the management of common-pool resources management: The Nile Basin.* Universidad Complutense de Madrid.

Tayia, A. (2019). Transboundary water conflict resolution mechanisms: Substitutes or complements. *Water, 11*(7), 1337. https://doi.org/10.3390/w11071337

Tayia, A., Allan, T., & Collins, A. (2019). *Water security in Egypt: Potential for and constraints on virtual-water decoupling in achieving food–water security.* Imperial College London.

Tayia, A., & Madani, K. (2017). *Resilient transboundary water management institutions* (pp. 618–627). American Society of Civil Engineers (ASCE). https://doi.org/10.1061/9780784480595.055

Tayia, A., Ramos Barrado, A., & Alonso Guinea, F. (2021). The evolution of the Nile regulatory regime: A history of cooperation and conflict. *Water History, 13*(3), 293–317. https://doi.org/10.1007/s12685-021-00287-3

UNDP. (2006). *Human Development Report 2006 – Beyond Scarcity: power, poverty and the global crisis.*

UNEP. (2011). Decoupling natural resource use and environmental impacts from economic growth. *A Report of the Working Group on Decoupling to the International Resource Panel.* http://www.unep.org/resourcepanel/decoupling/files/pdf/Decoupling_Report_English.pdf

UNEP. (2014) . *Decoupling 2 technologies, opportunities and policy options.*

UNEP. (2015) . *Options for decoupling economic growth from water use and water pollution.*

UNESCO. (2006). *The United Nations World Water Development Report 2: Water – A shared responsibility.*

UN-Water. (2008). Transboundary waters: Sharing benefits, sharing responsibilities. *Task Force on Transboundary Waters.* http://www.unwater.org/downloads/UNW_TRANSBOUNDARY.pdf

World Bank. (2020). World development indicators. *World Development Indicators.* https://databank.worldbank.org/data/reports.aspx?source=world-development-indicators

Unexpected bright spots: how the pandemic, climate change and biodiversity loss are shaping the evolution of the nexus

Nathanial Matthews, Bart Schoonbaert and Elizabeth Burlon

Introduction

The 2007–08 global financial crisis propelled the interlinkages of water, energy and food into the international spotlight. In 2009, United Nations Secretary General Ban Ki-moon focused attention on the nexus of water, energy, food and climate; and with the 2011 World Economic Forum (WEF, 2011) report, business and policymakers supported the need to focus on the security and links across these areas. Since then, the concept has been promoted by various organizations, including the UN which included the nexus approach as one of the main pillars for its Sustainable Development Goals (SDGs) set in 2015. By 2017, UN publications noted the 'most commonly discussed set of interactions' regarding the SDGs lie:

> in the nexus between food, water and energy, as reflected in the links between SDG 2 [food], SDG 6 [water] and SDG 7 [energy], with potential conflict in water use for energy production and generating hydropower with residential and industrial water use and for irrigation for food production. (Nikolova et al., 2017, p. 15)

The concept has been expanded to encompass many other fields including, but not limited to, ecosystems (De Strasser et al., 2016), climate and finance (Schmidt & Matthews, 2018), hydropower (Matthews & McCartney, 2018), and forestry (Melo et al., 2021). For water practitioners, the nexus shifts attention from a water-centric approach, akin to integrated water resources management (IWRM), to a multi-centric approach. Various formations of the nexus concept (whether it includes forests, climate, ecosystems or finance, etc.) draw specific attention to interlinkages, which has helped to elucidate system interactions and overlaps across water and its many Earth and people relationships; different manifestations have been applied depending on the context, but the focus on systems and the central role of water in these systems remains fundamental.

In this commentary, three of Professor Tony Allan's former students draw on his insights to follow the evolution of the nexus concept in the face of Covid-19, increasing impacts of climate change and biodiversity decline, and our accelerating hyperconnectivity in the Anthropocene. These pressures act against the backdrop of increasing global water use driven by population growth, economic development and changing consumption patterns – especially linked to food; water use is expected to increase to 20–30% above current levels by 2050 (UN, 2019).

Tony was acutely aware of myriad pressures on global water use and supply and its links to economies, politics and the environment. He saw particular value in how the nexus illuminates not only interdependencies but also choke points and trade-offs – helping us to understand the political and economic forces that drive decision-making and governance of natural resources. He also saw how the nexus could offer solutions – specifically how water's value could be usefully operationalized and how water was valued (or not valued) within supply chains (Allan & Matthews, 2016). Supply chains are complex networks of organizations, individuals, activities, technologies and resources where water, food and energy are traded and consumed globally and which have important implications for ecosystem stewardship: 80–90% of the world's freshwater is consumed by agriculture. Even so, water has been taken for granted, green water in particular. Tony's seminal work on virtual water identified how countries export or import water – often at the expense of the environment – and alerted society to the fact that farmers, at the beginning of the food supply chain, were key stewards of our freshwater.

Here, we consider three focal areas with links to the evolution of the nexus and the pressures upon it: (1) how the value of water is starting to be recognized in the energy sector; (2) how virtual water remains largely silent in the context of the food trade; and (3) how water is being revalued and recognized in agriculture and food as we face up to the challenges of climate change and biodiversity loss and the implications of the pandemic.

Water–energy: reimagining the value of water in the energy system

The long silence on the value of water is being broken by the rise of renewable energy and, in the wake of the Covid pandemic, a shift in the dialogue about the energy transition from cost to value. Until recently, energy policy and investments have been made in isolation, without real consideration of their implications for other sectors or the wider environment – including water. We can see this in fracking for natural gas in water-stressed areas of the United States: a politically popular policy of energy independence through domestically produced natural gas has led to unintended consequences when groundwater became contaminated or, in some cases, ran dry because of a process that was more water intensive than the water supplies could support.

Current geopolitics has put energy at the top of the political agenda, elbowing out the transformation required to shift the world onto a path to well below 2°C warming. From a purely cost point of view, fossil fuels are often still the least-cost option to meet burgeoning energy needs but power generation from coal, oil and natural gas is highly water intensive compared with renewables – and can have impacts on water quality: coal consumes 79–2100 m^3/TJ, oil 316–1830 m^3/TJ, and natural gas 76–1240 m^3/TJ, whereas wind power has a consumptive water footprint of 0.2–12 m^3/TJ and solar photovoltaic (PV) has a water footprint of 6.4–303 m^3/TJ (Mekonnen et al., 2015).

However, there is growing awareness of the water environment context and impact, as we see in the inclusion of water in policy and investment decision-making frameworks. This is part of a move from pure-cost assessments to wider value frameworks, for example, the WEF's System Value Framework considers 11 value dimensions beyond cost and investment competitiveness – explicitly incorporating the water footprint (WEF,

2020). Value-based frameworks like this are changing the way energy policy and investment options and decisions are communicated. Water is no longer taken for granted: it is incorporated explicitly when analysing options and considering the holistic value created by a transition to clean energy. The global response to climate risks and the reimagining of energy systems during the pandemic also open the field for a greater use of *nexus thinking* when engaging local stakeholders to understand the trade-offs in specific areas. Tony would have appreciated this shift in the way we communicate the value of water but he would have said it does not go far enough – for example, the WEF System Value framework considers only *blue water* and does not assign it an economic value.

A value-based assessment that includes water footprint could also help inform thinking and policies on the wider water impact associated with energy sources, for instance, in the context of biofuels. Biofuels provide renewable energy but they are water intensive; assessing investment in biofuels requires an examination of where the fuel can be sourced – if from an area with abundant water where production of biofuel would not impact food production or ecosystem services, then it may be appropriate. So, to the consideration of local water used in energy generation, we need to add consideration of the water imported through the sourcing of the fuel. That is to say, *virtual water*.

Water–food: virtual water remains largely silent

Tony introduced the concept of virtual water in the early 1990s; its significance was recognized by the award of the Stockholm Water Prize in 2008. It is as striking in its simplicity as in its profundity, and it remains an illuminating concept in any understanding of global water resource management and mismanagement. In the decades since Tony coined the phrase, great strides have been made in the appreciation of how completely water is embedded in the provision of goods and services, notably through the *water footprint* concept developed by the late Arjen Hoekstra. Virtual water can contribute to food and water security but, at the same time, detrimental impacts on the water environment can be outsourced or, rather, *off-shored* by importers and consumers to the producers and exporters. The extensive work undertaken to calculate and map international virtual-water flows demonstrates the embedded water that accompanies trade in agricultural products, which make up 80% of virtual water flows (Qu et al., 2018), all of which has raised public awareness of the impact that the consumption of food and goods can have on the water environment. Even so, and *pace* Hausmann (2022, in this issue), decisions on trade rarely consider the embedded water flows and their implications for water resource management at either end of the exchange. Allan et al. (2015, p. 6) noted:

> the scarcity values of water and energy embedded in food and manufactured commodities are not reflected in the prices paid by consumers for the goods they purchase in private-sector markets. Nor are the use values of water and energy captured. Because the exchange values along the supply chains have been very severely distorted by subsidies and taxes, the costs of degrading water and other ecosystem services have been invisible and until recently ignored.

Has anything changed due to the pandemic? Virtual water flows were perturbed by the interruption of trade. The Bank of England (2021, p. 1) reported, 'Covid-19 caused a significant decline in global trade in 2020, particularly in services,' and some have started to look specifically at the impact on virtual water flows of the resulting trade alterations or restrictions (Roidt et al., 2020). But virtual water thinking does not seem to have informed the immediate responses to the pandemic. Questions surfaced about the reliance on food imports and the need for greater food self-sufficiency, and several countries put restrictions on the export of agricultural products; Vietnam, for example, one of the world's biggest rice producers, temporarily halted rice exports (see also Woertz, 2022, in this issue). Debate about food self-sufficiency and food security is not new, but the questions get sharper at a time of crisis, just as they did in 2007–08 when there were significant price increases.

It is clear that the pandemic instigated a push for greater food self-sufficiency. In the related decision-making processes at the beginning of the pandemic, any implications for the management of water do not seem to have featured. Keulertz et al. (2020) noted, 'Recent reactions to the Covid-19 crisis, such as attempts to increase local food self-sufficiency, could irreversibly damage local ecosystems and the sustainability of current local production.' The issues are ongoing. At the end of October 2021, the World Bank, (2021, p. 5) starkly noted:

> An increasing number of countries are facing growing levels of acute food insecurity, reversing years of development gains. Even before COVID-19 reduced incomes and disrupted supply chains, chronic and acute hunger were on the rise due to various factors including conflict, socio-economic conditions, natural hazards, climate change and pests. COVID-19 impacts have led to severe and widespread increases in global food insecurity, affecting vulnerable households in almost every country, with impacts expected to continue through 2021, into 2022, and possibly beyond.

In November 2021, the Food and Agriculture Organization (2021, p. 17) commented:

> Just as many parts of the world have reopened their economies in the aftermath of the COVID-19 pandemic to stimulate economic growth – even though the pandemic remains unchecked – the current rise in food and fuel prices is highly regressive in the case of poor consumers, compounding economic stress and exerting a negative impact on their livelihoods.

Those observations preceded the tragic events that have unfolded in Ukraine. Food and water security are casualties of Russia's blockade of Ukraine's exports; global food prices in May 2022 were 30% higher than in May 2021 and UN Secretary General Antonio Guterres said the conflict, combined with the effects of climate change and the pandemic, threatened to 'tip tens of millions of people over the edge into food insecurity followed by malnutrition, mass hunger and famine'. In the middle of May, India, the world's second largest producer of wheat, banned all exports with immediate effect after a heatwave diminished the crop. Beyond the immediate impacts, the full effects, and implications of such major and far-reaching events, will play out over many years. Two things are clear: in the near to medium terms making or keeping food available and affordable will occupy policymakers much more than the impact this may have on water resources; and food security and food self-sufficiency seem likely to stay high up the policy agenda.

Water, food, climate change and biodiversity – where opportunities and solutions are emerging

Tony saw that the greatest challenges to sustainable freshwater were in food and agriculture (Allan & Dent, 2021; Allan & Matthews, 2016). He saw farmers as the principal custodians of our water, yet notably absent from many key discussions, and he always argued for their inclusion. Food water remains undervalued and under-recognized and the 'silence of water' in the context of food enables and helps perpetuate a perverse system of 'under-priced food for underpaid people' – *underpaid people* reflecting his strong views on social injustice and *under-priced food* reflecting the fact that the environmental impacts of farming are not accounted for in the price of food.

Now, the dynamics are changing. With the increasing attention given to the implications of climate change, agriculture and water are moving centre stage. There is growing awareness that water is the primary medium through which we will feel the effects of climate change; water supplies are becoming less predictable; the impacts and frequency of droughts, floods and storm surges due to sea-level rise will be amplified. Moreover, transforming agriculture is critical to meeting the Paris goals: by 2050, at the latest, agriculture must move from a net source to a net sink of carbon for us to have any chance of staying within a 2°C, let alone a 1.5°C rise in global temperature. A significant focus on supporting agricultural systems that improve soil health, biodiversity, water quality and quantity, and capture carbon is being driven by politicians through incentives, by businesses and investors, and by consumer demand for more sustainable food.

In the UK, for example, real change in farming systems is being driven by payments to reduce nitrogen and phosphorus leakage and to increase biodiversity, an emerging carbon market, and increasing consumer demand for organic and regenerative agriculture. Groundswell, a UK farming event that Tony always attended and that focuses on working with nature to grow food sustainably, has grown over the last six years from small beginnings to an international event attracting 6000 participants in 2022. And worldwide, despite almost complete absence of support by governments and schools of agriculture, conservation agriculture, which follows these principles and practices, has been adopted across more than 15% of the world's cropland (Kassam et al., 2018). These changes are not a panacea. Given the current commodity markets and higher farm gate prices, most farmers still find industrial farming for food production more attractive than protecting soil, water and biodiversity. However, Tony would have welcomed the new opportunities to transform farming systems and recognize the critical role of farmers as stewards of water, the biosphere and the landscape.

Conclusions

Early on, Tony Allan appreciated the detrimental consequences for society and the environment of undervaluing water. His mission was to see it properly valued and taken into account where it needs to be. He saw the nexus concept as a useful, powerful tool and contributed to its evolution and subsequent input into concepts such as integrated systems, intersectionality, wider value frameworks and resilience. He saw the greatest challenges in food systems, where the most freshwater is used and managed globally. He helped to shape policy and practice globally through deftly crafted, simple communication and a staunch focus on taking messages out of the ivory tower to decision-makers and practitioners. His body of work on the nexus was ground-breaking in its attention to systems thinking and clear articulation of the issues, challenges and solutions. Beyond all this, his stewardship of his students has critically shaped generations of academics and opinion leaders in the private sector, ensuring that his ideas, impact and legacy continue.

It is heart-breaking that he should have passed away just as the world is, finally, beginning to recognize the centrality of food and farming to many of our problems, but also to their solutions. Perhaps these changes are driven by the climate crisis rather than by any desire to confront the social injustice that Tony felt so keenly and captured in his mantra 'under-priced food for underpaid people', or than by a ubiquitous understanding of virtual water. Even so, the outcomes are leading towards what he wanted and the pandemic has both underscored and compounded the issues of accessibility and affordability of food which might still be seen – as they traditionally have been – as forces pulling in different directions: affordability in one direction and sustainability in the other. This this is a false dichotomy: the two are converging.

Positive changes are not confined to agriculture. Policymakers and businesses are incorporating water impacts into decision-making frameworks. This fresh way of looking at water and the value it brings moves away from a pure cost focus and, also, moves beyond the risk focus that has been the main lens through which water impacts have been considered – if at all. Water is finding a voice. It is increasingly recognized, appreciated, valued and taken into consideration. Virtual water remains well understood by relatively few, but it framed a different way of looking at the water system, kickstarted a movement and underpins a lot of progress. We are seeing the public at large increasingly appreciating the natural environment and shifting received wisdom in the direction Tony wanted it to go. He would have appreciated a lot of what is happening and, certainly, played a big part in it making it possible.

Acknowledgements

The authors thank Professor Tony Allan for his kindness, guidance and friendship. They also thank the editors for their initiative and attention.

Disclosure statement

No potential conflict of interest was reported by the authors.

ORCID

Nathanial Matthews http://orcid.org/0000-0002-1512-1142

References

Allan, J. A., & Dent, D. L. (2021). The cost of food. Consequences of not valuing soil and water and the people who manage them. In D. L. Dent & B. P. Boincean (Eds.), *Regenerative agriculture. What's missing? What else do we need to know?* (pp. 3–20). Springer Nature Switzerland.

Allan, J. A., Keulertz, M., & Woertz, E. (2015). The water–food–energy nexus: An introduction to nexus concepts and some conceptual and operational problems. *International Journal of Water Resources Development, 31*(3), 301–311. https://doi.org/10.1080/07900627.2015.1029118

Allan, J. A., & Matthews, N. (2016). The water, energy and food nexus and ecosystems: The political economy of food and non-food supply chains. In F. Dodds & J. Bartrum (Eds.), *The water, food, energy and climate nexus* (pp. 104–116). Routledge.

Bank of England. (2021). *How has covid affected global trade?* Retrieved December 28, 2021, from https://www.bankofengland.co.uk/bank-overground/2021/how-has-covid-affected-global-trade

De Strasser, L., Lipponen, A., Howells, M., Stec, S., & Bréthaut, C. (2016). A methodology to assess the water energy food ecosystems nexus in transboundary river basins. *Water, 8*(2), 59. https://doi.org/10.3390/w8020059

Food and Agriculture Organization. (2021). *Food outlook. Biannual report on global food markets.* Trade and Markets Division, Food and Agriculture Organization of the United Nations.

Hausmann, C. (2022). I remember Tony. *Water International, 47*(7), 1021–22. https://doi.org/10.1080/02508060.2022.2131814

Kassam, A., Friedrich, T., & Derpsch, R. (2018). Global spread of conservation agriculture. *International Journal of Environmental Studies, 76*(1), 29–51. https://doi.org/10.1080/00207233.2018.1494927

Keulertz, M., Mulligan, M., & Allan, J. A. (2020). The impact of COVID-19 on water and food systems: Flattening the much bigger curve ahead. *Water International, 45*(5), 430–434. https://doi.org/10.1080/02508060.2020.1779515

Matthews, N., & McCartney, M. (2018). Opportunities for building resilience and lessons for navigating risks: Dams and the water energy food nexus. *Environmental Progress & Sustainable Energy, 37*(1), 56–61. https://doi.org/10.1002/ep.12568

Mekonnen, M. M., Gerbens-Leenes, P. W., & Hoekstra, A. Y. (2015). The consumptive water footprint of electricity and heat: A global assessment. *Environmental Science: Water Research & Technology, 1*(3), 285–297.

Melo, F. P., Parry, L., Brancalion, P. H., Pinto, S. R. R., Freitas, J., Manhães, A. P., Meli, P., Ganade, G., & Chazdon, R. L. (2021). Adding forests to the water–energy–food nexus. *Nature Sustainability, 4*(2), 85–92. https://doi.org/10.1038/s41893-020-00608-z

Nikolova, A., Karazhanova, A., Schneider, N., & Weinberger, K. (2017). *Integrated approaches for sustainable development goals planning: The case of goal 6 on water and sanitation.* United Nations.

Qu, S., Liang, S., Konar, M., Zhu, Z., Chiu, A. S. F., Jia, X., & Xu, M. (2018). Virtual water scarcity risk to the global trade system. *Environmental Science & Technology, 52*(2), 673–683. https://doi.org/10.1021/acs.est.7b04309

Roidt, M., Chini, C. M., Stillwell, A. S., & Cominola, A. (2020). Unlocking the impacts of COVID-19 lockdowns: Changes in thermal electricity generation water footprint and virtual water trade in Europe. *Environmental Science & Technology Letters, 7*(9), 683–689. https://doi.org/10.1021/acs.estlett.0c00381

Schmidt, J. J., & Matthews, N. (2018). From state to system: Financialization and the water–energy–food–climate nexus. *Geoforum, 91*, 151–159. https://doi.org/10.1016/j.geoforum.2018.03.001

United Nations (UN). (2019). *Leaving no one behind. The UN world water development report.* Retrieved December 28, 2021, from https://www.unwater.org/publications/world-water-development-report-2019

Woertz, E. (2022). Virtual water, international relations and the new geopolitics of food. *Water International, 47*(7), 1108–1117.

World Bank. (2021). *Food security and Covid 19.* Retrieved December 28, 2021, from https://www.worldbank.org/en/topic/agriculture/brief/food-security-and-covid-19

World Economic Forum (WEF). (2011). *Water security: The water-food-energy-climate nexus.* Island Press.

World Economic Forum. (2020). *System value framework.* Retrieved December 28, 2021, from https://www3.weforum.org/docs/WEF_Overview_Path_to_Maximise_System_Value_2020.pdf

Tony Allan: a magic toolbox of theoretical frameworks, a never-ending story

Francesca Greco

Tony Allan was always seeking a more all-encompassing theoretical framework within which to explore the role of water in society: power relations, political economy, virtual water and in recent years, food-water along the food supply chain. According to Tony, virtual water passes through the entire food supply chain along three market modes (production, processing/trade and retail). Tony and I wanted to add 'power plus' lenses to this food supply chain, in order to let winners and losers clearly emerge. His most troubled thoughts were about communication of these issues, in particular to journalists: scholars should not follow them blindly but seek to inform them better.

A never-ending quest

Tony Allan never stood still. He was always seeking a more all-encompassing theoretical framework within which to explore and account for the fundamental role of water in the environment and society. And beyond the facts, his goal was their honest and disinterested communication.

In November 2019, he undertook his last viva as a supervisor – a study of virtual water from faraway and apparently insignificant tomato production in southern Sicily; it was my PhD. The initial idea revolved around the concept of *bad virtual water*; that is, when a vulnerable or non-renewable aquifer is used to sustain export-led irrigation (Greco, 2012). The theme links with several critical issues in political ecology and political economy such as water-justice, appropriation of natural capital and groundwater governance in water-scarce environments (Jakeman et al., 2016; Krausmann et al., 2013; Syvitski et al., 2009). Wada et al.'s (2010, 2012) global assessment of unsustainable use of groundwater was followed by Arjen Hoekstra coining 'the paradox of virtual water exports from water-scarce environments' (Hoekstra, 2014). In 2017, another global study ranked Italy among the top 10 exporters of vulnerable groundwater (Dalin et al., 2017) and, at that point, my case study in Sicily seemed to fit nicely in all its theoretical settings.

I was wrong. Tony had moved on to another development in his interests and, allow me to say it, passions. And I was ready to follow him. I had been with him in 2004 at the first conference on the Red Sea–Dead Sea Water Conveyance Project, where he was persistently ignored, if not ridiculed, by the Middle Eastern media and most of the audience: he was trying to explain the significance of *virtual water*. The Red–Dead kickoff agreement was signed in 2005. Only three years later, he was awarded the Stockholm World Water Prize and, by then, he was creating yet another new theoretical framework and I followed him once again.

We manage what we measure and Tony was convinced that 'accountants will save the world' or, alternatively, 'farmers will save the world' if only they were given the tools for the job – the job being stewardship of our most precious and inseparable resources: land and water. The food system – as the consumer of at least 90% of fresh water – its main actors, how food is produced and traded and consumed, the livelihoods of farmers, and public health became the focus of his attention. He saw farmers producing 'under-priced food for underpaid people' as victims of the system, and accountants and economists of various stripes guilty of sins of omission – they should be valuing water and educating society about its worth and fundamental role 'in the environment *per se*'.

At that point, my case study in Sicily pivoted towards the analysis of virtual water along the food supply chain, using the latest tool presented in *The Oxford Handbook of Food, Water and Society* (Allan et al., 2019) whereby water and, subsequently, virtual water passes through three market modes (production, processing and trade, and consumption) in which winners and losers clearly emerge. Myriad farmers cannot make a living without subsidies; at the other end of the chain, myriad consumers cannot pay for this food without food stamps or other subsidies; but the middlemen in trade and processing make the profits! Set within this framework, my little case study became more significant to me – and to Tony. It illustrated how export-led production was shaping local communities and local economic development, and how global market forces are responsible for the way water is used, perceived, monitored and valued (or not-valued) by society.

I was able to categorize the stakeholders along this specific food-supply chain, discovering whom were winners and whom losers. You may judge for yourselves where each of the following find themselves: migrant workers, farmers, labour organizations, local government, commercial intermediaries, criminal intermediaries, final consumers. Water scarcity and the vulnerability of the aquifer remained facts that corroborated the original *bad virtual water* categorization of my study, but now the facts of what Tony called 'poor farmers' and their 'failed market' emerged dramatically. I found farmers oppressed by: (1) the imposition of a farm-gate price by external traders, food industries and big supermarkets; (2) the imposition of a low prices by competition from abroad; (3) the imposition of a Mafia-tax in terms of choice of transportation or a percentage on the farmers' revenues; and (4) the presence of 75,000 'slaves' in the Italian agricultural and farming industry – by 2020 there were 180,000 (Cgil, 2015, 2020).

In light of the activities of the *Mafia*, Tony worried about me interviewing farmers in Sicily, but I downplayed the risks and embarked, farm by farm, interviewing farmers and key stakeholders with the help of local experts. Again, Tony was right: during my fieldwork, a tomato company belonging to the Pachino Tomato PGI consortium was under investigation by the Italian police for Mafia affiliation, and the journalist who investigated the case, Paolo Borrometi, received death threats (he is still under police protection; Borrometi, 2018). By the time of the Borrometi event, I had already collected my main information, although farmers were only willing to disclose their information anonymously. What emerged was not only that intermediaries were taking most of the revenues but also criminals were active in this segment of the supply chain.

Back from my fieldwork, I applied the food–water framework and noticed the semblance with what Sojamo and others had written on *virtual water hegemony* (Sojamo et al., 2012) – that is, an hourglass distribution of revenues along the food supply chain of Pachino tomatoes: myriad poor farmers, a thin neck of intermediaries and distributors and, finally, numerous hapless consumers who passively accepted the price demanded by the distributors. My PhD was a long and amazing journey with a beautiful mind and I am still wondering how much I can still dig into this gold mine.

My greatest joy has been that, despite being his most unproductive scholar, I had the honour of being free to follow him for 20 years to conferences, Stockholm Water Weeks, World Water Forum, etc., starting from transboundary groundwater issues in the Middle East (my master's thesis in 2004) to his latest conceptualizations. This commentary cannot possibly do Tony justice, but I will try to match his tenacity and perseverance in the face of received wisdom (Cascão et al., 2018). During these particularly difficult days, I would like to remind those who might be encountering Tony for the first time with this special issue that his most recurrent and troubling thoughts have always been related to the communication of science – in particular, to journalists. Tony always scolded his scholars against following journalists in their *water-wars* rhetoric. I can only appeal, again, against scaremongering. Water wars do not exist per se because 'fortunately there are substitutes for water' (Allan, 1993); water is usually one of several topics in negotiations and conflicts among states and critical hydro-political scholars know that water is part of the ideological narrative of some countries – although, apart from ideologies, there is the reality of water infrastructure being targeted in armed conflicts. This is precisely what Tony's scholars currently deal with on a daily basis and at the highest levels, working towards peace (Geneva Water Hub, 2021; Cascão et al., 2018, Mirumachi et al., 2021; Doudy, 2020). If journalists want to scream *war*, let them scream, and move on as Tony did.

Now, I have a new adventure. In the last months of his life, Tony asked me to review how the concept of virtual water has been used over 30 years across different fields. I promised I would do it 'as soon as I could'. That is a promise, but it is another story for another time.

Disclosure statement

No potential conflict of interest was reported by the author.

ORCID

Francesca Greco http://orcid.org/0000-0002-1696-3472

References

Allan, J. A. (1993). Fortunately, there are substitutes for water, otherwise our hydro-political futures would be impossible. In *Priorities for water resources allocation and management* (pp. 13–26). Overseas Development Administration.

Allan, J. A., Bromwich, B., Keulertz, M., & Colman, A. J. (Eds.). (2019). *The Oxford handbook of water, food and society*. Oxford University Press.

Borrometi, D. P. (2018). *The company of the Mafia boss Giuliani is a member of the Pomodora da Pachia protectioin consortium. La Spia* 27 Feb 2018. https://www.laspia.it/la-societa-del-capomafia-giuliano-socia-del-consorzio-tutela-igp-pomodoro-pachino/(Italian)

Cascão, A. E., Closa, A., Fantini, E., Gebreleul, G., Ide, T., Jobbins, G., Kinna, R., Loures, F. R., Magsig, B. O., Matthews, N., McIntyre, O., Menga, F., Mirumachi, N., Moynihan, R., Nicol, A., Oestigaard, T., Rieu-Clarke, A., Selby, J., Sujamo, S., & Zeitoun, M. (2018). Why are water wars back on the agenda? And why we think it's a bad idea! *Undisciplined Environments*, 22 March. https://undisciplinedenvironments.org/2018/03/22/why-are-water-wars-back-on-the-agenda-and-why-we-think-its-a-bad-idea/

Cgil. (2015). *Agromafie and gangsters. Dell'osservatorio Placido Rizzotto della Flai-Cgil.* Futura.

Cgil. (2020). *Agromafie and gangsters. Dell'osservatorio Placido Rizzotto della Flai-Cgil.* Futura.

Dalin, C., Wada, Y., Kastner, T., & Puma, M. (2017). Groundwater depletion embedded in international food trade. *Nature*, *543*(7647), 700–704. https://doi.org/10.1038/nature21403

Doudy, M. (2020). Water weaponization in the Syrian conflict: Strategies of domination and cooperation. *International Affairs*, *96*(5), 1347–1366. https://doi.org/10.1093/ia/iiaa131

Geneva Water Hub. (2021). *Zeitoun appointed as director general of geneva water hub*, 21st of November 2021. https://twitter.com/bluepeace/status/1457730210531618817

Greco, F. (2012). Introducing an historical perspective to virtual water flows: How hydro-metabolism explains the "Bad" virtual water flows from non-renewable/overexploited groundwater. *EGU Leonardo Topical Conference Series on the Hydrological Cycle*, 14–16 November 2012. Turin

Hoekstra, A. Y. (2014). *The water footprint of modern consumer society*. Routledge.

Jakeman, A. J., Barreteau, O., Randall, J., Rinaudo, J. D., & Ross, A. (Eds.). (2016). *Integrated groundwater management concepts, approaches and challenges*. Springer Open.

Krausmann, F., Erb, K.-H., Gingrich, S., Haberl, H., Bondeau, A., Gaube, V., Lauk, C., Plutzar, C., & Searchinger, T. D. (2013). Global human appropriation of net primary production doubled in the 20th century. *Proceedings of the National Academy of Science*, *110*(25), 10324–10329. https://doi.org/10.1073/pnas.1211349110

Mirumachi, N., Gregulska, J., & Smetek, J. (2021). *The human right to drinking water: Impact of large-scale agriculture and industry: In depth-analysis*. Directorate General for External Policies of the Union, European Parliament.

Sojamo, S., Keulertz, M., Warner, J., & Allan, J. A. (2012). Virtual water hegemony: The role of agribusiness in global water governance. *Water International*, *37*(2), 169–182. https://doi.org/10.1080/02508060.2012.662734

Syvitski, J., Kettner, A., Overeem, I., Hutton, E. W. H., Hannon, M. T., Brakenridge, G. R., Day, J., Vörösmarty, C., Saito, Y., Giosan, L., & Nicholls, R. J. (2009). Sinking deltas due to human activities. *Nature Geoscience*, *2*(10), 681–686. https://doi.org/10.1038/ngeo629

Wada, Y., van Beek, L. P. H., & Biekens, M. (2012). Non sustainable groundwater sustaining irrigation: A global assessment. *Water Resources Research*, *48*(6), W00L06. https://doi.org/10.1029/2011WR010562

Wada, Y., van Beek, L. P. H., van Kempen, C. M., Reckman, J. W. T. M., Vasak, S., & Bierkens, M. F. P. (2010). Global depletion of groundwater resources. *Geophysical Research Letters*, *37*(20), L20402. https://doi.org/10.1029/2010GL044571

Accountants will save the world!

Andrew Ross

Money matters

I met Tony Allan for the first time in 2013 and invited him to a meeting at the Institute of Chartered Accountants where Peter Bakker, chief executive officer of the World Business Council on Sustainable Development, asserted: 'Accountants will save the world!' (Bakker, 2013). The visionary mantra strongly appealed to Tony who knew that all things, not least water, have a cost – and if that cost is not accounted for in the ultimate price of goods, then all manner of decisions will be flawed. This is certainly the case in the food system, which consumes more than 90% of global freshwater. As Tony elegantly described the progress of water through farm production, food processing and retail, he noted that rational use of *green water*, *blue water* and *virtual water* requires accounting on a whole-system scale (Allan, 2019). And in the absence of such accounting, we have a grotesquely distorted market that squanders land and water resources.

So, why did accountants not account earlier for the externalities of the system, including water? That question had been largely confined to academic discussion (Mekonnen & Hoekstra, 2011) until the Great Financial Crash of 2007–08 made the concept of global systemic shocks an accounting reality – and governments woke up to the similarity with climate change. *Scope 1/2/3 carbon emissions*, which had been described since 2001 in the Greenhouse Gas Protocol (World Resources Institute, 2001) have since become mandatory for corporate reporting in the UK, and this accounting has been adopted worldwide. Simultaneously, central banks insisted that sustainability reporting could no longer be an appendix to corporate reports but should be a front-page matter. So, at the Paris 2017 Summit, they launched the Taskforce on Climate-related Financial Disclosure (TCFD) based on corporate carbon accounting but, as Peter Newborne observes in this issue (Newborne, 2022), water and biodiversity have struggled to identify appropriate metrics. Mark Carney then established the Network for Greening the Financial System encompassing 52 central banks (which at the instigation of the Joe Biden administration now includes the US Fed and China) and, in time for COP26, directed the Glasgow Financial Alliance for Net Zero linking asset managers and asset owners and insurers accounting for US$70 trillion of assets under management.

Climate change is now recognized as an existential threat to the global economy. Institutional investors such as pension funds are required by law and financial regulators to provide a secure return of income to enable pensions to be paid. They cannot do that if they are holding stranded assets such as fossil fuel reserves that are untouchable without breaching 1.5°C; they are required to transition their investments into sustainability. The EU Sustainable Finance Disclosure Regulation (SFDR), the phenomenal growth of environment social governance (ESG) corporate reporting, and fines for misreporting all reflect the urgent need that chief financial officers now have to explain to their institutional investors how they are providing true-cost accounting for sustainability. A total of 60% of the world's largest public companies have committed to reach net zero. But how are they going to do it? Where are the data on which to judge their performance?

The annual cost of restricting global heating to 1.5°C and adaptation to climate change that is already in train is estimated at US$300 billion (and cumulatively up to US$50 trillion). In response, the volume of green bonds issued globally is anticipated to top US $1 trillion by the end of 2021, and it is on an exponential trajectory. The availability of money is not a problem: the problem is the lack of propositions at institutional investment grade and scale of more than US$1 billion. These are very powerful transformations of a massively complex system, but they are, still, only dealing with the easy part: carbon accounting.

The difficult parts are yet to come: accounting for natural capital. It is axiomatic that global water resources are all of one with natural capital and a 40% shortfall by 2030 is projected. The key question, again, is: Where are the reliable data?

Thanks to the labours of Tony and his apostles over a quarter of a century, the financial world now realizes the consequences for capital assets and corporate supply chains: companies exposed to this risk face a downgraded credit rating – and an increase in their cost of capital – with potential losses to their insurers and investors. For insurers, the physical impacts of the climate crisis are obvious. The last two or three years have seen historic climate-related devastation from severe flooding and drought to unprecedented wildfires and heatwaves, and back-to-back hurricanes ravaging coastlines and inundating cities. What were, not so long ago, thought of as extreme weather events are now more frequent and more intense, costing thousands of lives and tens of billions of dollars in damage. Prolonged drought, shrinking glaciers and drying up of reservoirs dammed for hydroelectricity generation threaten entire economies; and *The Lancet* (Helldén et al., 2021) has reported the negative effects on health of the destruction of the natural world, floods and droughts; evidenced in malnutrition, disease and pandemics.

Natural catastrophe and credit-risk models are being adjusted for these climate effects by downscaling macro-economic data to the regional and corporate level, largely based on publicly available syndicated loan information rather than upscaling from creditworthy site data. The annual value-at-risk from physical climate impacts just the syndicated loan portfolios of major banks is approaching 10%: two-thirds of banks' physical risk comes from the indirect economic impacts of climate change, such as corporate supply chain disruptions and lower productivity; the biggest direct risks come from flooding of cities and infrastructure driven both by sea-level rise and stronger storms. Economic development in prosperous parts of the world, such as Florida and California, is becoming either uninsurable or unaffordable. The floods in Germany and Belgium, following a week of continuous heavy rain, caused second-order impacts with long-term social disruption. Similarly, countries such as Brazil and Argentina are seeing third-order impacts with their sovereign bond credit rating impaired for deforestation (and its disruptive effect on the jet stream) and the alteration of water availability impacting on the global food system.

The knowledge base of TCFD has, therefore, led to the creation of the Taskforce for Nature-related Financial Disclosure (TNFD). This aims to establish a framework for the provision of data enabling nature-based solutions to be incorporated into financial asset valuation, investment decisions and capital allocation. Given that the Dasgupta Review (HM Treasury, 2021) described how the global financial economy is a wholly dependent subset of the global ecosystem and its natural capital assets, a key financial focus of COP26 in Glasgow was on nature-based solutions.

Scientists, step up!

There are ways to mitigate both the physical impacts and the economic costs of the climate crisis. Insurers account for US$25 trillion of the US$100 trillion global economy – and they are searching for sustainable investment opportunities. It requires insurance-risk innovation to bridge the ever-widening *protection gap*. Fundamentally, data are required that enable insurers to provide risk-resilient insurance cover, and make risk-mitigation investments in sustainability. If you are to be of tangible value to insurance data analysts and credit rating agencies in the financial market, water economists, hydrologists, agronomists, natural-capital economists and scientists with research and development (R&D) projects at the landscape and river basin scale can, and should, be able to provide these data. You can do so in four steps:

1) MEASURE CLIMATE RISK

Assess all elements of climate risk and the opportunities for mitigation and adaptation at the river basin scale.
Measure and account for the current impact of natural disasters on the value of financial assets [insured and uninsured] at the river basin scale.
Project the future cost of climate change at the river basin scale.
Perform climate-stress tests on the exposure of local authorities, cities and corporates at the river basin scale.
Collect asset-level insurance data about their exposure and loss vulnerability.

2) TAKE ACTION

Build connections with external experts in insurance and finance.
Integrate climate risk into R&D project design and cost pricing.
Engage corporate and financial clients on their physical risks and exposure.

3) CAPITALIZE ON OPPORTUNITIES

Understand the changing banking and insurance landscape.
Focus on adaptation projects to mitigate credit risk.
Enable innovative financing solutions with the provision of tangible data.
Advocate smart financial regulatory and policy actions on adaptation.

4) MEET THE MOMENT

Set and disclose financing targets for sustainability at the river basin scale.

Help to develop and provide investable financial instruments such as Municipal Green Bonds for cities and their natural capital assets in their upstream and downstream hinterlands at the river basin scale.

The Water Group at KCL, established by Professor Tony Allan, has provided the background for major new R&D projects such as ReSET and ICATALIST, and his work has been influential in the recent synthesis of evidence in a key reference book (Allan, 2019) that has a comprehensive description connecting nature-based solutions with water-based security and the global insurance industry. To a significant extent, the Modes 1/2/3 of global supply chain infrastructure that Tony foreshadowed have been globally re-described in terms of tracking Scope 1/2/3 corporate carbon emissions – and now in terms of natural capital and biodiversity impacts, and by the very profession with which he most wanted positive engagement in order to 'save the world': accountants!

Disclosure statement

No potential conflict of interest was reported by the author.

References

Allan, J. A. (2019). Food, water and the consequences of not valuing the environment. In J. A. Allan, B. Bromwich, M. Keulertz, & A. J. Colman (Eds.), *The Oxford handbook of food, water and society*. Oxford University Press.

Bakker, P. (2013, March 3). Accountants will save the world. *Harvard Business Review*. https://hbr.org/2013/03/accountants-will-save-the-worl. See also *The Guardian*, London, March 5 2013.

Helldén, D., Andersson, C., Nilsson, M., Ebi, K. L., Friberg, P., & Alfvén, T. (2021). Climate change and child health: A scoping review and expanded concept. *The Lancet Planetary Health*, *5*(3), E164–E175. https://doi.org/10.1016/S2542-5196(20)30274-6

HM Treasury. (2021). *The economics of biodiversity: The Dasgupta review*. www.gov.uk/officialdocuments

Mekonnen, M. M., & Hoekstra, A. Y. (2011). *National water footprint accounts: The green, blue and grey water footprint of production and consumption* [Value of Water Research Report Series50]. UNESCO-IHE.

Newborne, P. (2022). Crossed wires: Public regulation and private action for water stewardship and sustainable farming. *Water International*, *47*(7), 1169–1186. https://doi.org/10.1080/02508060.2022.2133798

World Resources Institute. (2001). *The greenhouse gas protocol: A corporate accounting and reporting standard*. WRI.

OPEN ACCESS

Irrigated agriculture: more than 'big water' and 'accountants will [not] save the world'

Bruce A. Lankford

ABSTRACT

Two of Tony Allan's phrases – *big water* and *accountants will save the world* – invite me to argue that irrigation is poorly served when its hydrology is seen solely as *big* or via *accounts*. While *big* applies because irrigated areas deplete considerable volumes of water, irrigation systems contain many more water relations, behaviours and puzzles. In this problematic, environmental, social and governance (ESG) and water accountants and accounts will become a dominant force. This is worrying for the degree to which individual irrigation systems are rendered into catchment-level accounting abstractions, removing us from a more vital, multidisciplinary, cross-scale and action-oriented approach.

Introduction

Two of Tony Allan's phrases – *big water* and *accountants will save the world* – prompt me to voice concerns about the future of irrigation hydrology knowledge (IHK). I worry that IHK is moving towards *accounting–abstraction–absolutism* (AAA), by which I mean desk-based modelling, supported by remotely accessed data, leading to quantification abstractions that create blueprint irrigation policies and quasi-scientific paradigms and principles. With this shift, IHK is moving away from *system–action–transformation* (SAT) characterized by an action-oriented, grounded, field-based, farmer-partnering, system-specific, empirical approach that acts in the real world; testing, experimenting, observing and measuring. I fear that if current trends continue, accountants will swamp, rather than save, the world. That is, they and their methods will shape our thinking and policies for many years to come. This will bring some benefits but, on the whole, it will make the change management of sustainable catchment-fitting irrigation systems much more difficult. If my AAA fears come to pass, the future of IHK will not be recognizable as saving the world.[1]

The expression *big water* works because irrigated agriculture as a sector and total area consumes huge volumes of water, produces considerable amounts of food, feed and fibre, and creates sharp externalities and opportunity costs that fall on other sectors such as cities and the environment (Keulertz & Allan, 2019). *Big* also allows us to distinguish the difference between this water and non-food water (Allan, 2013), meaning water for other sectors. Into this problemscape, accountants, business consultants and economists and their hydrological accounts (note, not the financial accounts of irrigation systems, nor hydrologists with hydrological computations) are set to become a dominant force in the science and understanding of irrigated hydrology – as Tony predicted when he affirmed Peter Bakker's argument (Allan, 2019; Bakker, 2013), that 'accountants will save the world' (AWSTW) as part of Tony's prognosis that 'farmers, accountants and optimism' (FAO) will help meet global water challenges.

Just as with his *virtual water* concept (Allan, 2011), we can marvel at Tony's ability to capture a changing problem/waterscape in so few words. His use of expressions and handles encouraged debate, invited non-irrigation policymakers to the table and released funds for research. It is a skill that few of us have and more need. In the spirit of that invitation to debate, I now comment on the downside of irrigation accounting.

Irrigated agriculture is more than a *big* volume

The volume of water withdrawn into and consumed by global irrigated agriculture is *big*. Assuming the global area under irrigation in 2022 is 350 million ha, and the average depth equivalent of water consumed (not withdrawn) is 800 mm per annum (equivalent to one season of rice cropping or two non-rice crops), then approximately 2800 km^3/year of water is consumed. For comparison, this is about 50 times the daily domestic water consumption of, say, 20 litres per person per day for the world's 8 billion people. Furthermore, *big* continues to expand: the growth in global area irrigated exceeds global population growth (Puy, 2018).

So significant are these volumes that irrigation systems reshape the hydrology and society of the catchments they occupy (Damonte & Boelens, 2019). This reshaping brings water-governance problems such as how to share water between competing sectors (e.g., the environment and growing cities) whilst retaining desirable agricultural outcomes. Tony posited 'FAO' as the answer to these questions. I fully accept FAO as a partial answer, but believe it falls short of the mark. To follow my criticism requires some discussion of whether irrigation is more than *big*.

What is the puzzle (or problematic) of irrigation? Once we might have defined irrigation as the withdrawal and distribution of water from an intake via canals, pipes, outlets and fields to meet crop evapotranspiration needs (ET) not furnished by rainfall. This anodyne definition contains some of the volumetric accounting aspects of irrigation such as the deficit between rainfall and ET, and the difference between withdrawal and ET consumption.

However, irrigation is much more than a *big* volume or that is captured by 'in and out' accounts. In today's water-scarce/-variable catchments, the hydrological puzzle of irrigation is sharply drawn and is growing. In my humble attempt to mimic Tony's handles, irrigation is a puzzle because it is:

- *Turn water*: farmers on gravity systems[2] share, use and pass on flows of water known as leadstreams (*main d'eau*). This means farmers are socially and materially connected to each other and to how water cascades, bifurcates and distributes through an irrigation network.
- *Timely water*: in a soil profile with a limited storage capacity of 3–15 days of available water, irrigation must be applied on time. To time irrigation requires a gamut of design, operation and maintenance factors to work coherently. Low irrigation efficiency in canals and fields slows down this timing.

- *Fractal divisional water*: the continuous division of water happens from intakes on a river down to secondary and tertiary canals, and on down to farms, fields, parts of fields and individual plants. This sets up competing/equity relations between branches of those bifurcations not easily captured in water accounts (Lankford, 2012). Furthermore, unlike all-too-neat water accounts, which sees water clearly split into 'disposal fractions', the ever-cascading division of water flowing through an irrigation system is coupled, obscure, difficult to track, hard to measure and nearly impossible to manage separately and distinctively.
- *Conjunctive water*: farmers are adept at managing overlapping sources of water from soil water storage, direct rainfall, groundwater, rainfall runoff/harvesting, capillary rise from shallow water tables, and water from streams, rivers, storage and non-conventional sources, e.g. wastewater. This means it is difficult to ascribe changes in cropping patterns solely to one source of water or to one variable such as irrigation efficiency.
- *Common water*: via withdrawal and consumption, irrigation is part of a commons connecting people competitively and cooperatively at all scales, for example, farmers on canals, and cities to their neighbouring irrigation systems. Even the *water commons* can be further characterized, for example, via the non-consumption and release of freed-up water salvages, irrigation systems play their role in the catchment *paracommons* (Lankford, 2013).

Even these handles do not describe the 'more-than-big' and 'more-than-accounting' aspects of governing catchment-, climate- and economy-fitting irrigation.[3] In irrigation systems that are often inadequately designed, operated and maintained, water supplies have to match demand in the face of imperfect information in a highly socialized, spatially differentiated knowledge-scape: that of farmers and water-use groups. This imperfectly known environment is easily buffeted and shaped by an array of experts who decree that 'today's irrigation is inefficient' or 'is consumptive', but 'tomorrow's will be more efficient and less (or paradoxically more) consumptive' as long as 'this technology', 'that rule and institution', or 'these *xyz* methods of analysis' are adopted or that 'water depletion only should be managed'. Put simply, my comprehension of the puzzle of irrigation means we need SAT. Favouring such an approach, I question the rise of water accounts, accountants and accounting in IHK.

Accountants will save the world (AWSTW)

In emails dated 29 September 2015, I cross-checked with Tony what or whom he meant by accountants. In the light of this exchange, his writings and presentations, I surmise he meant people with accounting, business, financial and economic backgrounds who report on hydrological accounts of water use in irrigated systems and food supply chains. What he was driving at was that water should be properly valued and this value should be accounted for in the price of things water goes into, especially food (Allan, 2019; Allan & Dent, 2021). With AWSTW, Tony wanted accountants to shift their focus from corporate risk-measuring to being much more involved in wider monitoring, and in deeper ethical decisions about societal- and consumer-led trends in global value chains. Quoting from his emails:

> When speaking about the role of accountants in public, my purpose is to draw attention to their potential role in changing behaviour in private sector food supply chains and Better Accounting Rules, Responsible Private Investment and Sustainable Public Policy so that we have sustainable markets rather than blind and dangerous ones.

In keeping with his *virtual water* insight (Aldaya et al., 2010; Allan, 2011) he wanted society to bring light to 'blind' markets that cause crops to be irrigated in water-scarce catchments despite consuming much more water and bringing less income than, say, microchips or hotels.

So here we are in 2022, and Tony's prescience was spot on, and as evidence of that I draw attention to a growing need for non-financial environmental, social and governance (ESG) reporting (Markota Vukić et al., 2018), and the burgeoning ESG staff recruitment plan by the 'Big Four' (PwC, KPMG, EY, etc.) and other accounting firms to fill that growing need.[4]

However, much as I see the benefits of well researched water accounts, I am far from convinced about AWSTW. From now on, I characterize *accountants* by method and training rather than by career designation; in other words, an irrigation engineer can be an accountant if he or she intervenes in irrigated systems primarily via accounting methods. And an accountant by job title is an *irrigation transformer* when working with farmers applying multiple IHK methods. Also, the quality of engagement on both sides matters; irrigation systems can be informed and transformed by diligent accounting of the right metrics at the right scale, or be harmed either by accounting delivered to lax standards (Bennett & James, 2017) or by SAT that is not systematic and self-learning.[5]

Towards accounting–abstraction–absolutism (AAA): accountants swamping not saving

Let me be clear; if well conducted, resource accounting brings clearer science and policy oversight to economic, hydrological and environmental trajectories at the basin level (Delavar et al., 2020) plus visibility to inter-state/-nation water conflicts (Bassi et al., 2020). What I fear is that handing sole or major responsibility for steering these trajectories to accountants will substitute other expertise and greatly weaken our ability to deliver water systems that are equitable, productive and sustainable. Far from the forensic oversight that Tony sought, accounting reports are often incomplete, without a common framework and in need of reform (Davies et al., 2020). Worse still, they can be incurious, blind and biased, or narrowly descriptive, as demonstrated by Coca Cola's claim for replenishment that aims to 'safely return to communities and nature an amount of water equal to what we use in our finished beverages' (quoted in Rudebeck, 2019), which highlights the water consumed for manufacturing beverages but ignores the water consumed in irrigating sugarcane. We should be further concerned by the political economy and financialization of resource accounting (Rudebeck, 2022) and ask who pays for and referees ESG accountants and their methods and reports, which Tariq Fancy terms 'convenient fantasies' (Fancy, 2021).[6]

But even good ESG accounts will be insufficient to incorporate the puzzles I introduced above in order to steer complex, multi-scale irrigated systems towards desirable basin-level outcomes. The shortfall lies with the substantive way that accountants render and abstract the world into *accounts* (Sullivan & Hannis, 2017) which emphasize selected countable or quantifiable dimensions of the material world but demote or exclude other dimensions. An

Table 1. Irrigation hydrology knowledge (IHK): accounting–abstraction–absolutism (AAA) and system–action–transformation (SAT).

Future IHK pathways	AAA	SAT
Main foci and benefits	Higher scale outcomes and trajectories at the basin or system level over longer time periods. To guide decisions on basin allocation and sustainability	Lower scale, quicker time-span workings and underlying factors that create higher scale outcomes. To guide water management to achieve system performance and basin sustainability
Main location	Desk based	Field, system and farmer based
Knowledge evolution and ethical praxis	Towards accounting abstraction	Empirically triangulated by the material and human world
Quantitative calculations	Cost-curve methods, water accounting, water footprints	Bespoke calculations for the problem at hand
Views on irrigation efficiency (IE)	All traditional systems are inefficient and leaky. All modern systems are efficient but either save water or consume more water	Fluid, recursive, hybrid, heterogeneous and overlapping
Sources of IE information	The first few pages of returns from a web search engine	Multiple sources working at all scales
Professional training	Professional accounting qualifications plus MBAs that include environmental, social and governance (ESG)	Irrigation, soil science, agronomy, hydrology, engineering, social science, practical experience
Evidence for growth/ shrinkage	'Big Four' ESG recruitment. More than 2500 MBAs available in English globally[a]	At the time of writing, no dedicated MSc/ MEngs globally on irrigation

Note: [a]See https://find-mba.com/what-is-an-mba/.

irrigation system is a true system wherein many underlying, smaller, more numerous or seemingly inconsequential, qualitative, human or non-material dimensions affect other, larger or more material outcomes. Therefore, the change management of irrigation with its many factors cannot be guided by ESG accounts that report on meta- or basin-level water disposals. On the contrary, irrigation systems need the integrated thinking that I label SAT; partnering with irrigation systems and their farmers and engineers using an array of worldviews, skills and work modalities.

I have divided these two approaches to IHK in Table 1. On the left, AAA abstracts the complex world of irrigated systems into numbers and axioms. On the right, SAT puzzles out this world using heterodox methods and information sources.

The rise and rise of the ESG and water accounts and accountants

I predict the next decade will see AAA ascend and SAT decline. What is driving this?

- The need for international companies, including financial companies, involved in the food-value chain to identify risks whereby a lack of water might disrupt production and profitability (Rudebeck, 2022).
- The Big Four recruitment drive for ESG analysts responds to other growing forces: the need for organizations to report on environmental sustainability to shareholders and the wider public, or for investment (Davies et al., 2020) and regulatory purposes (Markota Vukić et al., 2018).
- The economics of ESG/AAA are attractive: desk studies are extremely cost-effective. A consulting company can charge considerable daily rates for environmental reporting, and if reports appear to show numeracy and competence, there is no need to add costs by visiting the field and meeting farmers (let alone measuring agrometeorology, water flows, start–stop times and areas irrigated).

- AAA methods, accounting aims and system comprehension co-create what appear to be credible change-management paradigms. For example, McKinsey devised the *Cost-curve Model for Decision-making* as a basis for investing in and modernizing irrigation systems (2030WRG, 2009), which eventually brought 2030WRG-led switches from canal to drip irrigation in Karnataka, India (Meland, 2021). As a contribution, it merits discussion, but as a universal blueprint, it ignores local geography, sensibilities and trajectories (Dyer & Counsell, 2010) and fails to identify who bears the costs of such changes (Lobina & Hall, 2009).
- Similarly, water research institutes need to sell water accounting services and thus claim these services span a very wide set of scales 'from field-to-basin' and serve 'communities'.[7]
- It is relatively easy for a water accounting person to advise how to improve irrigation systems when reproducing advice from internet-based sources. Over the years, many people (ecologists, economists, civil engineers, agronomists, etc.) have told me what constitutes better irrigation, despite no sign they have managed or systematically researched irrigation performance. Recommendations tend to be: line canals; irrigate at night; add soil moisture measurement; instal drip irrigation; laser-level fields; fix leaks; meter canal flows; train farmers on soil and water management; and create water user groups. It is not that these are individually wrong, but what troubles me is *how* they are dropped into the discussion. To me these shibboleths reveal that the adviser knows little about irrigation. There is no commensurate sense of how to think about these solutions such as: what the baseline performance might be; how effective they might be (for different purposes); their cost and who bears this; in what order of priority; their 'system' fit (to each other and to the local situation); how farmers might use and own these technologies or have equivalent solutions and different priorities;[8] or that many 'design-manageability' reforms never get mentioned such as the density of canal network per hectare, retrofitting of manageable hardware or how to design-in the water duty down to a tertiary scale (Lankford, 1992; Lankford & Gowing, 1998). And it appears there is little awareness that there is a debate (still incomplete in my view) on irrigation manageability (Horst, 1999; Plusquellec, 2002; Plusquellec et al., 1994).
- Similar to the previous point, it is relatively easy to confidently take up a position on one side of the *inefficient–efficient* binary. For example, one can assert that gravity irrigation schemes are inefficient, should save water and need to be modernized (for the critical discussion on this, see Lankford, 2004; or De Bont et al., 2019) or that modern irrigation systems are efficient but, paradoxically, raise water consumption (Food and Agriculture Organisation (FAO), 2017). But axiomatic pronouncements about the efficiency status of heterogeneous *peopled* irrigation systems in a state of flux do not equate to the delivery of future performance towards desired aims.
- Irrigation is no longer a senior or international career track. Unlike 20–30 years ago, there are globally no dedicated irrigation master's degrees that view the topic as more than, but includes, engineering. With very few exceptions, irrigated agriculture continues not to attract science research, university posts and teaching, careers and professional development.[9]

Conclusions

Within the space of IHK, the role of accountants in environmental reporting is on the increase. While this to be welcomed for the oversight brought to water governance, it should not substitute for other disciplines whose management modalities involve working with and alongside farmers and their knowledge, practices, irrigation systems and material resources. Plus, although we all need irrigation actors to be more quantitative, accountancy is not the only quantitative voice.

Beyond concerns of: (1) accountants numerically substituting field-based, multi-stakeholder-oriented irrigation actors; and (2) the quality of ESG reporting, there is also a risk of an ontological substitution. By this I mean the manner in which accountants abstract the material world into *high-level accounts* (Sullivan & Hannis, 2017), which, by definition, have removed other important information and, then, generalize from these speculations back down to factors and parameters operating at lower scales. Moreover, the risk is that non-accountants fail to see the significance of this elision and are unable to judge how AAA and SAT work together. Similarly, Wichelns (2011, p. 609), with reference to water footprint analyses, is also concerned about the weaknesses of water accounting abstractions:

> Estimates of water footprints do not contain sufficient information to enhance understanding of water's role in any of its many functions, or to guide policy makers wishing to motivate improvements in water management.

Notes

1. I take this phrase at face value and do not unpack *saving the world*. Also, I did not witness Tony suggesting accountants be seen as additive and complimentary to other irrigation world-views, expertises and skills. Furthermore, I only focus on the 'accountants will save the world' (AWSTW) part of 'farmers, accountants and optimism' (FAO).
2. Water-sharing and turn-waiting also occur in pressurized piped systems if they are poorly designed, operated and pressurized.
3. Many more labels cover the challenges of irrigation. For example, to source and distribute water, farmers manage different energy (e.g., gravity, solar, diesel) and information types (e.g., timing, depth equivalents, area covered, flows).
4. See https://www.reuters.com/business/sustainable-business/pwc-planning-hire-100000-over-five-years-major-esg-push-2021-06-15/; and Big Four accounting firms rush to join the ESG bandwagon
5. During my six years working on and researching irrigation systems in Swaziland, I observed many ad hoc and weak SAT-type attempts to improve irrigation performance in sugarcane systems that had little effect, such as HR Wallingford's failure to get their irrigation software developed for divisional canal systems on flat Asian floodplains to work on rotational canals located on the more undulating landscape of the Simunye system.
6. Fancy is ex-head of sustainable investing at BlackRock.
7. Such as expressed in two quotations from a 2021 job vacancy at a global water research institute for a Water Accounting and Hydrology Postdoc, who should be 'trained in spatial hydrology, remote sensing and water accounting developing tools for water accounting and water balance assessments applicable to a range of scales (from field to basin)'. The job had a remit 'to understand water availability and use, water risks, and water values, the aim is to enhance the ability of African governments, communities and companies to better manage their water resources'.

8. After I used the participatory tool 'the River Basin Game' with irrigators in Usangu, southern Tanzania, they asked not for water training but for bookkeeping training. They knew how to manage water, but wanted to do this more collectively and needed to be confident that their shared financial contributions to the operation and maintenance of irrigation systems could be tracked.
9. 27 May 2021; I checked with Joshua Newton who runs the excellent database at www.joshswaterjobs.com that approximately 1% of jobs posted in the last five years referred to irrigation or irrigated agriculture.

Acknowledgements

I thank the editors of this special issue for their invaluable comments, plus Alex Meland and Therese Rudebeck for insightful discussions.

Disclosure statement

No potential conflict of interest was reported by the author.

ORCID

Bruce A. Lankford http://orcid.org/0000-0001-5580-272X

References

2030WRG. (2009). *Charting our water future: Economic frameworks to inform decision-making.* https://www.mckinsey.com/~/media/mckinsey/dotcom/client_service/sustainability/pdfs/charting%20our%20water%20future/charting_our_water_future_full_report_.ashx

Aldaya, M. M., Allan, J. A., & Hoekstra, A. Y. (2010). Strategic importance of green water in international crop trade. *Ecological Economics*, *69*(4), 887–894. https://doi.org/10.1016/j.ecolecon.2009.11.001

Allan, T. (2011). *Virtual water: Tackling the threat to our planet's most precious resource.* I.B. Taurus & Co Ltd.

Allan, T. (2013). Food–water security: Beyond water resources and the water sector. In B. A. Lankford, K. Bakker, M. Zeitoun, & D. Conway (Eds.), *Water Security* (pp. 321–335). Routledge.

Allan, T. (2019). Food, water and the consequences of society not valuing the environment. In T. Allan, B. Bromwich, M. Keulertz, & A. Colman (Eds.), *The Oxford handbook of food, water and society* (pp. 859-878). Oxford University Press. https://doi.org/10.1093/oxfordhb/9780190669799.013.58

Allan, T., Dent, D. (2021). The cost of food: Consequences of not valuing soil and water and the people who manage them. In D. Dent & B. Boincean (Eds.), *Regenerative Agriculture* (pp. 3–19). Springer. https://doi.org/10.1007/978-3-030-72224-1_1

Bakker, P. (2013). Accountants will save the world. *Harvard Business Review.* March 5. https://hbr.org/2013/03/accountants-will-save-the-world

Bassi, N., Schmidt, G., & De Stefano, L. (2020). Water accounting for water management at the river basin scale in India: Approaches and gaps. *Water Policy*, *22*(5), 768–788. https://doi.org/10.2166/wp.2020.080

Bennett, M., & James, P. (2017). Key themes in environmental, social and sustainability performance evaluation and reporting. In M. Bennett, P. James, & L. Klinkers (Eds.), *Sustainable measures* (pp. 29–74). Routledge. https://doi.org/10.4324/9781351283007

Damonte, G., & Boelens, R. (2019). Hydrosocial territories, agro-export and water scarcity: Capitalist territorial transformations and water governance in Peru's coastal valleys. *Water International, 44*(2), 206–223. https://doi.org/10.1080/02508060.2018.1556869
Davies, P. A., Dudek, P. M., & Wyatt, K. S. (2020). Recent developments in ESG reporting. In D. C. Esty & T. Cort (Eds.), *Values at work: Sustainable investing and ESG reporting* (pp. 161–179). Springer International Publishing. https://doi.org/10.1007/978-3-030-55613-6_11
De Bont, C., Liebrand, J., Veldwisch, G. J., & Woodhouse, P. (2019). Modernisation and African farmer-led irrigation development: Ideology, policies and practices. *Water Alternatives, 12*(1), 107–128. https://www.water-alternatives.org/index.php/alldoc/articles/vol12/v12issue1/481-a12-1-7/file
Delavar, M., Morid, S., Morid, R., Farokhnia, A., Babaeian, F., Srinivasan, R., & Karimi, P. (2020). Basin-wide water accounting based on modified SWAT model and WA+ framework for better policy making. *Journal of Hydrology, 585*, 124762. https://doi.org/10.1016/j.jhydrol.2020.124762
Dyer, N., & Counsell, S. (2010). McREDD: How McKinsey 'cost curves' are distorting REDD. Policy Brief. Rainforest Foundation Climate and Forests. http://redd-monitor.org/wp-content/uploads/2010/11/McRedd-English.pdf
Fancy, T. (2021). *The secret diary of a 'sustainable investor'*. Online blog. https://medium.com/@sosofancy/the-secret-diary-of-a-sustainable-investor-part-1-70b6987fa139
Food and Agriculture Organisation (FAO). (2017). *Does improved irrigation technology save water? A review of the evidence. Discussion paper on irrigation and sustainable water resources management in the Near East and North Africa*. FAO.
Horst, L. (1999). The failure of adjustable irrigation technology, the options for change and the consequences for research. *Agricultural Water Management, 40*(1), 101–105. https://doi.org/10.1016/S0378-3774(98)00111-5
Keulertz, M., & Allan, T. (2019). *The water–energy–food nexus in the MENA region: Securities of the future (Routledge Handbook on Middle East Security*. Routledge.
Lankford, B. (1992). The use of measured water flows in furrow irrigation management – A case study in Swaziland. *Irrigation and Drainage Systems, 6*(2), 113–128. https://doi.org/10.1007/BF01102972
Lankford, B., & Gowing, J. W. (1998). Participatory research of water control on surface irrigation systems: Informing perceptions. In L. S. Pereira & J. W. Gowing (Eds.), *Water and the Environment: Innovative Issues in Irrigation and Drainage* (pp. 223–230). E & FN Spon.
Lankford, B. (2004). Irrigation improvement projects in Tanzania; scale impacts and policy implications. *Water Policy, 6*(2), 89–102. https://doi.org/10.2166/wp.2004.0006
Lankford, B. (2012). Fictions, fractions, factorials and fractures; on the framing of irrigation efficiency. *Agricultural Water Management, 108*, 27–38. https://doi.org/10.1016/j.agwat.2011.08.010
Lankford, B. (2013). *Resource efficiency complexity and the commons: The paracommons and paradoxes of natural resource losses, wastes and wastages*. Routledge.
Lobina, E., & Hall, D. (2009). *Thinking inside the box: Why the World Bank is not learning*. University of Greenwich: Public Services International Research Unit (PSIRU). https://gala.gre.ac.uk/id/eprint/1727/1/2009-03-W-wbank.pdf
Markota Vukić, N., Vuković, R., & Calace, D. (2018). Non-financial reporting as a new trend in sustainability accounting. *Journal of Accounting and Management, 7*(2), 13–26. https://hrcak.srce.hr/194750
Meland, A. J. (2021). *Outsourcing institutional entrepreneurship to the world economic forum. Mono-glocalizing 'water security' with the world bank group & government*. University of Bremen.
Plusquellec, H., Burt, C., & Wolter, H. W. (1994). *Modern water control in irrigation: Concepts, issues, and applications*. World Bank. https://documents1.worldbank.org/curated/ru/100291468765601092/pdf/multi-page.pdf
Plusquellec, H. (2002). *How design, management and policy affect the performance of irrigation projects*. Report. Food and Agriculture Organisation, FAO Regional Office for Asia and the Pacific, Bangkok, Thailand. http://citeseerx.ist.psu.edu/viewdoc/download?doi=10.1.1.132.636&rep=rep1&type=pdf

Puy, A. (2018). Irrigated areas grow faster than the population. *Ecological Applications*, *28*(6), 1413–1419. https://doi.org/10.1002/eap.1743

Rudebeck, T. (2019). *Corporations as custodians of the public good? Exploring the intersection of corporate water stewardship and global water governance*. Springer.

Rudebeck, T. (2022). Framing water as a financial risk: Reviewing the processes shaping a narrative. *WIREs Water*, e1596. https://doi.org/10.1002/wat2.1596

Sullivan, S., & Hannis, M. (2017). "Mathematics maybe, but not money": On balance sheets, numbers and nature in ecological accounting. *Accounting, Auditing & Accountability Journal*, *30*(7), 1459–1480. https://doi.org/10.1108/AAAJ-06-2017-2963

Wichelns, D. (2011). Assessing water footprints will not be helpful in improving water management or ensuring food security. *International Journal of Water Resources Development*, *27*(3), 607–619. https://doi.org/10.1080/07900627.2011.597833

Farmers will save the world!

Brian Chatterton

Australian dryland farmers measure water-use efficiency using their rain gauge and come up with kilos of grain per mm of rain in the growing season. In Libya, they also introduced farmer-to-farmer technology transfer with rotations including annual medick to maintain the traditional integration of crops and livestock, instead of green revolution monocultures. All of which went against the academic and bureaucratic grain – although Tony Allan was a rare supporter. Things are more complicated in irrigated agriculture, but it suffers from just the same bureaucratic handicaps, in particular the rift between ministries and mindsets of water and agriculture.

Dryland farming

My wife, Lynne, and I first met Tony Allan about 40 years ago because of our shared interest in Libya. We were at a conference organized by the Libyan Studies Association in Sheffield and kept in touch. Some years later, we were discussing or perhaps even arguing fiercely about the misleading concept of the water footprint that had been developed by others from Tony's initial work on virtual water. It was, perhaps, this argument that led Tony to develop his theme that farmers are the ultimate managers of water. He was surprised to learn that Australian dryland farmers, like myself, had been using the water footprint for decades to measure our water-use efficiency.

No one had noticed because we used a much simpler way to measure it. Every farmer in Australia has a rain gauge and they assess their performance as water managers by the kilos of grain they harvest per mm of rainfall during the growing season. It measures their farming success in terms of weed control, fertilizer application, all other aspects of farm management and, of course, as dryland farmers, a substantial proportion was also luck – when the rain falls is as important as the amount. The concept seemed easier to understand than cubic metres of water per kilo of a food product.

This discussion might have been just a passing interest if it had not been for the Libyan link. South Australia, which enjoys a Mediterranean climate, had been deeply involved in the Jabal al Akhdar region in the north-east of the country, inland from Benghazi (see Salem Maiar, 2022, first part of this double special issue). We had replaced the Green Revolution cereal monoculture recommended by the Food and Agriculture Organization (FAO), the World Bank and other international development agencies with a rotation of wheat and legume (annual medick) pasture. These Libyan farmers were able to produce more wheat on half their arable than they had previously produced on all of it. The other half was sown to productive pasture for the sheep, which provided more income than the wheat. Many international water experts found this confusing and asked us why we did not sow all the land to wheat and increase yields too: they found it difficult to grasp the idea of sustainable farming where the pasture restored the land degraded by cereal production. In other words, they struggled to see the connection between farming constraints and water use.

The Libyans had blown up the orthodox development model. Not just modified it, but completely destroyed it. They had not set up the usual complex bureaucratic system of trickle down by employing Australian experts to come and talk with Libyan experts and then set up hundreds of trial plots. From this kind of work, the conventional model produces a recipe of seeding rates, seeding depths and a host of other detailed recommendations. But the agronomists lack the skills to implement them; they could not adjust the seeding-rate mechanisms on the seed drills or the tine pressure on the scarifiers. The Libyans had seen Australian farmers at work and decided to leapfrog this tortious and precarious path – not only because it takes so long and they were in a hurry but also because it usually fails to transfer the practical skills that make the new practices work. The agronomists produced the perfect recipe but, like cooking, you need technique as well. The South Australian farmers had a recipe. It might have been a rough at the edges but it was oven-ready; and they had the technique too and could take it directly to the Libyan farmers. They transformed not just food production but farmer prosperity. They increased the water productivity four-fold. Agronomists tidied up the loose edges afterwards.

I like to think that the Libyan example inspired other themes that Tony tried to popularize, such as under-priced food for under-paid people, and regenerative farming. In the world of water theory, water is converted into food. Many in the world of water policy think it is a pity that it is done by farmers: I am sure they would prefer robots because the farmers need an income to support themselves and their families. In the case of Libya, the government had subsidized the price of wheat to more than double the world price; and the price of local sheep breeds for meat was quite incredible by Australian standards – without any need for government support.

It is not surprising that the Libyan model has not been imitated. The bureaucrats and academics have fought back and won against a disorganized group of farmers who were doing this work because they enjoyed it as much as for the money they earned. The idea that technology should be transferred directly – farmer to farmer – would undermine the justification for their existence. They were deeply shocked that people without university degrees should play such an important role – and even worse, an independent role – in technology transfer. They determinedly opposed the concept and were triumphant.

Irrigation

Things are more complicated in irrigated farming (Lynne was bought up on an irrigated farm on the banks of the Murray River in South Australia, and worked as an irrigation farmer for many years). First, we have to discard the fake improvements in water-use efficiency from drip irrigation and other improved methods of delivering the water to the plants. That is not easy, as Australians have demonstrated in their muddled management of the Murray–Darling river system. The next difficulty is that irrigation farmers, like Lynne, did not want to increase their output. This may seem strange but they were growing fruit, grapes and vegetables where the monetary return and the output were often inversely related. It was easy to produce more grapes but they had less sugar and fetched a lower price. Bigger peaches were easy too, but the cannery rejected them as they did not fit their machinery.

Tony did not find this farming experience very relevant to the big water picture he was working on, but later, when Lynne and I (on Tony's recommendation) advised the Iranian government on water management, we became involved in the more widely grown irrigated field crops such as wheat, oil seeds and sugar beet. Tony saw this as vindication of his contention that farmers were the real managers of water: in contrast to the current water managers who are, in fact, managers of water delivery. He pleaded for farmers to be included in discussions on water management. His plea largely failed, which disappointed him greatly, but he realized that he was up against the academic and bureaucratic club which is tough for non-members to gate-crash. However, he managed to get Lynne and I into many conferences and seminars in disguise; he wanted us there as farmers but we were granted entry based on our credentials as a former Rural Policy Adviser to the South Australian Premier and as a former Minister of Agriculture in South Australia.

The bureaucratic division of water between ministries, which is almost universal, is not just frustrating. It is a fundamental cause of much of the current water crisis; it is worse than the cereal/livestock divisions in the Ministry of Agriculture. The big stuff is handled by the Ministry of Water; the trivia of actual water use is passed on to the Ministry of Agriculture; and they have a hard boundary – *the farm boundary*. The water people handle the big picture of the dams and the canals, and make a big splash in the media when they announce expensive infrastructure projects: the agriculture people work with farmers, unnoticed, converting this water into food in the most efficient manner. The two ministries do not cooperate – they fight – over boundaries, prestige and funding. Water usually wins. There is more political credit generated from opening a new dam than a 5% improvement in water-use efficiency.

Tony Allan's legacy

Bureaucratic divisions plague us in other areas too. Tony was not just supporting the role of farmers as water managers but, much more generally, as managers of the environment as a whole. He was a strong supporter of the regenerative farming movement that is currently squeezed in the middle of agricultural policymaking. The policy is divided into two territories with environment having an environment area and agriculture having an industrialized farming area for food production. The environmentally sustainable groundswell farmers, whom Tony strongly supported, fall between these two power grabs. They are farmers who want to produce food, but in an environmentally compatible manner. They do not want to become nature reserves or tourist theme parks – but the national agricultural policies do not allow for shades of grey between the rival ministries.

It is not for me to say what Tony Allan's international legacy will be, but what I saw him trying to do was to bridge the gap between academic studies of water and the real world of farming, and he was not afraid to break with the sanctioned discourse.

Disclosure statement

No potential conflict of interest was reported by the author(s).

Reference

Maiar, S. (2022). Professor Tony Allan and Libya. *Water International*, *47*(6), 887–889. https://doi.org/10.1080/02508060.2022.2118335.

Crossed wires: public regulation and private action for water stewardship and sustainable farming

Peter Newborne

Introduction

The rivers of England are in a mess. According to the House of Commons Environmental Audit Committee (2022, p. 5), amongst a 'chemical cocktail' of pollutants:

> farm slurry and fertiliser run off is choking rivers with damaging algal blooms. [...] The build-up of excess nutrients like phosphorous and nitrogen from animal waste and sewage is reducing oxygen levels in rivers and in severe cases can cause fish kills. [...] Biodiversity has not been priced adequately into economic decision making.

The committee notes that only 14% of rivers in England have good ecological status, and the UK is not on track to meet the European Union (EU) Water Framework Directive target (subsequently transposed into UK law) for all rivers to reach good status by 2027. The committee highlights agriculture and rural land management among the 'reasons for not achieving good status' (p. 12): there needs to be 'a step change in action' to tackle this problem of river water quality, including 'cross-catchment collaboration with farmers'.

As for water supply, Emma Howard Boyd, Chair of the Environment Agency, asserts:

> England's water is not as secure as people believe. Parts of the country are over-abstracted, which is requiring proposed changes to abstraction licenses. [...] During the Covid pandemic, few people were going on foreign holidays, so the usual 10% drop in demand from July to September didn't happen. Thousands of South East Water's customers found themselves without tap water. (Boyd, 2021)

She adds that a higher risk of drought, caused by hotter and drier summers, combined with a growing population, is 'an operational challenge for UK plc'. Lower river flows mean lesser dilution of pollutants.

Purpose of this study

That growing population must be fed. Agriculture consumes most of global fresh water and farmers make the day-to-day decisions about soil and water management, but they do not act alone – the food system includes food traders, manufacturers and processors; retailers selling to end-consumers; governments that set the legal and policy framework; and civil society organizations advocating and monitoring progress against environmental and social goals. Here we consider how farmers might be supported in producing food sustainably, which depends on how soil and water are managed – seeing that soil is the main reservoir and water regulator (Dent, 2019). Currently, primary producers receive only 8% of the value added in the entire UK food chain (Lang, 2020) and, while farmers can collectively influence political leaders, their individual bargaining position with food processors and retailers is weak except for big corporate farms (Allan & Dent, 2021).

The UK retailers considered here are: Tesco, Sainsbury's, Asda, Morrisons, Aldi, Co-op, Lidl, Waitrose, Iceland, Ocado and M&S; the first 10 account for more than 90% of UK food sales (UK Food Retail Statistics, 2021). These 11 retailers are the same as those studied by *Which?* magazine (2022),[1] which looked at three aspects of sustainability: greenhouse gas emissions, plastics use and food waste. It did *not* look at water use, nor deforestation or pollution, although those issues were noted as having 'big environmental impacts'. Water management is accordingly picked up here, in conjunction with support to farmers in their efforts to shift to more sustainable practices. James Rebanks, author of *English Pastoral – An Inheritance* (2020), calls this 'nature-friendly farming', otherwise known as *conservation agriculture* or *regenerative agriculture*. Conservation agriculture is described as 'a farming system that promotes minimum soil disturbance' (i.e., no tillage) and 'maintenance of a permanent soil cover, and diversification of plant species' (i.e., a diverse crop rotation or intercropping; Vanelph & Benites, 2001). Defining regenerative agriculture, John Cherry, co-creator of the Groundswell Regenerative Agriculture Show and Conference, says: 'It is any form of farming which, at the same time, improves the environment. This primarily means regenerating the soil. It's a direction of travel, not an absolute' (Cherry, 2021).

Farmers wanting to shift in this direction are looking for allies. Interviewed at his farm in the Lake District, Rebanks talks about the rewards that farmers would like to see:

> You need [. . .] wholesalers and retailers to [. . .] ask more, and more searching, questions of their suppliers. How are they producing this stuff? [. . .] You can't ask farmers [like me] to farm in a way that is nature-friendly and appropriate for this landscape and then put my stuff on the shelf without any help, next to the stuff from a totally unsustainable system. That's not going to work. (Rebanks, 2021)

Part of his inheritance has been building a grass-fed herd of beef cattle. Currently, at least two-thirds of the UK countryside, including highly productive arable land, services livestock production (Lang, 2020). A shift back to feeding cattle on grass and home-grown forage, together with a change in consumer diets to eat less meat, would soften the environmental impact of farming. A plant-based diet contributes only one-third of the greenhouse gas emissions of an average diet, and one-quarter of those from high-income countries (Springmann, 2019). Moreover, the water footprint of animal products is generally[2] greater than plant products; the water footprint of beef per gram of protein is six times greater than for pulses (Mekonnen & Hoekstra, 2010).

As defined by the Alliance for Water Stewardship (AWS) (2019), water stewardship brings together both individual and collective actions: 'Good water stewards understand their own water use, *catchment* context and shared [water-related] concerns and then engage in meaningful individual and *collective* actions' (emphasis added). In this context, the questions posed here are as follows:

- Do these major UK food retailers – and food processors and manufacturers who supply them – intend to use their position to support improvements in water management in the food supply chain, including sustainable soil and water stewardship by farmers?
- If so, how? What initiatives or activities are they engaged in to help farmers farm sustainably? In particular, what pricing or other contractual incentives are they offering?
- And why? What do the retailers say (if anything) about their motivations?
- What are the gaps in the regulatory system in terms of supporting the shift towards sustainability?

Scope and methodology

The first section of this article considers a case study highlighted by the House of Commons Environmental Audit Committee in its (2022) report on *Water Quality in Rivers* – the section relating to phosphate pollution from poultry production in the Wye Valley. The efforts of a range of actors to find a solution to the problem are described, as led by a local non-governmental organization (NGO), including the role of a food company, and a retailer, Tesco.

This subsequent section makes preliminary observations of the stances of the 11 UK food retailers. Each asserts aspiration to be a 'responsible' business – in other words, setting out its stall in terms of sustainability. But what is it that differentiates them in terms of:

- The way they are constituted as companies or other business entities?
- Their commercial niche and corporate culture?
- The language of sustainability they use?
- Their commitments relating to water management and sustainable farming practices?

For this review, the author read the reports and materials available on the websites of all 11 retailers without, in the time available, seeking to talk to representatives of those companies.[3] Interviews were conducted with a few farmers and representatives of farmers' groups, as well as NGOs. No scoring of the retailers is proposed here, including on how they are performing against their commitments. In the absence of data, that would be difficult – the *Which?* (2022) report noted that 'the supermarkets don't yet report comprehensively on all of these [water, deforestation and pollution] measures for the products they stock'.

Review

Experience in the Wye Valley

The Wye is the fourth-longest river in the UK, with a catchment of 4285 km^2. The river flows 250 km from its source in mid-Wales, through Hay-on-Wye, Hereford, Ross-on-Wye, Monmouth and Tintern, joining the Severn estuary below Chepstow. The last 26 km form the border between England and Wales. Both the upper and lower reaches include Sites of Special Scientific Interest (SSSI) for their geology and topography, flora, mammals, invertebrates, fish and birdlife; as well as a Special Area of Conservation (as an important wildlife corridor and breeding area for nationally and internationally important species); and much of the lower valley is an Area of Outstanding Natural Beauty (AONB). However, the river is overwhelmed by phosphate pollution. Efforts to combat this pollution have been led by an NGO, the Wye and Usk Foundation (WUF), working with other stakeholders.

The public authorities on both the English and Welsh sides of the valley have permitted the establishment of many poultry farms. Free-range chickens deposit their phosphate-rich excrement on the ground; chicken manure is also sold as fertilizer. Both sources contribute to pollution of the river, which has become increasingly severe as more and more poultry farms have been established. The associated algal blooms have been appearing further upstream earlier in the year and lasting longer, and the decomposing material takes oxygen out of the water, suffocating the fish. The river sometimes looks like 'pea soup' (McKie, 2020).

By mapping supply chains in the catchment, WUF identified Avara Foods as an important interlocutor, owning several poultry farms in the catchment and also processing poultry from other local farms. When Tesco proposed a major new contract for Avara's supply of chicken pieces, Avara made a substantial investment to increase production capacity at its Hereford processing plant by 30%.

WUF affirms that it is not against free-range farms. The issue is their location, how many and how they operate. It contends that planning permission should not be granted for farms in places where their runoff pollutes the river. An independent, peer-reviewed study led by the University of Lancaster[4] found that some 70% of the river's phosphorus inputs are of agricultural origin. The phosphorus surplus, 60% greater than the national average, is driven by locally produced manure[5] – more than the land can absorb, more than the grassland or crops need to grow, an excess of over 2000 tonne of phosphorus every year. Having previously denied that it was part of the problem, Avara responded to the authoritative independent study and the pressure exerted by stakeholders brought together by WUF: in October 2021, Avara's Head of Agriculture announced that the company accepted a level of responsibility (Hartland, 2021) and committed to a 40% cut in phosphates from its own and supplying farms by 2025 – which will require significant new infrastructure and logistics.

Local authorities and national environmental regulators have a legal responsibility to protect Special Areas of Conservation such as the Wye. The local planning authorities that permitted[6] the multiplication of poultry farms in the valley seem to have considered applications farm by farm, without taking into account their cumulative impact on the catchment as a whole.

As a national body, the Environment Agency coordinates the catchment-based approach called by its acronym, CaBA, in river catchments across England and cross-border with Wales, which aims to 'embed collaborative working at a river catchment scale, delivering a range of environmental, social and economic benefits and protecting our precious water environments for the benefit of us all' (CaBA, 2022). The CaBA programme provided annual funding to WUF to host the Wye catchment partnership.

According to its *regulatory* function, the Environment Agency is responsible for a wide range of environmental issues, including applying existing laws on water quality and water resources.[7] The Environmental Audit Committee concluded in its January 2022 report that the national Farming Rules for Water[8] were insufficiently enforced. These rules require farmers to plan and apply manures and fertilizers to meet, but not exceed, soil and crop nutrient levels.[9] The Chief Executive of the Environment Agency conceded that 'because of the reduction in our grant [from the UK Treasury . . .] we have been able to do fewer and fewer farm inspections over the last several years'. The committee also heard that many farmers were unaware of the Rules. It recommended that the Environment Agency with the responsible ministry, the Department for Environment, Food & Rural Affairs (Defra), focus more attention on the Rules, with more funding to support their application. The committee also referred to the new Environmental Land Management Scheme (ELMS),[10] which is designed to provide farmers with public funding for delivering environmental outcomes, including on water quality. (Note: at the time of proof-reading this article, ELMS has been suspended, pending a review by government.) The Committee noted (House of Commons Environmental Audit Committee, 2022, p. 121) that ELMS could be a source of financial help to farmers in assessing the existing phosphorus and nitrogen status of their land before spreading farmyard manure. Meanwhile, the committee recommended that the Environment Agency should intensify inspections (p. 48).

In the Wye catchment, Defra has now provided additional funding to the Environment Agency to carry out this compliance and enforcement, retrospectively strengthening regulatory environmental oversight. There remains, however, a disconnect between the two key parts of public regulation: the planning regime and environmental control are operating separately – and in the Wye case, those crossed wires were a primary cause of the problem.

The Wye Nutrient Management Board argues that the food industry should shoulder its share of responsibility for the state of the river and be proactive, for instance, in finding ways for farmers to earn their living other than by environmentally damaging practices. Avara has stated that instead of selling chicken manure as fertilizer, it will investigate its potential as a fuel source. Other businesses in the catchment also need to take action and WUF aims to push them and support them in changing farm practices to cut pollution. Simon Evans, Chief Executive of WUF, describes the process as 'de-risking' the agricultural supply chain, focusing on a key driver of farmers' and food processors' behaviour, namely contracts from buyers.

In this case, the identification of Avara as an interlocutor was significant: as both owner of poultry farms and poultry processor, it operates at two levels in the supply chain. The question arises: Could Avara have shifted its position earlier? From a purely commercial perspective, as a private for-profit company, Avara had invested substantially in free-range poultry, presumably looking to the positive image free-range has for many consumers. To take up the new contract with Tesco, Avara had increased capacity at its Hereford plant. Having done that, the company would not wish to incur the substantial cost of modifying the farm infrastructure and would be inclined to resist statements that it bore responsibility for polluting the river (it may not have believed them). Meanwhile, selling chicken manure to other farms was bringing in additional revenue.

Previously, however, Avara had signed up to the Courtauld initiative, expressing its interest in water stewardship according to the aim of the Courtauld initiative as then defined, namely that 'by 2025 all business signatories are participating in collective action to improve the quality and availability of water in key sourcing areas'.[11]

In Wye & Usk – one of the project areas identified by the Courtauld initiative as 'being critical areas for sourcing key foods such as fresh produce, dairy and crops', all 'suffering water stress' – the focus was on the problem of pollution from poultry farming. So, why did this not bring Avara to embrace its stewardship role sooner? The voluntary commitment by companies to the Courtauld goal brings with it an expectation that they will act according to considerations beyond profits for their shareholders. This kind of enlargement of corporate purposes is discussed further below, as is the uptake of water stewardship in the food supply chain. In the Wye case, the decisive event was the University of Lancaster report that closed the information gap. At this point, in the face of the pressure from WUF and others, Avara conceded its part of blame.

What of the role of the retailer? Tesco was also a signatory to the Courtauld initiative and had provided some initial funding to support WUF's work in Wye & Usk. What was Tesco's motive? Was its investment in WUF's capacity to identify problems and find a solution an indication that Tesco was looking to identify shared water challenges[12] in the catchment? Or, like Avara, did Tesco react retrospectively to bad publicity?

It seems unlikely that, without the work of WUF, the situation would be moving towards a solution. WUF has been an energetic leader for many years on the Wye pollution issue, including identifying the need for the scientific expertise of the University of Lancaster. Looking ahead to future prospects for a shift to more sustainable farming, the question follows: Are the leaders of catchment partnerships in other places going to have to work so long and hard to push food processors and retailers to move – rather than those key food-chain players actively demonstrating a voluntary shift towards sustainability? The following review of the 11 UK food retailers picks up this issue.

Sustainability stances of major UK food retailers

What, how and why are the retailers proposing in terms of support to improvements in water and soil management in the food supply chain, including help to farmers' efforts to shift to more sustainable practices? The same question is relevant to food processors who supply the retailers – for example, Avara Foods.

Constitutions

How far businesses have a responsibility to society is contested. Company lawyers in different jurisdictions as well as financial commentators[13] disagree on whether the purpose of a corporation is to serve the interests of shareholders to the *exclusion* of others (as long as the company operates within the law), or if company directors should, or can, take account of other stakeholders (Newborne, 2012). Under Section 172 sub-section (1) of the 2006 Companies Act which applies to companies registered in England,[14] company directors have a duty 'to promote the success of the company for the benefit of its members' [i.e., shareholders] and in doing so:

> have regard . . . to:
>
> (a) the likely consequences of any decision in the long term
> (b) the interests of the company's employees

(c) the need to foster the company's *business relationships with suppliers*, customers and others
(d) the impact of the company's operations on *the community and the environment*
(e) the desirability of the company maintaining a *reputation for high standards of business conduct*
(f) the need to act fairly as between members of the company. (emphasis added)

When consulting company reports and statements, the reader can try to ascertain how far company directors have regard to the community and the environment (as per S. 172 (1) (d) above) and at what level of *priority* – as well as to the other considerations in (a)–(f) above, especially (c) and (e). Section 172 (1) was drafted in such a way – the deliberate choice by the legislature of the words 'have regard' – as to allow directors to *choose* when (and whether) they put profits before those other purposes listed, including the social and environmental aspects (Newborne & Mason, 2012).

Of the 11 retailers, nine (Tesco, Sainsbury's, Asda, Morrisons, Aldi, Lidl, M&S, Iceland and Ocado) are limited companies in England; Tesco, M&S, Sainsbury's and Ocado are listed on the stock market, with those shares traded publicly. Companies may, however, opt under Section 172 sub-section (2) to re-define their corporate aims to *expressly* include other purposes (other than the benefit of the members/shareholders). They can, for example, opt to become Certified B Corporations (B Corps), which formalizes a company's alignment with the B Corp Movement's values and 'embeds a stakeholder-focused mindset that separates B Corps from other businesses': B Corp status is designed to exclude any company which only 'dabbles in a bit of corporate social responsibility, but instead to legitimise companies whose purpose is to have a material positive overall impact on society and the environment' (UK B Corporation Movement, 2022).

None of the limited companies in this list has chosen to redefine its constitution in this way – none has marked itself out as *specially* motivated by purposes other than profit.

As for Waitrose, the constitution of the John Lewis Partnership (of which it is part) is constructed differently; it is an 'employee-owned' business. As co-owners, the employees (called partners) have a share in the business. John Lewis states on its website: 'As we're not owned by shareholders, we can share this success with our Partners [. . .] customers, suppliers and communities.' To do that, the company must generate profits and, in good profit-making times, employees receive an annual bonus. The statement by the Chairman in the 2020/21 progress report[15] made it clear that John Lewis is positioning itself as a 'purpose-led and sustainable business'. To this end, its constitution, with its governance duty to the *internal* community of employees, may be of some assistance as compared with limited companies whose shares are owned by *external* shareholders (e.g., as publicly traded on stock markets). But the question remains, when and whether any bonuses will be accorded to *suppliers* (premium prices or other special contract terms) after the employees have received theirs.

The Co-op is owned by UK consumers and several other UK cooperatives, making the business a hybrid of a primary consumers' cooperative and a cooperative federation:

> We're one of the world's largest consumer co-operatives, owned by more than 5 million members' [. . .] and 'we manage our business for their benefit.

> The Co-operative Movement was founded to serve a *social* purpose, as well as a commercial one.
>
> We continue to be guided by the long-established co-operative values of self-help, self-responsibility, democracy, equality, equity and solidarity, alongside our commitment to ecological sustainability and social responsibility. (emphasis added)

This distinguishes the Co-op also from the limited companies.

Commercial niche and corporate culture

Based on the retailers' reports and statements online, the following are observations as to how the 11 retailers seek to portray themselves. There are clear differences in terms of niche in the food retail sector, reflecting corporate culture.

Asda identifies itself as a discounter: 'From our beginnings as the first UK discounter, offering our customers the very best value has been our driving force.' Lidl says it takes pride in providing its customers with the highest quality products 'at the lowest possible prices'. Likewise, Aldi states: 'Sourcing the best products for the very best price is at the heart of what we do'; adding: 'Our lower operating costs and the unique relationships we have with our suppliers means we can put top-quality products on our shelves for a price no one else can match.'

Both Lidl and Aldi offer substantially fewer items in a store as compared with the other retailers. The other retailers compete (according to what they say) on a slightly different basis as, for example, stated by Tesco: 'sourcing great quality' and 'affordable' products. Sainsbury's emphasizes its international reach, sourcing of 'over 12,000 products from over 70 countries'. Morrisons highlights its niche as both 'a food maker' and 'shop-keeper', which makes it 'different from competitors. Uniquely, we source and process half of the fresh food that we sell in our own manufacturing facilities and stores' which gives it 'close control over provenance and quality'. That would seem to indicate that Morrisons could, if it chose, apply that leverage to encourage its suppliers to shift their farming practices. Morrisons additionally reports its range of 'For Farmers' products launched in 2015 and, so:

> became the first supermarket to give part of the retail price directly back to farmers – £15m in 2019 and £18.5 in 2020. The range, which now includes milk, cream, cheese and butter, has so far given back £18.5m directly to British farmers. In 2020 we added eggs to our range with farmers receiving 1p for every 'For Farmers Chuckle Eggs' medium egg sold. The money raised is intended to support farmers to create better free-range farmland to create a healthier environment for happier hens, who naturally tend to lay more eggs.

So, a special pricing deal.

M&S underlines its customers' trust in the good quality of its products: 'We are proud to be known as a retailer that customers can trust and to sell products that are sourced and made with care.' Iceland is a specialist in frozen foods (though not exclusively). Ocado has its niche as an 'on-line smart platform'. As for Waitrose and the Co-op, each defines its culture in the terms of its mission. The Waitrose chairman describes its aspiration to become a leading sustainable business as *work in progress* – in other words depending on its future performance against its commitments. The Co-op, as noted above, aims to pursue its social purpose, as well as a commercial one.

The information available on corporate websites and at Companies House indicates strong representation of women in the leadership teams[16] at Waitrose and the Co-op, with a balance of women and men at M&S, Sainsbury's, Asda and Ocado, as contrasted with executive teams led predominantly by men at Iceland, Lidl, Aldi, Morrisons and Tesco. The latter five companies are, it seems, lagging behind what is, according to Deloitte's (2022) report on boardroom gender diversity,[17] the UK average of 30% of board seats held by women – with UK boardrooms generally lagging behind European counterparts. A separate study could seek to compare attitudes to sustainability of the 11 companies' leaders – female and male.[18]

Language

Alongside sustainability, the language employed by both Waitrose and the Co-op includes reference to ethics. With its 'Ethics and Sustainability Strategy', Waitrose 'takes an integrated approach to being a more sustainable, ethical business'. M&S refers to sourcing 'ethically' as well as 'responsibly' having positioned itself as a leader on sustainability over more than a decade through its 'Plan A' (recently reset on the basis that the company knows 'we need to do more'). Morrisons refers to buying from 'sustainable, ethical and responsible supply chains'. Iceland posits: 'If you act ethically and do the right thing, people will root for you.' For Ocado: 'enacting a culture of ethical procurement will be a key focus in 2022'. Some retailers (Sainsbury's, Aldi, Asda) prefer to refer to Fairtrade or principles of fairness, for example, Asda has made 'efforts to engage fairly with suppliers'.

The question arises what engaging 'fairly' means? In a market, prices are set by supply and demand: there is no such thing as a 'just price' (Sandell, 2009, p. 4). Lang concurs, in that (in his words) fairness is a 'social judgment'.

> True cost is about whether the price fully reflects the costs which have actually been incurred in creating a product or getting it to the consumer. Studies by environmental economists show that prices do not cover real costs. Price signals need to align more fully with health and the environment. (Lang, 2020, p. 208)

Some retailers highlight their British credentials. Lidl says it is committed to supporting British producers with 'up to two thirds of core products sourced from UK suppliers'. The Co-op states that all the fresh meat sold in its stores in the UK is from British producers. Aldi's website shows features on family farming businesses in the UK and the company says it is 'the market leader in the amount of fresh produce [it sells] which is British' including its 'entire core range of fresh meat, shell eggs and milk'. Morrisons is, it says, 'British farming's biggest customer', while Asda announced in 2020 its intention to:

> source a higher-proportion of products from UK-based suppliers, covering a range of products such as chicken, dairy, wheat and potatoes' – to reassure its customers that, despite the major US equity stake in the company, it is still committed to supporting British suppliers.

Where retailers devote space to stories of relationships with individual farmers, for example, with videos (on the Sainsbury's website), the question arises how far the company has advanced to planning systematically. Fully developed sustainability strategies will surely list specific commitments in defined timelines.

Commitments

What, then, are the retailers setting out as *targets* in relation to improvements in soil and water management in the food supply chain and support to farmers in their efforts to shift to more sustainable farming practices?

The retailers make commitments regarding water consumption in their own premises – a basic requirement of good corporate water management. For example, Sainsbury's says it will 'minimise use of water in its own operations'. Similarly, Asda will reduce water consumption in its own premises by fixing leaks and adding rainwater harvesting at some depots and 'wants to make efficient water use a fundamental focus in new stores'. Water stewardship, according to its accepted meaning, goes *further*, to take account of water challenges beyond the factory fence.

In the environmental, social and governance (ESG) prioritization exercise that Asda conducted with its customers and experts to identify the sustainability issues which they considered important, water stewardship (Asda employs the term) came out as the lowest of three categories in terms of importance to stakeholders and the capacity of Asda to support positive impact. Their priorities were carbon emissions (in operations and the supply chain), 'healthy food' and 'sustainable dietary choices', 'food and operational waste' and 'plastics'. Asda accordingly indicates a future date, 2025, for development of a water strategy. Similarly, Ocado set a future date for producing a strategy on water. Whether those future strategies distinguish the plans that the companies have beyond the factory fence will indicate the extent to which those retailers will expect particular businesses from which they source food products to engage in water stewardship. In those two cases, there is, evidently, a planning gap at present.

As a preliminary step, several retailers refer to analysis of water risks. Morrisons says it is mapping water scarcity in its own-brand supply chains, on which basis it will develop a water stewardship plan (the question remains as to what that plan will entail). Tesco reports it has analysed a number of its UK supply chains (it is not stated which) 'to understand their exposure to water risk, including their overlap with environmentally sensitive river basins'. Tesco goes on:

> Starting with some of our fresh produce supply chains, which rely heavily on natural sources of freshwater, we are putting in place improvement action plans. These include setting local reduction targets for water-stressed areas and establishing on-the-ground collaboration with other stakeholders.

The latter is, presumably, a reference to Tesco's participation in the Courtauld initiative, of which M&S, Asda, the Co-op and Sainsbury's are also partners. The water component of the Courtauld initiative is designed to promote collective action in recognition that 'Water supplies are at risk in many production regions, and water users need to work together to safeguard this shared resource'. The water targets under the Courtauld Commitment 2030, reflecting both water efficiency and broader water stewardship issues, are that:

> businesses will continue to increase water use efficiency in their own operations; and that, by 2030, the UK food & drink industry will have helped to achieve sustainable water management to improve the quality and availability of water at catchment scale in the top 20 most important product & ingredient sourcing areas in the UK and overseas.

The target for 2030 is that '50% of fresh food is sourced from areas with sustainable water management'. The focus in the UK is on East Anglia, the Medway, the Tamar and the Wye & Usk. That specific, time-bound target is a commitment entered into collectively by five of the 11 retailers (the detailed method of measurement may be observed as the project continues).

Waitrose's agricultural strategy aims to introduce 'regenerative farming techniques', including soil improvements and 'improving the water cycle'. Part of more sustainable farming practices – it is noted on the Waitrose website – will be a reduction of dietary reliance on meat, to embrace 'more plant based alternatives'. It adds significantly:

> Currently there are no clear plans for how livestock farmers can adapt to these production changes with existing business models. Food producers must urgently tackle these obstacles to truly nature-friendly food systems. Our Waitrose Agricultural Strategy will enable us to *start* facing a number of these challenges head-on and enable us to help create a more sustainable food system. (emphasis added).

The note of realism is useful. At the same time, the reference to existing business models seems to refer to those of the food producers. How Waitrose itself will help to create a different food system is not clear.

M&S similarly refers to the 'Farming with Nature' programme that it has developed 'with suppliers and external stakeholders' to support a 'nature-friendly farming' approach to sourcing its fresh British products, including helping 'our UK farmers become more resilient to the biggest environmental challenges – such as climate change, biodiversity, soil health and water usage'. In collaboration with the Wildlife Trusts and the Farming & Wildlife Advisory Group, M&S is working specifically with 17 Indicator and Innovation Farms: 'spanning fresh produce, dairy, beef, lamb, pork, turkey, chicken and eggs' to trial 'new ways of farming in harmony with nature and measuring environmental improvements and benefits over time'. M&S says that learnings will be shared with other M&S Select Farmers and the wider industry. The question arises: where will the 17 Indicator and Innovation Farms go in terms of water management? Sainsbury's talks about guidance to its customers in making food choices. It emphasizes the potential health benefits and positive environmental impacts of plant-based diets: '31% less GHG emissions; 34% less land use; 17% less water use'. It notes that 'our customers are all *on a journey* and seek our help as they tell us that it can sometimes be confusing to know what is both a healthy and sustainable choice' (emphasis added).

These statements of Waitrose, M&S and Sainsbury's are framed as general approaches that each intends to consider, rather than as specific targets. The support to farming referred to by the retailers also includes dialogue with farmers, training and trialling new farming methods:

- Tesco's Sustainable Farming Groups provide 'a forum to discuss sustainable production'.

- Morrisons announced in 2019 the creation of a fund for training farmers – supporting farm employers with the fees at local agricultural colleges for young people aged 19 or over.
- Morrisons and Aldi provide funding to the Prince's Farm Resilience Programme to 'help family farms improve their skills and knowledge of business and environmental management'.
- Asda is working with the Cambridge Institute for Sustainability Leadership on soil testing, 'identifying risks and exploring ways to improve current practices for better soil health'.

Asda is also supporting a project on water-sensitive farming in East Anglia:

> in 2019, focused on protecting water quality in the River Ouse by helping farmers to build earth barriers slowing the movement of water from their fields into the water course, reducing sediment runoff. Such interventions proved to be important as the River Ouse flooded extensively at the end of 2020.

Meanwhile, the Co-op and Iceland support projects on peatlands 'as one of the UK's key ecosystems making up around 12% of the UK's land area' (Co-op) and in Wales in collaboration with the Wildlife Trusts (Iceland).

Tesco, in collaboration with WWF-UK, has devised an innovative approach, expressed in a specific, time-bound target (by 2030), to 'halve the environmental impact of the average UK food shopping basket'. The 'Sustainable Basket Metric' will measure seven key environmental issues in the food system:

> 'Climate change' (weighted at 25%), 'Deforestation' (20%), 'Sustainable diets' (15%), 'Sustainable agriculture' (12%), 'Marine sustainability' (10%), 'Food waste' (10%) and 'Packaging waste' (8%).

Water features alongside soil health and biodiversity as one of four '*sub*-metrics' under the sustainable agriculture criterion (at 4% of the total weighting) to be measured according to the 'percentage of fresh produce from regions with sustainable water management'. Tesco accepts that the most effective means to understand whether water resources are being managed sustainably is to assess their use above and beyond individual farms or sites. The question then becomes: for these purposes, what is a 'region' with sustainable water management? For the water metric to be meaningful, the company will need to take account of water volumes which it sources at the level of urban municipalities or rural counties or catchments, not provinces or whole river basins. In relation to water, the end result of Tesco's Sustainable Basket Metric is similar to that of Asda's ESG prioritization exercise – a very low level of priority. However, the combination of Sustainable Agriculture and Sustainable Diets, weighted at 27% together, does put food system issues on the company radar. And agriculture is a major emitter of greenhouse gases, which brings the expansion of nature-friendly farming practices into an important position in this composite index.

Indeed, most of the retailers in this selection set specific, time-bound targets for carbon emissions:

- Waitrose: net zero greenhouse gas emissions from its UK farms net zero by 2035.

- Co-op: net zero in terms of carbon emissions from its products by 2040.
- M&S: net zero across its entire value supply chain by 2040.
- Tesco: net zero for emissions in its supply chain and products by 2050.
- Morrisons: net zero agriculture by 2030.
- Iceland: net zero for all products sold in the UK by 2040.

How emissions will be measured in farms and products across those supply chains is not discussed in the documents consulted during this review.

As for the standards that retailers look to farmers to apply, the LEAF (Linking Environment and Farming) Marque is most commonly cited, for example, by Waitrose, Tesco, Co-op, M&S, Sainsbury's and Lidl. The section of LEAF relating to water focuses on water-use efficiency within *farm units*: it does not consider what is happening beyond the farm boundary. Neither catchment, nor watershed, nor river basin is mentioned in the 52 pages of text on the criteria against which compliance with LEAF is verified. That makes LEAF a farm-assurance tool, not a water–system–assurance tool. A separate LEAF advisory document includes reference to catchment-sensitive farming (p. 16), but that does not form part of the control criteria of the LEAF Standard.[19]

Perry (2018, 2019) describes how technical improvements in irrigation efficiency at the farm level often *increase* water consumption. He argues compellingly for a combination of water pricing *and* rationing 'which farmers may resist and their elected representatives like to avoid' (Perry, 2019, p. 87). But 'quotas will inevitably be required to ensure that demand is constrained to sustainable levels', and those quotas must be based on sound water *accounting* clarifying how much water is available, with what reliability, and where that water goes. That is the equivalent, in relation to water quantity, of provision of clear, authoritative information on water quality – as highlighted on the Wye.

The AWS Standard 'stakeholder-inclusive' process of water stewardship (AWS, 2019, p. 4) recognizes the role of public bodies as well as private businesses. As part of gathering information on what is happening in the relevant catchment, Step 1 (criterion 5) requires water users to identify the existence of 'water-related public policies and major publicly-led initiatives' so as to inform themselves of 'possible opportunities for water stewardship collective action' (AWS, 2019, p. 10). The 11 retailers are not currently proposing to farmers or suppliers that they apply the AWS Standard, which suggests they are not taking account of water risks in the supply chain in the rigorous manner that the AWS process identifies. The broader catchment vision incorporated in the AWS Standard includes companies being aware of upstream dependencies (sources of supply and factors that threaten to disrupt those supplies) as well as downstream water-related impacts (e.g., polluting discharges). The case of Avara Foods in the Wye, with its substantial presence on the ground (as compared with disparately distributed retailers' outlets), shows what could have been achieved, sooner, had the company followed the AWS Standard.

What support may retailers provide? Information from AWS[20] shows that 85% of companies that have registered and certified sites under the AWS Standard are applying the Standard to sites that they own. Companies cannot exercise the same control over sites owned and operated by others and only a minority of companies are currently taking on the additional challenge of supporting the implementation of the AWS

Standard down the food supply chain. AWS suggests that the next stage of uptake of water stewardship in the food & beverage sector can usefully focus on geographical hubs where suppliers of particular categories of product are located, with a pooling of effort and financial investment in those locations. Retailers and food processors are supplied with fresh fruit and vegetables from sources that are clustered and can be identified by mapping of supply chains.

The issue then becomes how that combination of funding and effort is to be organized. One model is the Courtauld initiative, coordinated by an NGO, WRAP, which leads projects within the UK and beyond. The Courtauld-supported project in the Wye valley receives funding from both the private companies that are partners in the Courtauld Initiative, and the Environment Agency as a public agency.

Part of wholesalers and retailers helping farmers to farm in a nature-friendly way will be offering good prices and contracts for pasture-fed beef. The organization Pasture for Life[21] promotes the 'restorative power of grazing animals on pasture', offers certification to help consumers identify meat and dairy produced from grass, and aims to help farmers identify routes to market. It emphasizes the need for clear and accurate labelling to distinguish pasture-fed from grain-fed, and the business niche requires the ability to hold to the pasture-fed sourcing. Its website shows butchers' shops, farmers' markets and farm shops in the UK that are selling beef and lamb purely from pasture – the list of which, it notes, is growing. A farmer interviewed for this article tells that 'a *high-end* wholesaler supplying restaurants and hotels is paying a premium for [their farm's] wholly pasture-fed beef' (emphasis added). Pasture for Life further reports that 'a retailer[22] has shown interest in buying certified pasture-fed beef as a premium offering to its customers'. The uplift could be as much as £1 per kg on hooks (as opposed to on the hoof). The retailer is question is, similarly, *high-end* which suggests that there is room for the pasture-fed product at the more affluent end of the market. The challenge – the farmer reported – will be for the producers to guarantee 'a sufficient number of carcasses per week to meet the demand indicated by the retailer'.

Conclusions

Farmers have a responsibility for sustainable soil and water management but they need support. The question posed here is whether major UK food retailers intend to use their power to contribute to that support. On the basis of the information gathered here, it is not clear that they do. They say they want to be environmentally responsible but, for nine out of the 11, their position is, fundamentally, a function of their mission (as set out in their corporate constitutions) as well as their niche in the retail market. The duty of directors under the conventional form of limited company in England, combined with the commercial culture of low prices for their customers, puts shareholders and end-consumers ahead of suppliers, including farmers.

The two exceptions are Waitrose (as part of the John Lewis Partnership) and the Co-op, each with its form of organization different from the limited company. The constitution of John Lewis, with its governance focus on the internal community of employees, will presumably afford the directors of Waitrose room for manoeuvre in deciding whether any premium prices are granted to suppliers – after bonuses paid to the employees. The Co-op group is also a special case, guided by cooperative values.

Another level of prioritization by the retailers is the choice of which sustainability issues to address. When they look to their environmental impact, the retailers first think of carbon emissions: water issues are towards the foot of the list. When they think water, they think of the *efficiency* of use in their stores and warehouses and their supplier farms, without looking beyond to what the aggregate of individual farms' actions means for the amount and quality of water in the catchment. Compared with their targets for reduction of carbon emissions in the supply chain, the language employed to express the undertakings relating to water stewardship (beyond own operations) and sustainable farming practices lacks specific, time-bound commitments. The retailers seem to be uncertain as to what to do, choosing activities and circumstances that they can control, and which are visible (presumably good for public relations). The Courtauld initiative with its specific and time-bound water target is an exception. That it is a group undertaking may allow those retailers a feeling that they have some individual room for manoeuvre. Future progress towards the 2030 Courtauld target will reveal their level of commitment.

As for food processers, Avara Foods is a signatory to the Courtauld initiative. After some delay, the company accepted its part of responsibility for phosphate pollution in the Wye valley. A key test of corporate commitment to water stewardship is the readiness of businesses to *anticipate* water problems. The first step in the Alliance for Water Stewardship Standard is for aspiring water stewards to gather information to enable them to understand water-related challenges in the relevant catchment(s), including identifying the existence of 'water-related public policies and major publicly-led initiatives' that may reveal opportunities for linking public initiatives and private action.

The Environment Agency, as regulator, has been inadequately funded to do its job of inspecting farms for compliance with the Farming Rules for Water: an example of a government preference for soft or self-regulation (Lang, 2020). That put the onus on voluntary action by private businesses – which have been slow to respond – a case of crossed wires between regulation and incentives. For their part, farmers are looking to see whether the Environmental Land Management Scheme will be implemented to provide funding for improving soils and water management – on terms they can handle.

Another instance of crossed wires is the planning regime exercised by local authorities, which is operating separately from the (national) Environment Agency – a key cause of the problem which arose in the River Wye. Experience in other jurisdictions points to the need to firmly anchor environmental goals so that they are not undermined (e.g., Beasley, 2021, writing about the Murray–Darling basin in Australia). It should not be up to civil society organizations like the Wye and Usk Foundation – charities with an environmental mission – to shoulder such a large part of the task of defending the public interest.

The retailers would be well advised to show leadership by supporting suppliers like Avara in adopting the AWS Standard. That could be part of the next stage of water stewardship which AWS proposes with a focus on key sourcing locations as geographical hubs for pooling effort and financial support.

As for contractual incentives for farmers to farm in a nature-friendly manner, the goal of alignment of price signals with environment – as advocated by Lang (2020) – seems remote, at least on the basis of this preliminary review. In the competitive UK retail market, let alone the current cost-of-living crisis, how much room is there for increasing the prices paid to farmers? As in the US and Canada (Beck, 2019), the promotion of sustainable healthy food in the UK is, currently, a niche market of relatively affluent consumers, rather than a challenge to the structure of the UK food system.

Notes

1. The *Which*? study collected data 'both from their published reports and by asking the companies directly'.
2. The precise footprint of a piece of beef depends on the production system and the composition and origin of the stockfeed (The Water Footprint Network n.d.).
3. All quotes from the 11 retailers in this article are taken from their websites (in alphabetical order):
   Aldi: www.aldi.co.uk/about-aldi; www.aldi.co.uk/corporate-responsibility#fairer
   Asda: www.asda.com; http//:corporate.asda.com/newsroom/2021/05/06/asda-launches-first-esg-report
   Co-op: www.coop.co.uk; www.coop.co.uk/climate
   Iceland: http://about.iceland.co.uk; http://about.iceland.co.uk/doing-it-right/
   Lidl: www.lidl.co.uk; http://corporate.lidl.co.uk/sustainability?_ga=2.47549671.1100339008.1645025232-1224021255.1645025232
   M&S: http://corporate.marksandspencer.com; http://corporate.marksandspencer.com/sustainability
   Morrisons: www.morrisons-corporate.com/about-us/; www.morrisons-corporate.com/morrisons-sustainability/planet/
   Ocado: ocadoretail.com; www.ocadogroup.com/our-responsible-business/corporate-responsibility
   Sainsbury's: http//:about.sainsburys.co.uk/sustainability; http://about.sainsburys.co.uk/~/media/Files/S/Sainsburys/CRS%20Policies%20and%20Reports/Plan%20For%20Better%20Report%20-%202021-22%20Half%20one.pdf
   Tesco: www.tesco.com; www.tescoplc.com/blog/tesco-little-helps-plan-corporate-responsibility/
   Waitrose: www.johnlewispartnership.co.uk; www.johnlewispartnership.co.uk/csr/reporting/progress-report.html
4. As part of the RePhOKUs project, Resilience Phosphorus UK – Re-focusing Phosphorus Use in the UK Food System at Lancaster University. Evidence submitted to the House of Commons Environmental Audit Committee, http://committees.parliament.uk/writtenevidence/40668/pdf/.
5. According to the evidence presented to the Environmental Audit Committee by Professor Paul Withers of Lancaster University (House of Commons Environmental Audit Committee 2022, p. 36).
6. For example, in Powys, from July 2015 to June 2018, there were, according to the Campaign for the Protection of Rural Wales, 116 applications for poultry units, including 83 for free-range eggs, with only two being refused.
7. As well as regulation of major industry and waste, treatment of contaminated land, fisheries, river, estuary and harbour navigation, and conservation and ecology. The agency is also responsible for managing the risk of flooding.
8. Rules for farmers and land managers to prevent water pollution (www.gov.uk).
9. The rules also include minimum storage distances of organic manures from water bodies and minimum distances for the application of organic manures and manufactured fertilizers, and to require farmers to assess weather and soil conditions to reduce the risk of runoff and soil erosion. The rules additionally require farmers to manage livestock by protecting land within 5 m of water.
10. Environmental Land Management schemes: overview – GOV.UK, March 2021

11. The aim of the Courtauld initiative, as re-defined, is expressed in the Courtauld Commitment 2030, http://wrap.org.uk/taking-action/food-drink/initiatives/courtauld-commitment.
12. Identifying 'shared water challenges' is part of the process undertaken by water users who apply the AWS Standard. Under Step 1.6, aspiring water stewards 'understand current (and future) shared water challenges in the catchment' by following the stakeholder-inclusive process of the Standard.
13. For example, Lex Column, *Financial Times*, 25 May 2013: 'Companies exist to please the people who own them [i.e., the shareholders]. The freedom to be as avaricious as possible within the law keeps companies hard, fast and efficient, which is how we like them.'
14. In the case of Aldi and Lidl, the English companies are part of corporate groups in Germany. Asda is an English company in which the US retailer, Walmart, retains an equity stake.
15. Chairmans-statement.pdf (johnlewispartnership.co.uk)
16. Those with day-to-day leadership responsibilities such as, for example, the executive/highest level of management at the Co-op at four women to one man, as compared with the Board of the Co-op which has five non-executive directors out of a total of 13 board members of which (according to the company website, consulted in April 2022) six are women and seven are men.
17. Women in the Boardroom: A global perspective published in February 2022, according to which, women leaders on the continent boast the largest share of boardroom seats, led by France (43%), Norway (42%) and Italy (36%).
18. With each also representing and expressing her/his own cultural, ethnic, social and socio-economic perspective.
19. In relation to the risk of polluting discharges from farms into watercourses, LEAF requires farmers to look beyond the farm fence, for example, farm workers are instructed not to spray/mix pesticides and other chemicals close to water bodies and similarly farms should manage slurry/silage storage to avoid pollution (as per the Farming Rules for Water).
20. The presentation by Scott McCready of AWS at the AWS Forum on 19 May 2022.
21. Pasture for Life, http://www.pastureforlife.org/.
22. At the time of writing, the name of the retailer cannot be disclosed.

Acknowledgements

With thanks to Tony Allan for his inspirational leadership and generous support.

Disclosure statement

At the time of writing, the author was contracted by AWS as a co-trainer for training sessions introducing the AWS Standard to businesses.

References

Allan, J. A., Dent, D. L. (2021). The cost of food: Consequences of not valuing soil and water and the people who manage them. In D. L. Dent & B. P. Boincean (Eds.), *Regenerative agriculture* (pp. 3–19). Springer Nature Switzerland.

AWS. (2019). *The AWS international water stewardship standard*, (version 2.0), March. Edinburgh: Alliance for Water Stewardship. https://a4ws.org/

Beasley, R. (2021). *Dead in the water: A very angry book about our greatest environmental catastrophe: The death of the Murray–Darling Basin*. Allen & Unwin.

Beck, T. (2019). Disruptive food supply chains: Bringing it all back home. In J. A. Allan, B. Bromwich, M. Keulertz, & A. Colman (Eds.), *The Oxford handbook of food, water and society* (pp. 827–841). Oxford University Press.

Boyd, E. H. & Environment Agency. (2021). Water security in a changing environment [Speech], 30 June (https://www.gov.uk/government/speeches/water-security-in-a-changing-environment).
CaBA. (2022). Catchment-based approach in England. Website. http://www.catchmentbasedapproach.org/
Cherry, J. (2021). What is regenerative agriculture? Weston Park Farms. https://www.westonparkfarms.co.uk/post/what-is-regenerative-agriculture
Deloitte (2022). Women in the boardroom: A global perspective. Report, 7th edn. February 2022. https://www2.deloitte.com/global/en/pages/risk/articles/women-in-the-boardroom-seventh-edition.html
Dent, D. L. (2019). *Green Water* used by plants and managed by farmers. 29–44. In J. A. Allan et al (Ed.), *The Oxford handbook of food, water and society.* Oxford University Press.
Hartland, N. (2021). Company 'admits' chicken manure is polluting Wye. *The Ross Gazette*, 22 October. https://www.rossgazette.com/news/company-admits-chickenmanure-is-polluting-wye-325131/.
House of Commons Environmental Audit Committee. (2022). *Water quality in rivers.* Fourth report of session 2021–22, Water Quality in Rivers (parliament.uk).
Lang, T. (2020). *Feeding Britain: Our food problems and how to fix them.* Pelican Books.
McKie, R. (2020). 'It's like pea soup': poultry farms turn Wye into wildlife deathtrap. The Guardian, 20 June. https://www.theguardian.com/environment/2020/jun/20/its-like-pea-soup-poultry-farms-turn-wye-into-wildlife-death-trap/.
Mekonnen, M. K., & Hoekstra, A. Y. (2010). *The green, blue and grey water footprint of farm animals and animal products.* Vol 1, Main Report for UNESCO-IHE, December 2010. UNESCO/IHE Institute for Water Education. Report-48-WaterFootprint-AnimalProducts-Vol1.doc. https://cdn.odi.org/media/documents/8099.pdf
Newborne, P. (2012). *Private sector investment in water management – Company forms and partnership models for inclusive development.* Overseas Development Institute. ODI Briefing Paper 78.
Newborne, P., & Mason, N. (2012). The private sector's contribution to water management: Re-examining corporate purposes and company roles. *Water Alternatives*, *5*(3), 603–618. http://www.water-alternatives.org/index.php/alldoc/articles/vol5/v5issue3/188-a5-3-4/file
Perry, C. (2018). *Improving irrigation management in conditions of scarcity: Myth vs truth.* Improving irrigation management in conditions of scarcity: Myth vs Truth. Global Water Forum.
Perry, C. (2019). *Will irrigation technology, pricing, or quotas ensure sustainable water use*?. In J. A. Allan, B. Bromwich, M. Keulertz, & A. Colman (Eds.), *The Oxford handbook of food, water and society* (pp. 76–96). OUP.
Rebanks, J. (2020). *English pastoral: An inheritance.* Allen Lane.
Rebanks, J. (2021). *Nature-friendly farming: conversation with James Rebanks.* Short 10-minute film shot at the Rebanks farm in the Lake District, UK, in October 2021, filmed and edited by Gardar Thor Thorkelsson and produced by Peter Newborne, as part of the Environment Agency project: *Building collaboration in catchments using system-based approach.* CaBA.
Sandell, M. J. (2009). *Justice – What's the right thing to do?* Penguin Books.
Springmann, M. (2019). *Modelling report on alternative production and consumption scenarios for the UK.* Report to the Food, Farming and Countryside Commission. Oxford Martin School
UK B Corporation Movement. (2022). Website. https://bcorporation.uk/
UK Food Retail Statistics 2021: UK Food Retail Statistics 2021.
Vanelph, S., & Benites, J. (2001). *First world congress on conservation agriculture: A world-wide challenge: Report.* FAO/ECAF.
The Water Footprint Network (n.d.). *What is a water footprint?* (https://waterfootprint.org/en/water-footprint/what-is-water-footprint/).
Which? (2022). *News.* February.

How decisions are made by politicians through the advocacy of peer reviewed research: the lens of advocacy coalition theory

Anthony J. Colman

Tony Allan was a leading academic in this field and I was privileged to work with him in the period 2014–21, as both a water academic and a UK politician. A Luddite Norfolk farmer I met in September 2021 said to me: 'How did the change to ELMs [environmental land management] happen?' I answered: 'Tony Allan and his friends.'

Science and belief

An eminent water academic, on his retirement in 2012, commented to me that he wished he had made sure his research had been delivered direct to the politicians, rather than being left unpublished through peer-review pressure, or watered down in academic journals that never made their way to those who could take action. But politicians want certainty. The Danish Environment Minister at COP15 asked her climate academics for certainty: they gave her models, likelihoods, but no certainty!

So, how do academics influence politicians to change course and act on what may be good research outcomes, but yet are not conclusive? What is the process of change in decisions about water governance and other realms of natural capital? What lens should we use? In part, political economy and in part, for water, political ecology. Ultimately, who has voice and agency to present, with support, a coherent policy to the decision-makers could lie in the mechanisms of Advocacy Coalition Theory (ACT) (Sabatier & Jenkins-Smith, 1993; Sabatier & Weible, 2019).

ACT codifies this complex policy–consensus–formation process into a system-based approach, with power moving across, top-down and bottom-up. It understands the black box of decision-making as being explained by changes in the beliefs of the participants over a period of time. Underlying ACT are three belief levels: (1) *deep core beliefs*, the normal or standard attitude and behaviour of society; (2) *policy core beliefs*, which can be changed by evidence and can lead to coalition formation; and (3) *secondary beliefs*, narrower and subject to change over time, leading to empirical fine tuning of reforms (Sabatier, 1991). From these beliefs, come the subsystems to support change to a new coalition in support of a policy.

Deep core beliefs are those that have long been used to provide an overarching narrative for 'intractable difficult decisions, such as budget decisions or environmental stand-offs, where there is no consensual advocacy coalition [. . .] to provide a solution to the potentially intractable problem'. This narrative can come from 'stories from individuals or groups, which together make a way forward for the policy makers, which has broad support from the community' (Roe, 1991, p. 296). Such a story-telling approach

> brings the social sciences and humanities together to address real-world problems of public policy – particularly issues characterized by extreme uncertainty, complexity, and polarization – such as climate change which, if not more effectively managed now, will plague us well into the next century. (Roe, 1991, p. 296)

Moving opinions

The presentation of ACT has evolved beyond the simple three-belief system to involve linear movement of opinion (and feedback, when new evidence comes forward). Figure 1 shows in dynamic form how advocacy coalitions are formed and can be re-formed. The left-hand side, named *Relatively Stable Parameters*, represents the deep core belief systems (1). Here, this moves to the right to establish *Long-term Coalition Opportunity Structures* that, in turn, feed into the *Policy Subsystem*. This latter right-hand decision box explains the movements between coalitions A and B arising from potentially new,

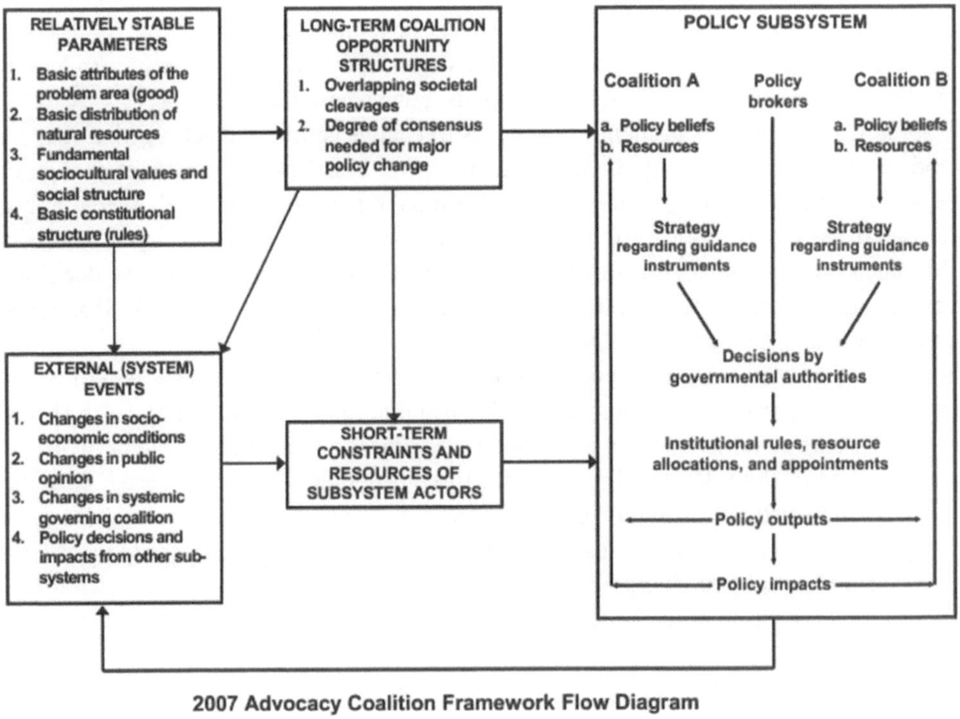

Figure 1. The advocacy coalition framework. Source: Sabatier & Weible (2019), reproduced with permission.

evidence-based, changeable policy core beliefs (2) mediated by policy brokers, the experts who have access to the secondary beliefs (3). The feedback loop from the *Policy subsystem* box, bottom left, takes account of *External (System) Events* that could impact on the *Relatively Stable Parameters* (1), subject to *Short-term Constraints* that could lead to the formation of new coalitions.

The logic assumes 'well organised interest groups, mission-oriented agencies, weak political parties, multiple decision-making venues, and the need for super-majorities to enact and implement policy change' (Sabatier & Weible, 2019, p. 199; see also Weible & Ingold, 2018) – seen as applicable to situations in the United States but not necessarily elsewhere (Weible & Ingold, 2018). Therefore, a new category of variables, known as *Long-term coalition opportunity structures* was developed referring to factors such as resource constraints that affect the behaviour of advocacy coalitions (Weible & Ingold, 2018, p. 3). This revision of ACT contained two additional sets of variables: the degree of consensus needed for major policy change, and the openness of the relevant political system.

Outside the United States, Westminster Rules adopted by the 54 Commonwealth democracies require a simple majority for change. However, the development of parliamentary select committees, mirroring the decision-making ministries, has enabled coalition-building for change and allows academics and non-governmental organizations (NGOs) to directly influence and shape legislation.[1]

ACT has been used to analyse water reform processes in a number of cases. A key example is in Denver in the United States, where ACT explains the positions of federal, state and local institutions in Colorado and their movement towards a policy on water for the city region (Ellison, 1998). After acrimony in the local community, the Lake Tahoe water quality analysis saw, through an ACT lens, the coming together of a consensual coalition on the use of the lake water (Weible, 2007). The 2021 difficulties with the allocation policy for Colorado River water was presaged by the work of an advocacy coalition in 2014 (Kendy et al., 2017). A Ghanaian example of the use of ACT explains the evolution of a common, agreed policy on private sector participation in water delivery where, previously, there had been deep divisions over the way forward (Ainuson, 2009). In Botswana, ACT can be used to analyse the movement to universal provision of clean drinking water (Colman, 2013).

Research in Spain shows a water-policy advocacy coalition to have replaced a hydrological paradigm with an environmentally based AC of politicians together with civil society organizations working through a democratic process, creating a new coalition based on the recognized need for water reform (Bukowski, 2007). Moves of the Chinese government towards a renewable-energy and climate-change policy after 2007 have been seen as a change that can be explained by an advocacy coalition (Stensdal, 2012). The evolution of climate policy in the United States, despite the Donald Trump negation 2017–21, was built around an AC (Kukkonen et al., 2017) which provided broad support for President Biden to stay in the United Nations Framework Convention on Climate Change (UNFCCC) Paris Agreement. Likewise, the push back on fracking for fossil fuels was an advocacy coalition approach (Ingold et al., 2017).

There are many other examples of ACT in action; the most recent has been, and still is, the issue of soil and water health, where big farming and the agrochemical industry are pitted against the knowledge and deep beliefs of farmers and academics espousing conservation agriculture or regenerative agriculture. Tony was a leading academic in this field and I was privileged to work with him between 2014 and 2021, as both a water academic and UK politician, in expanding his ideas in meetings and seminars to allcomers across Europe and the United States. His manifesto (Allan, 2019) was the summation of the work of academics, NGOs, private-sector experts and, most of all, farmers – whom he correctly saw as leading the move to a new coalition of belief. The success of his work was in the influence he had with UK Department for Environment, Food and Rural Affairs (DEFRA) with its move from a paradigm of the inherited European Union's Common Agricultural Policy (CAP), largely paid to farmers under the majority Pillar One according to size of the farm to one based primarily on environmental land management (ELM) embracing action to moderate droughts and floods, reverse carbon emissions, and maintain biodiversity. The Conservative Party Manifesto for the December 2019 UK General Election, won with a big majority, supported the move. The full endorsement of this change by the UK Cabinet Minister and Secretary of State for DEFRA, RH George Eustice, came in a speech at the Groundswell Farmers' Conference in June 2021, a month after Tony's death[2] and showed his success in being a key part of a successful advocacy coalition for change.

Conclusions

ACT can be seen as a way for coalitions of policymakers outside government to make it safe for politicians to endorse a changed paradigm with new scientific information underpinned by returning to deep core beliefs. In the case of the management of soil and water, the recipe for creating a coalition for change is to have science and scholarship-learning from the deep core beliefs of those involved, in our case farmers and custodians of the soil and water; then to test the work against audiences of stakeholders across the world; and, finally, to work with politicians and their advisors of all hues to get political change.

Notes

1. A Procedure Select Committee, set up in 1976, recommended this in 1978. Premier Margaret Thatcher introduced the innovation in 1980 and it was adopted in all Commonwealth countries.
2. See https://www.youtube.com/watch?v=Lf8btbyJo0k&list=PL5ZXKRVhhiD1WZ92D8rJesmw7smI2G5It&index=4/.

Disclosure statement

No potential conflict of interest was reported by the author.

References

Ainuson, K. (2009). An advocacy coalition approach to water policy change in Ghana: A look at belief systems and policy-oriented learning. *Journal of African Studies and Development*, *1*(2), 16–27.

Allan, J. A. (2019). Food, water and the consequences of society not valuing the environment. In J. A. Allan, B. Bromwich, M. Keulertz, & A. J. Colman (Eds.), *The Oxford handbook on food, water and society* (pp. 859–878). Oxford University Press.

Bukowski, J. (2007). Spanish water policy and the national hydrological plan: An advocacy coalition approach to policy change. *South European Society and Politics*, *12*(1), 39–57. https://doi.org/10.1080/13608740601151673

Colman, A. J. (2013). *The political economy and coalitions in Botswana's water sector reform 2009–13: To what extent can the process of reform be understood?* PhD thesis. University of East Anglia: Norwich. Retrieved August 17, 2021, from https://ueaeprints.uea.ac.uk/4876

Ellison, B. (1998). Intergovernmental relations and the advocacy coalition framework: The operation of federalism in Denver water politics. *Publius: The Journal of Federalism*, *28*(4), 35. https://doi.org/10.1093/oxfordjournals.pubjof.a030000

Ingold, K., Fischer, M., & Cairney, P. (2017). Drivers for policy agreement in nascent subsystems: An application of the advocacy coalition framework to fracking policy in Switzerland and the UK. *Policy Studies Journal*, *45*(3), 442–463. https://doi.org/10.1111/psj.12173

Kendy, E., Flessa, K. W., Schlatter, K. J., de la Parra, C. A., Hinojosa Huerta, O. M., Carrillo-Guerrero, Y. K., & Guillen, E. (2017). Leveraging environmental flows to reform water management policy: Lessons learned from the 2014 Colorado River Delta pulse flow. *Ecological Engineering*, *106*, 683–694. https://doi.org/10.1016/j.ecoleng.2017.02.012

Kukkonen, A., Ylä-Anttila, T., & Broadbent, J. (2017). Advocacy coalitions, beliefs and climate change policy in the United States. *Public Administration*, *95*(3), 713–729. https://doi.org/10.1111/padm.12321

Roe, E. M. (1991). Development narratives, or making the best of blueprint development. *World Development*, *19*(4), 287–300. https://doi.org/10.1016/0305-750X(91)90177-J

Sabatier, P. A. (1991). Toward better theories of the policy process. *PS, Political Science & Politics*, *24*(2), 147–156. https://doi.org/10.2307/419923

Sabatier, P. A., & Jenkins-Smith, H. (1993). *Policy change and learning: An advocacy coalition approach*. Westview Press.

Sabatier, P. A., & Weible, C. M. (2019). The advocacy coalition framework: Innovations and clarifications. In P. A. Sabatier & C. M. Weible (Eds.), *Theories of the policy process* (pp. 189–220). Routledge.

Stensdal, I. (2012, September). *China's climate change policy 1988–2011: From zero to hero*. Fridtjof Nansen Institute Report.

Weible, C. M. (2007). Stakeholder perceptions of scientists: Lake Tahoe environmental policy from 1984 to 2001. *Environmental Management*, *40*(6), 853–865. https://doi.org/10.1007/s00267-007-9005-2

Weible, C. M., & Ingold, K. (2018). Why advocacy coalitions matter and practical insights about them. *Policy & Politics*, *46*(2), 325–343. https://doi.org/10.1332/030557318X15230061739399

The cost of food. Consequences of not valuing soil and water and the people who manage them

Tony Allan and David Dent

ABSTRACT

Affordable food is a political imperative: there is nothing more expensive. The food system delivers cheap food but takes no account of the damage done along the way; it operates in three modes. In *Mode 1*, farmers manage land and water to produce crops and livestock; incidentally, they manage the landscape and carbon capture from the atmosphere by crops and its recycling through soil organic matter. Soil organic matter matters: it fuels life in the soil that breaks down wastes and toxins, regenerates crop nutrients, and controls pests and diseases; it maintains soil structure that enables infiltration and storage of rainfall; and it holds more carbon than the atmosphere and all standing vegetation put together. If these services are considered at all, they're taken for granted. Nowadays, Mode 1 accounts for only one-tenth of the value added in the food system; farm-gate prices have been driven down relentlessly so farms have had to get bigger or get out. Production certainly benefits from economies of scale and the gifts of technology (ever-more-powerful machines, new crop varieties, fertilisers, pesticides, and irrigation) but these gifts exact a cost by exposing topsoil to the elements, turning it upside down with every pass of the plough, and raiding soil organic matter. So, more power and more chemicals are needed; soil erosion, floods, and droughts are exacerbated; groundwater isn't recharged; and agriculture burns up more energy than it harvests. None of this is accounted in the cost of food but, over millennia, agriculture has amassed rights and vested interests that have resisted reforms that could take these costs into account. In the *second mode*, food is traded, processed, and retailed. These activities now account for most of the value added in the food system, and comprehensive legal foundations and accounting rules that have been impossible to install in primary production have proven politically feasible in trade, processing, and retail. In the *third mode,* food consumption, public policy again overrides the market to enable poor people to get food that is still too costly for them. The system is unsustainable – ecologically, environmentally, and economically. It will only be sustainable economically if power relations across the food system can be put on a more equal footing; if consumers paid the costs of maintaining environmental services, farmers in a viable commercial system might be able to remedy some of the ecological ills.

Food-water, soil, and society: competing claims and their consequences

The food system is a political economy, not an economy

Most people assume that the food system is an efficient, seamless conveyor of affordable food from farm to fork. Not so. It does deliver cheap food but takes no account of the damage done along the way. On top of the legacy of the previous 40 centuries, 25 per cent of all the land has been degraded in the last 40 years (Bai *et al.* 2025), and agriculture consumes more than 90 per cent of all the water we use. IPCC (2019) estimates that land use generates 23 per cent of manmade greenhouse gas emissions – but this is just from above ground. If we take account of the loss of soil organic matter, one-third of emissions is nearer the mark. Food supply chains are at risk and there is no certainty that they will be able to meet future needs. Julian of Norwich affirmed: 'All shall be well, and all shall be well, and all manner of thing shall be well' (Serenus de Cressy 1670). And politicians of all stripes promote this reassuring myth – just imagine what would happen if people were to think there won't be food on the shelves next week!

The political economy of the food system is a composite of three quite different modes (Allan 2019). In the first, which has been operating for more than 4000 years, there are no effective accounting rules. The political imperative – *affordable food* – has imposed a system in which farmers deliver food at well below its real cost. The financial difference is made up by farm subsidies but, as we explain, these come nowhere near the real cost. In Mode 2, food is traded, processed, and delivered to consumers in an effective market where profits are made and taxes paid. Shelf prices do capture most of the costs of inputs beyond the farm gate but not the enormous public health costs of poor diet. In Mode 3, where food is consumed, governments have to intervene again to ensure access to food that may still be unaffordable. The system is unsustainable: first, because farmers are trashing soil and water resources – whether knowingly or not. Second, it's a secure market system only so long as governments are willing and able to fund public subsidies – to farmers at the beginning of the supply chain – and at the end of the chain to enable the employed and unemployed poor to obtain food that is, otherwise, still too costly. Food prices are misleading. Farm-gate prices don't include all the costs of farm labour; they certainly don't reflect the environmental costs of mining the soil and polluting water resources, of floods and droughts, or of the climate crisis. So, growers subsidise consumers and exporting countries subsidise importers that receive this food at much less than its real cost.

Ghettos of the mind

For centuries, farmers and Society lived with volatile food prices. In England, prices fluctuated wildly but progressively increased until the repeal of the Corn Laws in 1846 (Figure 1). This was a decisive shift from a protected market to a version of free trade that has driven down the international floor price ever since (Figure 2). This is remarkable when, over the same period, the demand for food has grown by an order of magnitude.

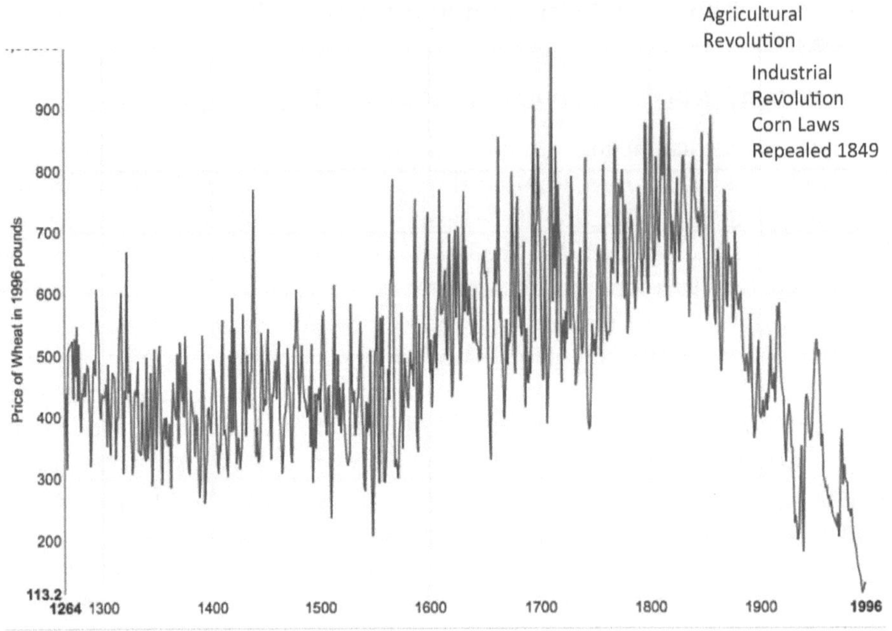

Figure 1 Wheat prices 1264–1998 in constant 1996 UK pounds. Rosner (1997)

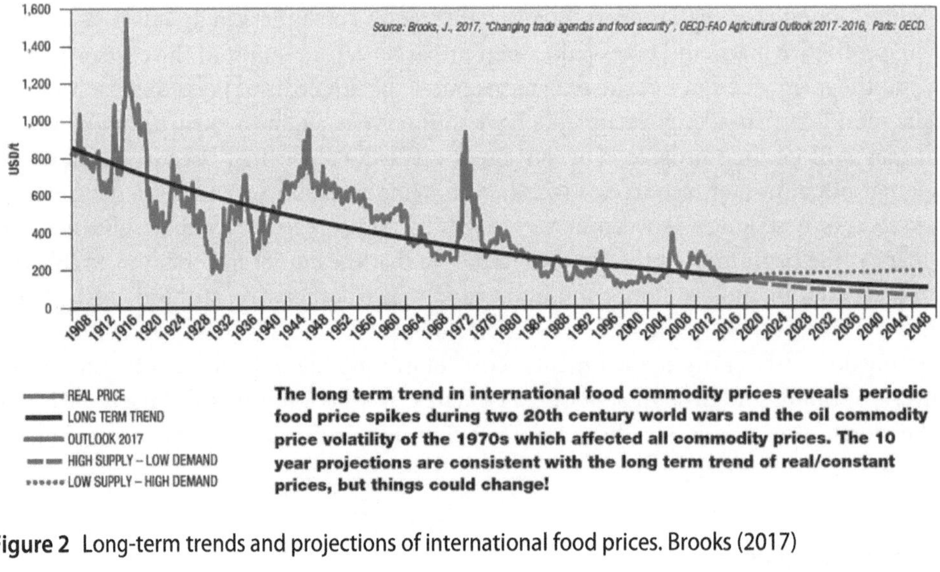

Figure 2 Long-term trends and projections of international food prices. Brooks (2017)

The profound effects of Mode 1 markets on farmers

Comparing the distribution of profits between three sectors of the food system in the USA – the farm sector, manufacture of the means of production (machinery, fertilisers, *etc.*), and the market sector that connects producers and consumers – the share of profits accruing to the farm sector shrank from 41 per cent in 1910 to 9 per cent in 1990, the industrial sector increased its share from 15 per cent to 24 per cent, the market sector increased its share from 44 per cent to 67 per cent (Smith 1991). In OECD countries, the market sector now accounts for 90 per cent of the value added in the food system and, by 2008, the share of the farm sector had shrunk to only 3 per cent (Eurostat 2009).

Farmers are price-takers, not the price setters, so farms must either get bigger or go out of business. Untold millions of small farms have gone bankrupt. Big farms have got bigger but have shed labour and adopted simplified farming systems that depend on ever-more-powerful machinery, more - potent pesticides and fertilisers, smarter irrigation, and new crop varieties that can take advantage of the new technology. Farmers are trapped in a system that forces them to rely on costly inputs promoted by powerful purveyors of these inputs. They are struggling for survival; they believe they are under siege; this is a lethal ghetto of the mind, revealed by the extraordinary suicide rate across the farming community (Weingarten 2017, USDA 2017). Meanwhile, the environmental services they render are unrecognised and unrewarded. Sustainability needs nothing short of a revolution. Only farmers can deliver it – but it's hard to be green when you're in the red.

The second Ghetto: weakness of regulation and absence of accounting rules

Much has changed over four millennia but some things don't change: production and supply of food is still driven by *who owns what, who does what, who gets what, and what they do with it.* Ownership and access to fertile land are eternally contested; there has never been any shortage of laws but, even in good times, there are disputes about private rights *v* communal rights, private *v* public interests, water rights tied to land rights… Now, as we push against planetary boundaries (Rockström *et al.* 2009), we have been caught off-guard by the absence of any legal protection or effective accounting rules that capture the value of environmental services like the supply of fresh water – on which food security depends – or carbon capture – on which the stability of climate depends. Water has never been valued or accounted for in Mode 1 markets; carbon capture has hardly been valued or accounted for anywhere.

Third Ghetto: complacency

This mindset has become firmly embedded as consumers have become distanced from the risks faced by farmers, and the stewardship needed to maintain essential services is taken for granted. In the absence of effective accounting and regulation, consumers are blithely unaware that their choices are destroying the ecosystems on which food-and-water security depends. Legislators are unwilling to break out of these ghettos of the mind. The unwritten contract that governments have struck with Society is that everyone is entitled to affordable food. The outcome is a political economy kept in place by direct payments to farmers and to consumers. Mobilising these payments is politically feasible: the politics of legislating for effective accounting rules that would reveal and capture the costs to the environment and the price of stewardship is not feasible – at present, without radical change. Unfortunately, these legislative and accounting measures are prerequisites of reforms that would create effective Mode 1 markets and attract urgently needed investment. Radical change calls for a shift in power relations between Mode 1 and Mode 2, and Society needs to give legislators the political space that frees them from their own ghetto of the mind – walled in by the imperative of delivering cheap food.

Discord: how we produce food confronts the ghettos of the mind

Growing crops the way we do trashes the land (Crews *et al.* 2018). Every year, we root them out, turn the soil upside down, and start again. Bare soil invites invasion by weeds. Rain splash turns bare soil into mud – mud that clogs the pores so that rainwater ponds or runs off the surface carrying the soil with it. When the rain stops, the pulverised surface sets as a crust that yields immediate runoff from the next rainstorm. Bare soil bakes in the sun – so do earthworms and myriad smaller creatures that should be maintaining soil permeability. And bare soil is carried off by the wind – three quarters of the topsoil and three and a half million people left the Dust Bowl of the American Plains States in the 1930s. None of this is accounted in the price of food.

Arable farming is not as productive as natural vegetation, crops are carried off, and separation of crops and livestock has cut the supply of manure – so farming is running down soil organic matter. This matters because soil organic matter is a bigger carbon pool than the atmosphere and all the standing vegetation put together (IPCC 2010). It is a brake on global heating; it is a bank of plant nutrients and fuel for soil life that cycles nutrients, disposes of wastes and toxins, and controls weeds, pests, and diseases; and it stabilises the architecture of the pore space that receives rainfall, releases water to plant roots, and drains any surplus to streams and groundwater. So, droughts and floods are exacerbated, agriculture burns up more energy than it harvests and emits clouds of greenhouse gases. None of this is accounted for in the cost of food. Sooner or later, a tipping point is reached when the soil itself leaves the stage. Then it's too late to keep better accounts.

Food production accounts for more than 90 per cent of the water consumed in the food chain (Mekonnen & Hoekstra 2011). But only a little of the water taken up by plants is used to make new plant material; most of it is transpired back to the atmosphere – the more green leaves, the more transpiration. Let us call this *green water.* Runoff and water entering the soil over-and-above its water-holding capacity drains to streams and groundwater. Let us call this *blue water.* The distinction is important: green water makes up two-thirds of all fresh water, but blue water gets all the attention because it can be tapped for many other uses. Setting aside water locked up in ice caps and glaciers, all fresh water is delivered by the soil. There are two critical junctures: the first when rain hits the ground, where it may infiltrate or run off carrying the soil with it; the second in the soil itself – soil water may evaporate from a bare surface, be taken up by plant roots or drain to streams and groundwater. Both junctures are managed by farmers. Rainfed farming cannot use last year's rain or anticipate next year's, but irrigators can draw on groundwater stored over centuries. They consume 70 per cent of water withdrawn from streams and aquifers but this is hardly ever regulated; groundwater is overdrawn wherever irrigation is practised, with the exception of Israel (Wada *et al.* 2012, Perry 2019).

Figure 3 depicts how water flows through the food system from the farm to the consumer. Our analytical framework introduces a further abstraction beyond green and blue water, to consider the water embedded in commodities as it moves along the food chain, commonly to the other side of the world. Let us call this *virtual water.*

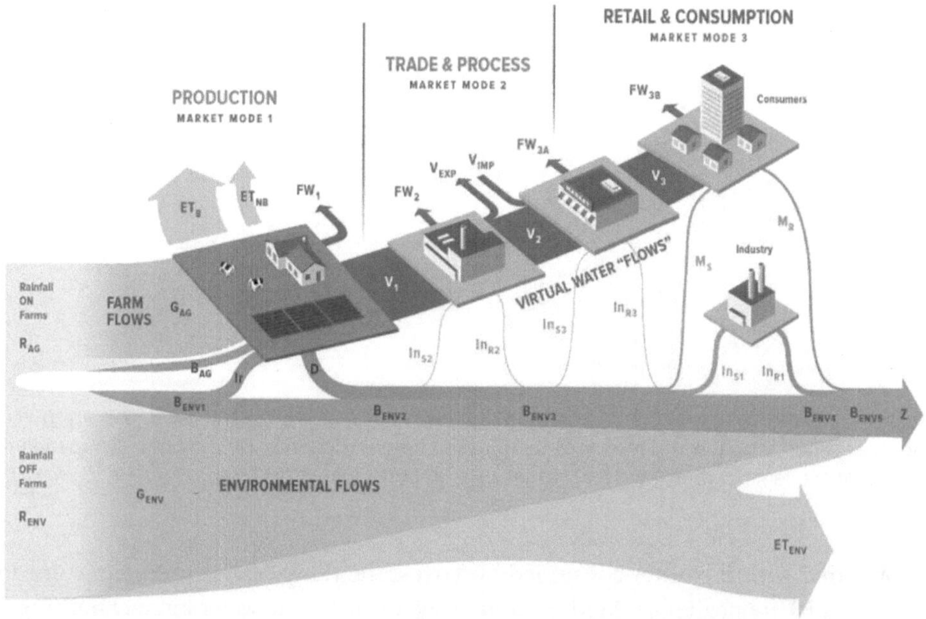

Water Category	Flows as labelled on the model
Water in the atmosphere: rainfall and evaporation	R_{AG} rain that falls on farms. (This either infiltrates and is available as green water on the farm G_{AG}; or runs off into channels and ponds as Blue water, B_{ENV}. If not collected and used this water may become polluted farm drainage D.) R_{ENV} rain that falls away farms in the environment. (This either infiltrates becoming green water in the environment G_{ENV}; or runoff becoming Blue water in the environment B_{ENV}.)
Water in the soil: "green water"	G_{AG} soil water on the farm G_{ENV} soil water off the farm
Surface water and aquifers: "blue water"	B_{AG} freshwater on the farm in ponds, streams, aquifers B_{ENV} Blue water off the farm. (This water assimilates pollution flows at points B_{ENV2}, B_{ENV3}, B_{ENV4}, B_{ENV5}.) IR. Blue water diverted to the farm for irrigation In_S Water supply to Industry. (IN_{S1} to non-food industry; IN_{S2} to market mode 2; IN_{S3} to market mode 3). M_S Water supply for Municipal use. Z Outflow to the ocean.
Polluted water	D. Farm drainage IN_S Water supply to Industry. (IN_{S1} to non-food industry; IN_{S2} to market mode 2; IN_{S3} to market mode 3). M_S Water supply for Municipal use.
Virtual water	V_1. 'Outflow' from market mode 1 to market mode 2. V_2. 'Flow' within market mode 2 V_3. 'Flow' in market mode 3 - food purchased for consumption. FW_1, FW_2, FW_3 Virtual water in food waste from market modes 1,2,3 FW_{3a}, FW_{3b} Virtual water in food waste in retailers and domestically V_{EXP} Virtual water 'exports' in market mode 2 V_{IMP} Virtual water 'imports' in market mode 2

Figure 3 Green, blue, brown, and virtual water flows in the three market modes of the food system. Allan (2019)

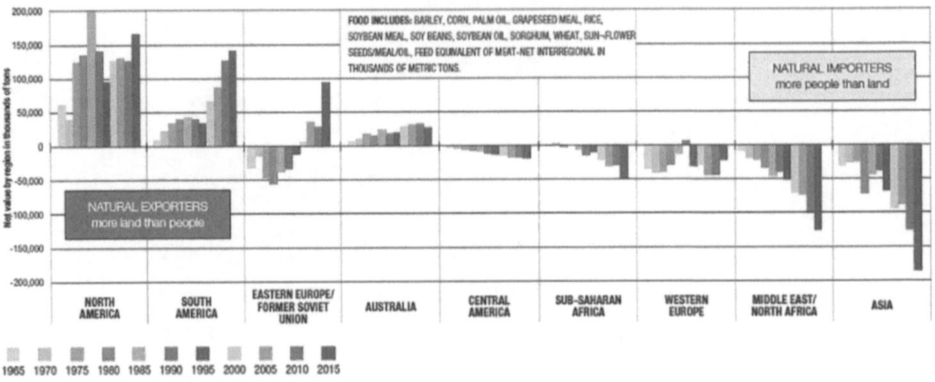

Figure 4 Global trade in major food commodities 1965–2015 showing, on the left, the seven economies that export virtual water resources to 160 or so other economies, on the right, grouped in five regions that are dependent on food and virtual water imports. Cargill 2017, reproduced in *World Energy* 46, 2020

Compared with the water consumed on farms, the flows of real water consumed in Modes 2 and 3 are negligible. Modes 2 and 3 pay for their blue water but pay nothing for the unaccounted virtual water embedded in the commodities they deal in. This raises a question about the international food trade (Figure 4). *Food exporters are exporting their environment.* Is this good news? And for whom?

Power relations in the food system

Power shifts but farmers, farm labour, and the environment are always weak

Farmers and graziers manage land and water but the food system dictates *how* they are managed. For a long time, farming was the mainstay of nearly every economy and the livelihood of many farmers. In rich countries, farmers are now few and governments have to ensure *cheap* food for the many. Figure 5 shows how the proportions of the value added in the first two modes of the food system have changed since the onset of industrialisation: in OECD economies, farmers and growers now account for less than 10 per cent of the value added; food traders, processors, and retailers add 90 per cent and generate lots of jobs, especially in retailing and hospitality.

Nowadays, in high-income countries, food accounts for only a fraction of household expenditure and of Gross Domestic Product (Figure 6). For example, in the USA in 2017, the farm sector accounted for 0.9 per cent of GDP; the whole food system accounted for about 5 per cent (CIA 2018). The economic transition from low to high income is marked by an increase in the value added in trading, manufacturing, retailing, and food services, matched by an increase in the proportion of jobs, from 9 to 79 per cent. At the same time, jobs in the farm sector shrank from 91 to 21 per cent; in the case of the UK 8 per cent (DEFRA 2015).

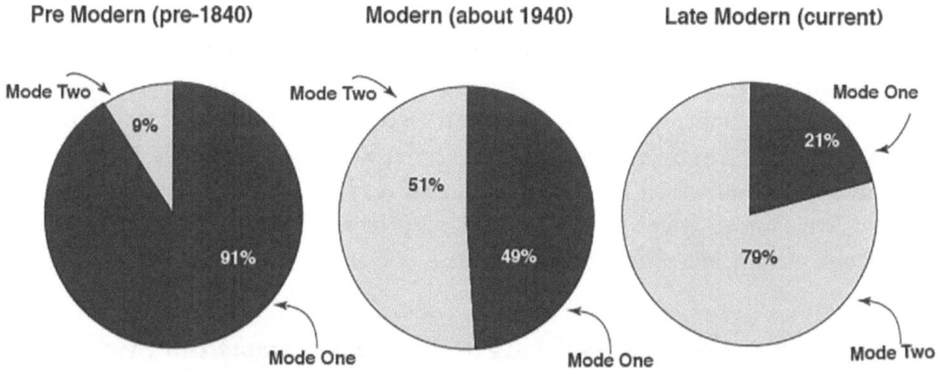

Figure 5 Value-added in the food chain in Market Modes 1 and 2

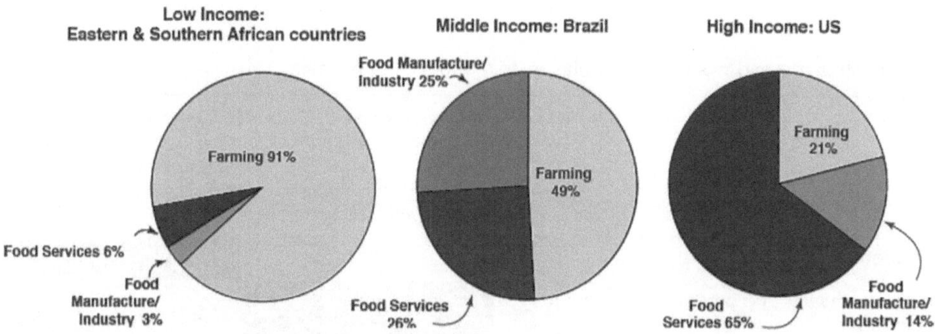

Figure 6 Jobs in the food systems of low-, middle-, and high-income countries. World Bank (2017)

The role of the state in Mode 1 markets

Governments have not established sound legal and accounting regimes in Mode 1 of the supply chain. Instead, they subsidise farmers. They do this to support livelihoods, maintain rural economies, and guarantee a degree of food self-sufficiency. These interventions have provided some protection for landscapes and ecosystem services, reinforced more recently by compliance conditions and agri-environment payments. However, the bar has never been set very high and this is an area of policy making where the state is learning on the hoof. Producer support from governments ranges from nil to more than 50 per cent with a mean value of about 20 per cent; OECD countries have cut back producer support since the early 1990s when the EU changed its Common Agricultural Policy from the payments for production that created a butter mountain and a wine lake to area-based payments; New Zealand and Australia have all but eliminated farm payments; but farm payments have been rising in emerging economies.

The first instinct of politicians is to pay more attention to people's livelihoods than to environmental services; food prices and food security are existential issues for politicians as well as consumers. In England, protection of landed interests has a long history but, since the early 19th century, governments have had to grapple with the competing claims of farming, manufacturing, and popular interests. The repeal of the Corn Laws, which opened the gate to food imports from North America, was a victory for popular interests. Ever since, Society has been comfortably deluded that the food system works; it delivers cheap food to consumers but pain to farmers and farm workers and serious harm to the environment at home and in food-exporting economies. Governments nurture the delusion, hide the harm by making payments to farmers, and, also, hide the erosion of the natural capital on which the system depends. *And the system doesn't deliver cheap food* – it delivers under-priced food. The real price is being paid by under-paid labour, the taxpayer, and our ultimate life-support system.

What needs to be done if we are to have a sustainable, commercially viable food system – and who needs to do it?

Table 1 lists the main societal players in the food system. The right-hand column highlights those that need to be strengthened or that should exert responsible influence:

Table 1 Strengths and weaknesses of players in the food supply chain

Main players	Players that are currently strong (in black)	Players & functions that Society needs to strengthen (in black)
Mode 1 supply chain market		
1. Farmers		Remunerate properly
Operators of water and other natural-resource infrastructure		
Seed, fertiliser, chemical, & equipment corporates		
Mode 2 supply chain market		
Food-commodity traders	Operate within an effective legal and accounting regime	Contracts with farmers should reflect stewardship costs
Food & beverage corporates		
Big supermarkets		
Food and water NGOs		
Multilateral and bilateral lenders and aid agencies		
Mode 3 supply chain market		
5. Society by adopting sound ideas and behaviour	Consumers	Badly informed, should eat responsibly
Functions that impact all 3 Modes		
2. The State – through law and regulation	The main player especially in Mode 1	
3. Accountants, reporting rules and regulations	Ignore soil and water costs & farming impacts	Fix legal regimes & accounting rules in Mode1
4. Investors	Few invest responsibly	Invest responsibly

This crude analysis indicates the need to (1) enable farmers to be good stewards of the land as well as growing food, fibre, and fuel, which means higher farm-gate prices or payments for environmental services; (2) introduce regulations to safeguard land and water resources and environmental services, which will require effective audit and accounting of these assets; (3) promote responsible investment; (4) convince consumers to change their behaviour. Five main players need to do things differently: farmers, accountants, the State, investors, consumers and Society. Corporate institutions in Mode 2 are important but they are already relatively well regulated. Some food and beverage companies are moving towards ecosystem-aware contracts with farmers but their language suggests it is possible to achieve parallel financial and societal/environmental goals with no trade-offs – a kind of market environmentalism (Bakker 2014, Rudebeck 2019); the rarity of such transactions reflects the balance of power. The silence of the corporate providers of hardware and chemicals is deafening. They have the capacity to watch and wait, content to keep wrecking systems in place.

Considering the main players, what do Society's consumers and its legislators need to do to make the food system sustainable?

1. ***Farmers*** are and will remain key players. The issue is this: farmers could be effective producers and stewards if they were properly paid – *but they are not.* Food is too cheap but are consumers willing to pay more? Most, probably, are not and there will always be many who cannot. As individuals, farmers are weak although some corporate farms operate at a scale enjoyed by the big food manufacturers and retailers. Collectively, farmers can exert political influence beyond their numbers, for example, in Europe because of the complex political economy involving arcane land ownership, under-paid farm labour, and cultural/emotional lock-ins.

   In many countries, farmers receive direct and indirect payments that enable them to provide well-understood production services but do not pay for vital but invisible environmental services. The path of least resistance appears to be direct payments for environmental services and audited compliance with specified conditions of stewardship. For instance, carbon credits may be paid for by imposing cap-and-trade legislation on emitters of greenhouse gases. In the case of water supply, non-farm water users will pay for their modest consumption but, ultimately, irrigators will have their water consumption capped. This will be contentious and many wetlands and water resources will be lost because regulation will have come too late. Whichever path is taken, credible accounts will be needed.

2. ***The State*** *has not enacted a legal framework and accounting rules that internalise all the costs of on-farm food production.* Instead, it subsidises farmers. Even then, poor people may need further support to buy food. The State is the most influential player in the food system. It has had to ensure that food is cheap and always available. And it is the only player that can nudge into place the regulations, accounting rules and an investment regime that would make farming commercially and environmentally sustainable. Politicians need Society to give them political space to implement reforms that will properly price food and conserve the natural resources on which food production depends. *But Society has not signalled that this is what it wants.*

3. ***Accounting and reporting rules:*** *Information is material if omitting it or misstating it could influence decisions that users make on the basis of financial information.* In the food system, this basic principle of accounting is more honoured in the breach than the observance. Emmanuel Faber, CEO of Danone, remarked in an interview for *Der Spiegel* on 8 June 2019:

   There is no cheap food. Every consumer must realise that if food has a low price, someone else is paying the real price. It's either the farmer, the degraded soil, or the consumers themselves with their health. At the end of the day, someone is paying the bill for cheap food.

   Effective markets depend on effective regulations and there is no better example of what happens without effective regulation and accounting rules than Mode 1 food production. Peter Bakker, an exceptional accountant and President of the World Business Council for Sustainable Development, has argued that accountants would save the world (Bakker 2013). Unfortunately, his perspective is not shared by the accounting profession – which has not advanced ways and means to deal with environmental capital. Accountants are no less expert than sustainability departments of the food-supply-chain corporates in keeping abreast of the discourse but, when it comes to the actual protection of soil, water, and biodiversity, they ignore information that is material. The financial system needs to be rebooted (Allan *et al.* 2015, Bakker 2017). Even then, accounting for intangibles like soil carbon or groundwater status won't change behaviour unless taxes, subsidies, and trade restrictions are managed effectively.

4. ***Investors.*** Investors also have immense power. If they choose to use it, they could change the behaviour of corporate institutions that contract with food producers. But investors, in general, are blind to the value of natural capital and the rights of those who produce and prepare food. Investors who only prioritise profit and return on capital (currently the fiduciary duty of a CEO of a Mode 2 food corporate) should be an endangered species – *but they are not.*

   The CEOs of some food manufacturing corporates *have* got the message. Emmanuel Faber has said:

   The food industry is going nowhere. Big companies have disconnected people from their sustenance. Consumers, especially millennials, are sceptics about industrial-scale food production. Even sellers of healthy products, such as mineral water, spread harm – just look at the billions of their plastic bottles that choke the ocean … A revolution and the end of globalisation are nigh.

   (*Economist* 2018)

These people know all about investment and operational risks; they know and are familiar with the reasons for market failure; they curse the darkness – but where is the candle to illuminate the reason why we have the current accounting rules and legal regime in Mode 1? Danone aims to convert from the conventional company approach to a *BCorp approach.* BCorp certification provides information on for-profit companies that demonstrate high standards of social and environmental performance; the idea is to drive a cultural shift to redefine business success and to provide an alternative legal norm – but it is a voluntary system as compared with a mandatory one established in law/by regulation. The predominant private-sector institution, the conventional for-profit company, remains a serious environmental risk (Newborne 2012). The revolution has yet to come and, without it, evolution is slow.

5. ***Consumers and Society.*** *Food consumers could significantly influence food security; they could be the voluntary regulators of the food system by consuming responsibly.* At present, their food consumption choices impair their own health and that of the planet. It is widely estimated that they waste about one-third of all the food purchased; unhealthy food choices and unnecessary consumption add to the volume of wasted and degraded resources. Quite clearly, consumers could significantly influence food and water security; they could be the voluntary regulators – *but they are not, yet.*

   We have not focussed on corporate food traders, processors and manufacturers, and the big supermarkets because they operate in markets that have rules and auditing systems that account for almost all of their inputs. All of the very little water they consume – about 1 per cent of food-water – is accounted for; they have adopted more efficient and more environmentally responsible practices that have halved their own water consumption during the past two decades, and they have reduced their costs. Their accountants and their shareholders are content; it has all made commercial sense. Perversely, there has been no equivalent enlightenment of farmers who actually manage and, on occasion, mismanage nearly all of Society's food-water consumption.

   How has Society adapted? Consumers have not starved. Farmers, mostly, make a living but it is troubling that suicide is an extreme component of the adaptation of the part of Society involved in Model food production. That Society as a whole has hardly begun to adapt to the Laws of Nature via the food system should also be troubling. The State and its taxpayers mitigate, to some extent, the market failures by providing subsidies to farmers but the State can only try to fix Nature's degraded ecosystems with unwelcome regulation – as opposed to effective incentives. As yet, there is no political will to fix the failures of natural resource mismanagement: they are denied or hushed up.

Conclusions

Governments dare not meddle with the certainty that affordable food will be available – they remain in power only as long as this certainty is in place – and practical policy advice concerns itself with piecemeal improvements (Lipsey 2007). It might be argued that the food system is a miracle of which governments and the private sector can be proud. Public payments to farmers and consumers have kept in place a political economy that delivers a version of food and water security. For two centuries, it has delivered ever-cheaper food to a burgeoning population. For all its failings, in the arcane calculus of political economy, it is cost-effective. Governments of industrialised economies will always be able to make the payments that fix the availability and affordability of food – although paying farm labour properly will remain an untidy, politicised issue and one of the lethal challenges faced by those who farm.

But because the cost of food is not reflected in the price of food, it impossible to introduce best, as opposed to second-best or worse, solutions to the problems of allocating and managing land and water. Whether current and future food systems will meet future demands for food and, therefore, increased food-water depends on Society, its expectations and its political systems. Society determines demand – it determines global population and population hotspots; and it is forcing global heating with all its uncertainties. *The capacity to fix the availability and affordability of food for low-income economies of Africa heading for a population explosion is not certain.* And what the existing food system certainly does not do is protect and conserve the land, water, and ecosystems; stabilise the climate; and protect public health. Society needs to recognise that if farmers are to do the right thing, legislators need to be given political space to install an effective legal regime and effective accounting and investing rules in Mode 1 of the food system. Can it be that something so prosaic will control our destiny?

Afterword on the Covid-19 crisis that disrupted global systems in 2020/1

Global systems are periodically tested. We introduced our analysis with divine revelation and asked you to imagine what would happen if people were to think there won't be food on the shelves next week. We have now seen what happens even when shelves are heroically replenished.

We also identified the major economic crises of the past half century; the 1974–79 oil crisis and the financial crisis of 2008–11 triggered exceptional commodity price spikes. We concluded that the food system – dysfunctional economically, environmentally, and socially according to economists, ecologists, and social scientists – is able to cope with exceptional market conditions. It has repeatedly adapted to volatility and resumed the long-term trend of falling international food commodity prices, remarkably quickly on each occasion. International agencies and analysts predicted, emphatically, that the 2008–11 financial crisis had shifted the food system to a new normal of volatile and higher prices: OECD did not recant until 2017 (Brooks 2017).

What will be the impact of the Covid-19 crisis on the global food system?[1] Scientists analyse the past impressively. They have made a few spectacular predictions about the future but have a poor record in predicting political economy outcomes, and the future political economy of the food system is certainly difficult to call. The first few months of the Covid-19 crisis have highlighted important differences between global food supply chains and other global systems. The food system and its supply chains have proven remarkably flexible, responsive, and pragmatic: there was no stressful politics associated with keeping food stores open, nor with the contradiction of allowing access to food stores despite lockdowns that kept most people isolated at home. The demand side was disrupted by the closure of pubs and restaurants but the system adjusted within a week or

[1] Essentially as published in 2021. History has, indeed repeated itself. Although the Covid-019 food spike was amplified by Russia's invasion and blockade of Ukraine 2022-4, a major food exporter, the downward trend of international food prices and farm gate prices reasserted itself.

so. Supermarkets, corner shops, and food supply chains adapted quickly. This contrasts with the situation in energy services, public transport, entertainment, hospitality and tourism, manufacturing, health, and education. Some of these ceased to function and many companies will fail.

Part of the adjustment has come via modest public funding but the sums are negligible compared with the trillions devoted to non-food sector corporations via subsidies and unprecedented social security payments. On the supply side, the food system was already encountering labour problems exacerbated by restrictions on the international movement of farm labour. Legislators have yet to come to terms with the pressure from northern hemisphere farmers wanting labour for vegetable production, vineyards, and orchards. Meanwhile, farm gate prices are falling, especially in the OECD economies.

What will be the impact of Covid-19 crisis on local food supply chains? Only 15–20 per cent of food is traded internationally: most of the food consumed worldwide is produced by farmers for themselves and their national markets. Family farms in Asia and Africa remain essential providers of food but they operate under the heel of low farm-gate prices brought about by rich countries exporting under-priced food. This means that farmers on small farms cannot invest to increase efficiency and make profits. The Covid-19 crisis might nudge up farm-gate prices in low-income economies but, if history repeats itself, the governments of these countries will still find it easier to import under-priced staples than to face volatile food politics played out on their streets.

Business-as-usual is, unfortunately, unsustainable. That the Covid-19 crisis will not be a food crisis is welcome amidst the chaos in other sectors. The system will keep consumers fed 'affordably', and keep farmers and farm labour in place, dependent on subsidies and direct payments. But it remains unsustainable and uninvestable – that is to say, uninvestable for private investors. Higher farm-gate prices would, amongst other things, make private investment feasible in Mode 1 food production but this is not yet politically feasible. Would-be reformers should continue to struggle with the contradictions; at the same time, they must recognise why the dysfunctional system stays stubbornly in place. Will we all remain complicit in complying with the imperative of under-priced food for the underpaid?

References

Allan JA 2019 Food, water and the consequences of society not valuing the environment. In JA Allan, B Bromwich, M Keulertz & AJ Colman (editors) *The Oxford handbook of food, water and society.* Oxford University Press, New York

Allan JA, T Colman & M Keulertz 2015 *The complexity and urgency of water: time for the accountancy profession to step up*. International Federation of Accountants, New York

Bai ZG, JD Russ, KF Mayr, & DL Dent 2025 How is Gaia doing? Trends in global land degradation and improvement. *Ambio*, 10.1067/s 1320-25-021791

Bakker K 2014 The business of water: Market environmentalism in the water sector. *Annual Review of Environment and Resources* 39, 469–494

Bakker P 2013 Accountants will save the world. *Harvard Business Review*, 3 March https://hbr.org/2013/03/accountants-will-save-the-worl (accessed 31 May 2018) see also *The Guardian*, London, 5 March 2013

Bakker P 2017 *We must reboot the financial system to deliver the SDGs.* WBCSD, Geneva https://www.wbcsd.org/Overview/News-Insights/Insights-from-the-President/We-must-reboot-the-financial-system-to-deliver-the-SDGs (accessed 8 August 2018)

Brooks J 2017 *Changing trade agendas and food security*. Chatham House. A Sustainable Food Future: Technology, Resource Use and Resilience, 27/8 Nov 2017. London. jonathan.brooks@oecd.org

CIA 2018 *World Factbook. 2018 Factbook archive*. Washington DC,

Crews TE, W Carton & L Olsson 2018 Is the future of agriculture perennial? Imperatives and opportunities to reinvent agriculture by shifting from annual monocultures to perennial polycultures. *Global Sustainability* 1. https//doi.org/10.1017/sus.2018.11

DEFRA 2015 *Annual business survey of the UK Organisation of National Statistics and Aggregate Agricultural Accounts*. Dept of Environment, Food and Rural Affairs, London

Economist 2018 Danone rethinks the idea of the firm: choosing Plan B. *The Economist* 9 August 2018

Eurostat 2009 *Dynamics of the food supply chain*. European Commission, Brussels

IPCC 2010 *The carbon cycle*. Intergovernmental Panel on Climate Change http://www.climate science.gov/Library/stratplan2003/final/annexfigure7-1.htm

IPCC 2019 Climate change and land. An IPCC special report on climate change, desertification, land degradation, sustainable land management, food security, and greenhouse gas fluxes in terrestrial ecosystems. Intergovernmental Panel on Climate Change, Incheon https://www.ipcc.ch/site/assets/uploads/2019/11/SRCCL-Full-Report-Compiled-191128.pdf

Julian of Norwich 1342–1416 *Revelations of divine love*. Translated by John, Serenus de Cressy 1670

Lipsey RG 2007 Reflections on the general theory of the second best at its golden jubilee. *International Tax and Public Finance* 14, 349–364. https//doi.org/10.1007/s10797-007-9036-x

Mekonnen MM & AY Hoekstra 2011 *National water footprint accounts: the green, blue and grey water footprint of production and consumption*. Value of Water Research Report Series 50, UNESCO-IHE, Delft

Newborne P 2012 *Private sector investment in water management – company forms and partnership models for inclusive development*, ODI Briefing Paper, December 2012

Newborne P 2012 The private sector's contribution to water management: re-examining corporate purposes and company roles. *Water Alternatives* 5 (3), 603–618

Perry C 2019 The water crisis: which solutions will work? In JA Allan B Bromwich, M Keulertz & AJ Colman (editors) *The Oxford handbook of food, water and society*. Oxford University Press, New York

Rockström J, W Steffen, K Noone & others 2009 Planetary boundaries: exploring the safe operating space for humanity. *Ecology and Society* 14 (2), 32

Rosner M 1997 Data source Makridakis, Wheelwright and Hyndman, https://ourworldindata.org/food-prices/

Rudebeck T 2019 *Corporations as custodians of the public good? Exploring the intersection of corporate water stewardship and global water governance*. Springer Nature, Cham

Smith S 1991 *Is there farming in agriculture's future? The impact of biotechnology*. College of Agriculture and Life Sciences lecture series, November 14. University of Vermont and State Agricultural College, Burlington VT

USDA 2017 *Rural America at a glance*. Washington DC

Wada Y, LPH van Beek & MFP Bierkens 2012 Nonsustainable groundwater sustaining irrigation: a global assessment, *Water Resources Research* 48 (6). https://doi.org/10.1029/2011wr010562

Weingarten D 2017 Why are farmers killing themselves? *The Guardian*, 6 December

World Bank 2017 Future of food. Shaping the food system to deliver jobs. World Bank Working Paper, Washington DC 114394-WB-PUB

Index

Note: **Bold** page numbers refer to tables; *italic* page numbers refer to figures and page numbers followed by "n" denote endnotes.

Abyei area 136, 142–143
accelerated agricultural growth 169
accountants 7, 292, 295, 297–302, 305, 346–347
accountants will save the world (AWSTW) 301–302
accounting: rules 339, 345–346; for water 122, 196
accounting-abstraction-absolutism (AAA) **303**
achieved/achieving: food 262–265, 267, 269, 271, 273, 275, 277, 279, 283; food security 74, 78
Aconcagua, River 26, *26*, 28
actual evapotranspiration (AET) 20, 230–231, 233–239, 242–246; agricultural 248; change in 235, 242, 246; global 239; global terrestrial 233, 247; national 238–239; for natural vegetation 246, 248
advocacy coalitions 332–334; approach 333; framework *332*
Advocacy Coalition Theory (ACT) 331
AET *see* actual evapotranspiration (AET)
affordability/affordable of food 289, 337, 347–348
African Development Bank 70
agriculture/agricultural 34–35, 49, 56–57, 66, 74–75, *77*, 78, 165, 194, 223–224, 231, 246–248, 285, 288–289, 311; accounts 253, 258; areas 165, 193; commodities 6, 48, 204, 213; demands 66; expansion 56–57; exports 57, 66; land 119; land covers 231, 241; measures 68–69; production 56, 68, 116–117, 119, 210, 224, 232, 280; products 56, 74, *75*, 286–287; research 169; sector 57, 68, 75; subsidies 194; trade 220, 258
Aldaco-Manner, L. 167
Aldi 314, 319, 320
Allan, Tony 2–8, 11–13, 16, 18, 20, 22, 24, 26, 34–44, 48–50, 60, 62, 64, 66, 68, 70, 72, 74, 76, 78, 84, 86, 88, 90, 92, 104, 106, 108, 110, 112, 116–118, 128, 130, 132, 134, 136, 138, 140, 142, 154, 156–157, 165–167, 171, 176, 178, 180, 182, 184, 186, 188, 194, 204, 206, 208, 210, 212, 214, 216, 218, 220, 222, 224, 228–229, 232, 234, 236, 238, 240, 242, 244, 246, 248, 252–254, 258–259, 262–264, 266, 268, 270, 272, 274, 276, 278, 280, 285–286, 288–289, 291–293, 295–296, 298–302, 308–312, 314, 316, 318, 320, 322, 324, 326, 336–338, 340, 342, 344, 346, 348; contributions 5, 148–149, 151, 153, 155, 157; legacy 1–8
Alliance for Water Stewardship 314
alternative policy interventions 87
American farmers 254, 257
analytical model 84, 86–87
annual use limits (AUL) 197
Arachet dam 105–106, 108–109
arc-min grid cells 209
Area of Outstanding Natural Beauty (AONB) 316
armed conflict 52–53, 293
ASME model water balance **66**
Aswan High Dam Reservoir *63*, 90–92
available water resources 27, 59, 165, 195
average price 121
AWS Standard 325, 327

Bacharach, S. B. 149
bad virtual water 154, 291
Bahr al-Arab 127, 130, 135, 137–138, 142–143
Bahr al-Ghazal 137, 142
Bahr region 137
Bakker, Peter 295
bare soil 340
bargaining power 48, 149, 151, 155, 277
baseline/baselining 5, 72, 76, 82–83, 85–86, 88–89, 91, 93, 233–236, 242, 245; AET 242, 245; conditions 83, 89; cover 235; land cover 236, 242, 244; process 83, 93; use 236
Beasley, R. 195
benefit analysis 82, 84, 86
big water 7, 194, 299, 301, 303

Blue Nile 56–57, 72, 76, 89, 155
blue water 14, 24, 35, 81, 116–117, 224, 286, 295, 340; resources 252, 258
boundaries 4, 43–44, 52, 127, 129–137, 139–140, 143, 156, 311; definition 140; river 127, 129, 144
Braverman, R. 83
British bouncing bomb attacks 79n7
Brooks, J. *338*
business models 6, 171–172, 182, 323
business systems 189–190

carbon: emissions 295, 322, 324–325, 327; footprints 230
Carmona-Moreno, C. 167
catchment: boundaries 101, 182; level 103, 181, 183, 185, 189; partnerships 186–187, **187**, 318; planning 111, 113
cattle troughs 106, 108, 110
Central Bank Yearly Statistics 117–120, 123
changes: in cropped area 221; management 299, 303; power relations 44, 46; in surface water irrigation 216, 224; in virtual water trade 213–214
cheap food 8, 337, 339, 343, 346
chief executive officers (CEOs) 182
Chobe, River 129, 135, 141, 143
climate: change 7, 54, 76–78, 85–87, 167–168, 234, 284–285, 287–288, 295–297, 323–324, 332–333; crisis 182, 289, 296–297, 337; effects of, change 7, 86, 287–288; nexus 7; risks 286, 297
Climate Hazard Group InfraRed Precipitation Stations (CHIRPS) 16
Climate-related Financial Disclosure (TCFD) 295
coexisting conflict 5, 153–154
collaborative approach 186, 189
collaborative management 184, *189*
colonial powers 130–131, 143
Common Agricultural Policy (CAP) 334
Commons Environmental Audit Committee 313, 315, 317
complex: problems 5, 185; problemsheds 99, 101, 103, 105, 107, 109, 111–112, 115; river boundary delimitations 131, 134; systems 183–184, 186, 296
conceptual framework 92, 263
conflict and cooperation 5, 48, 153
conservation agriculture 288, 314
consistent water balances 15, 27
consumed food 264, 268–270
consumptive use 16, 19–20, 26–27, 61, 65, 69, 231
consumptive water: footprint 285; use 245, 258
contested baselines 5, 82–83, 85–87, 89, 91, 93
cooperation 5, 41, 48–49, 51, 53, 71–72, 110, 152–155, 257
cooperative filling strategy 91–92
Cooperative Framework Agreement 155–156
core beliefs 8, 331–332, 334
corporate social responsibility (CSR) 171
corporate water accounting 231
cost-effectiveness analysis 84
costs 5, 7, 70, 82–87, 109, 117–120, 195, 224–225, 230–231, 257–258, 270, 285–286, 295–297, 303–304, 321, 337, 347; of food 8, 336, 339–341, 343, 345, 347
countries 34–35, 41–42, 48–49, 56, 71–73, 75–76, 118, 154–155, 168–170, 177–178, 214–215, 219–221, 229, 233, 247, 253–256, 258, 262–264, 287, 349
Courtauld initiative 317–318, 322, 326–327
coverage ratios *271*
croplands 19, 215, 221, 224, 230, 232–233, 235–236, 238–240, 242, 246–247; intensive nature of 242, 247
cropped area 65–66, 209, 215, 221
crops 54, 61–62, 65–68, 74, 165, 168–169, 208–209, 211–213, 221–222, 230–231, 235, 248, 271–272, 277–278, 340; growth 229; prices 215, 217; production 209, 211–212, 219, 221, 224; rotations **67**; structure 265, 281; water requirements/demands 18, 61, 66, 68, 197, 209

Daher, B. 167
Dahl, R. 149
Dalin, C. 154
dam: caretaker 108; construction 69, 111, 156, 206
Daoudy, M. 149
Dare, A. 167
day-to-day leadership responsibilities 329n16
decoupling 229, 262–263, 266, 270, 277
demands **60**
Department for Environment, Food and Rural Affairs (DEFRA) 334
Department for International Development 53, 103
desalination/desalinated: plants 119–120, 122; water 119–121
desert: agriculture 122–124; farming 121
developing countries 193, 254–255, 277
Development Policy Management 170
discourse 4–5, 43–48, 151–152, 155, 194, 346
district officials 108, 110
domestic consumption 119, 272, *275*, 276; of soybeans 276
domestic food 264; production 77, 120, 267–269
domestic physical water 276–277
domestic production 73, 78, 264, 266, 281
downstream 23, 152, 196–197, 230–232, 247, 325; context 231, 233, 241, 249; states 53, 150–151; water intakes 233, 247
drinking water 5, 157, 196, 225
drip irrigation 246, 304, 310

Dublin Principles 6, 192–195, 197, 199–200, 203
dynamic: baseline 5, 81–88, *85*, 91–93; model 61, 63, 67

Earth observations 4, 13, 15, 17, 19, 24, 27, 33
economies/economic 59, 75, 78, 155, 198, 201, 264, 277, 280–281, 285, 296, 337, 342; activity 78, 264; benefits 92, 316; capacity 149, 155; capital 45; conditions 214, 280; crisis 57, 267; decisions 224–225, 313; development 34, 41, 126, 278, 281, 284, 296; diversification 271, 280; efficiency 193, 197, 199–200; growth 73, 85, 270–271, 280–281, 287; mobility 6, 199; model/modelling 209, 213, 224–225; power 102, 111, 113, 255; reforms 35, 268, 270–271; responses 224; security 121–122; statecraft 255; use 34, 69; value 67, 193–194, 196, 201, 286
economists 82–83, 195, 257, 292, 299, 304, 346, 348
ecosystem services 20, 196, 286, 343
effective accounting rules 337, 339
Egypt/Egyptian 5–7, 49, 52–54, 56–57, 59–62, 66–78, 89–93, 154–156, 165, 252–253, 262–272, *269*, 270, 276–278, 280–281; economy 263, 267, 270, 280; and Ethiopia 53, 91–92; food 277–278; governments 266, 269, 277, 280; hegemony 69; perspective 91–92; socio-political contexts 276; and Sudan 5, 71–73, 89; water balancing act 5, 61, 63, 65, 67, 69, 71, 73, 75, 77
embedded water 48–49, 121, 286
employment 54, 215, 217, 220–221, 224, 226; in agriculture 224
endogenous dynamic baseline 87
energy 6–7, 167–168, 171–172, 184–185, 190, 194–196, 200, 234, 253, 255, 284–286, 340; and food 168, 172, 284; policy 285–286; prices 78; production 284; supplies 74, 171, 195; systems 285–286
Energy Information Administration (EIA) 170
Environment Agency (EA) 35, 179, 186, 313, 316–317, 326–327
Environmental Audit Committee 317
environmental land management (ELM) 334
Environmental Land Management Scheme (ELMS) 317
environmental, social and governance (ESG) 322
environment/environmental 7–8, 41–42, 104–105, 181, 185–186, 190, 193, 198, 247–248, 285, 299–301, 311, 317, 319, 342–343; baseline 82–83; costs 70, 337; flows 62, 196–197; impacts of farming 288, 314; non-governmental organizations 152, 183; outcomes 189, 317; perspective 121, 182; policy 181; risks 346; services 35, 339, 343, 345; sustainability 303
Ethiopia 53, 57, 69, 71–73, 76, 89–93, 102, 110, 152, 154–156
European Centre for Development Policy Management (ECDPM) 170
Evans, Simon 317
evaporative footprints 247–248; water 230–231, 243, 245–247

farmers 7–8, 11–12, 66–69, 157, 188, 192, 194, 197–198, 200, 288, 292–293, 300–304, 309–311, 313–315, 317–318, 325-327, 334, 337–340, 342–343, 345–349
farm/farming: labour 337, 342, 345, 349; sector 338–339, 342; systems 198, 288, 314
financial crisis 348
financial markets 255, 297
first virtual-water decoupling policy 271, 280
flood: irrigation 236, 248; partnership 187, **187**
food 6–8, 35, 52, 54, 116, 154, 168–169, 171–172, 230, 255–259, 261–263, 270–272, 278, 280–281, 284–289, 292–293, 310–311, 336–337, 339–343, 345–349; and agriculture 288; aid 254, 259; availability and affordability of 347–348; boycotts 122; chains 6, 8, 167, 340; commodities 42, 224; consumers 35, 347; consumption 213, 264, 269, 277, 286; crisis 7, 258, 349; and energy 171, 285; exporters 123, 342; growing 149, 263, 266; imports 4–6, 35, 52, 117, 120–121, 124, 252–253, 255–256, 259, 268, 270–271, 287, 343; independence 117, 122–123; insecurity 276, 287; net importers 256–257; nexus 6–7, 57, 195, 253; prices 211, 213, 218, 224, 258, 276, 288, 337, 340, 343, 347; processing 7, 255, 295; processors 314–315, 317–318, 326; producers 323, 346; producing 264, 269, 314; production 14, 165, 167, 170, 224, 230, 262–263, 278, 284, 288, 310–311, 345–346, 349; self-sufficiency 117, 121–123, 287; supplies 7, 54, 169–170, 224, 277, 339; supply chain 13, 170, 285, 291–293, 301, 315, 318, 322, 326, 337, **344**, 348; systems 3–8, 169, 254, 292, 295, 323–324, 337–340, 342, *343*, 344–345, 347–348; water 81, 166, 194, 269, 288; and water security 7, 35, 286–287
Food and Agriculture Organization (FAO) 16, 25, 65, 74, 200, 233, 235, 264–268, 277–278, 287, 300, 304, 309
food-importing countries 204, 253
food-insecure countries 254, 258
food security 6–7, 35, 48, 57, 73–75, 78, 194, 253, 277, 287, 343; in Egypt 75, 78; indicators *76*
food trade/traders 154–155, 229, 252–255, 257–259, 264, 285, 313, 342; politicization 257–258
free water, evaporation 236, 242, 248
fresh water 7–8, 292, 339–340; resources 262, 264
Frontier, River 138
functional water markets 199–200

Gardezi, N. Z. 167
gate prices 337, 349
Geneva Water Hub 293
Gilmont, M. 264
global: assessment 291; averages 242, 244; crop production 207, 213–214; economic hubs 255; economy 5, 39, 295, 297; financial crisis 271, 284; hydrological models 22; land cover 239; loss of tree cover 236; markets 52, 54, 225, 256; production 214, 223, 255; support 149; systems 348; trade 48, 52, 214, 287, *342*
global crop-suitable areas 242, 244; conversion of 242, 244
global food: crisis 253, 259; markets 195, 253; prices 54, 253, 287; production 6, 257; system 169, 192, 253–256, 263, 297, 348; trade 2, 6, 82, 154, 156, 257
Global Positioning System (GPS) 18
Global Precipitation Mission (GPM) 16
global water: consumption 213, 253; crisis 192; intakes 233, 247; resources 296; use 285
Government of Sudan 137
Gramsci, A. 149
Grand Ethiopian Renaissance Dam (GERD) 69, 71–73, 78, 82, 89, 91–93, 154–156
Greater Manchester Natural Capital Fund 189
Greco, F. 154
Green Revolution 253–254, 309
green water 24, 35, 42, 81, 116, 209, 220, 258, 285, 295, 340; pixels 23–24
grid cells 207–214, 225
grid-group model *184*
gross national income 74, *75*
groundwater 21, 49, 204–206, 208–209, 211, 213–215, 220, 225–226, 285, 291, 340; abstraction 21, 26, 204, 207–208; bucket 20–21, 23; consumption 204, 215; governance 224, 291; resources 18, 24, 49, 225; restrictions 211, 213, 220; status 207, 346; storage 18, 21, 28; supply 20, 24; and surface water 208, 211; sustainability policy 207, 214, 225; sustainability restrictions 211, 213–214, 220; systems 24, 205; use 208; withdrawals 25, 209

Hajer, M. A. 151
Hausmann, Carl 6
hegemonic power 43, 149–150
help/helping farmers 193, 324, 326
herbaceous/herb cover 235–236, 242, 245, 247–248
high-efficiency irrigation 236, 242
Hoekstra, A. Y. 13
human resources 35, 168, 177
hydro-hegemony 149, 151–152, 154
hydrology/hydrological 4, 7, 13, 19, 89, 300; capacity 149; challenges 230; models 207; processes 15, 19–20, 248
hydropolitics 39, 53
hydropower 57, 151, 200, 284; generating 200, 284

ideational power 102, 111, 150–151, 157n1
import/imported/importing 6–7, 56–57, 69, 74–75, 78, 214, 220, 229, 262–272, 276–278, 280–281; food 2, 26, 69, 74, 117; virtual water 49, 57, 264, 277; of virtual water 204, 252
income: countries 342; economies 348–349
individual farmers 199, 211, 321
industrial sectors 68, 74–75, 270, 338
integrated water resources management (IWRM) 5, 74, 101, 104, 151, 194, 284
intellectual friendship 37, 39
interception 20–21, 234, 246; losses 230, 236, 247–248
Intergovernmental Panel on Climate Change 86
internal: community of employees 319, 326; water resources 265
international: agreements 195, 257; boundaries 5, 126–127, 129–130, 132, 135, 140, 144; community 35, 89; conference 38, 193; corporations 7, 254; economic contexts 270–271, 278, 280–281; food prices 277; food trade 248, 270, 342; land boundaries 126, 129; law 127, 132, 135, 143–144; markets 271, 281; organization 256; players 178–179; political economy 255; post-oil crisis conditions 270, 280; relations 149, 252, 255, 257, 259, 261; river boundaries 127, 143; supply 281; systems 49; trade 35, 211, 213–214, 224, 267; transboundary water arrangements 154, 156; virtual-water market 276–278, 280, 281; watercourses 71, 142, 150; water law 71, 88, 93, 150
International Conference on Water and the Environment 193
International Court of Justice 129–130, 135–136, 141
International Finance Corporation (IFC) 178
International Policy Council (IPC) 165
International Renewable Energy Agency (IRENA) 167
interviews 103, 315, 345
intrinsic resource price 197–198
irrigation hydrology knowledge (IHK) 299, **303**
irrigation/irrigated 5–7, 24, 116–119, 121, 208–209, 220–222, 230, 236, 300–301, 303, 340; agriculture 6, 19, 194, 198, 215, 221, 232–233, 236, 239–240, 242, 248, 299–301, 303–304; areas 62, 87, 118, 198, 215–216, 233, 239; cropped area 215; cropping 235, 243; crop production 210, 212; deficits 89, 91; development 93, 111; drylands 6, 231, 233, 235, 237, 239, 241, 243, 245, 247, 251; efficiency 27, 301, 325; farmers 310–311;

intensity 213, 224; production 215, 217, 219; runoff 208; in Saudi Arabia 116; schemes 23, 109, 119; systems 24, 205, 299, 301–305; water 67, 104, 121, 197, 206, 209

Jevons' paradox 198–201
Jordan 5, 12, 41, 49, 127, 135, 140, 143, 148, 155, 220
Jordan, River 8, 134, 140

Kaabong district 101, 109–110
Karamoja region 5, 99, 101–105, 107–113, 115
Kasikili/Sedudu Island 129, 136, 141, 143
Kerry, John 170
Ki-moon, Ban 284
Kinchy, A. 86

Lancaster University 316, 318
land: boundary 130; cover 205, 232–233, 235–236, 239, 247–248; use 19, 21, 27, 218, 223–225, 231, 233, 235–236, 323, 337; and water management 12, 234
LandScanTM Global Population Database 233
land-use land cover (LULC) 19
Lawler, E. J. 149
Lekshmi, S. 17
Lewis, John 319
Libya/Libyan 4, 34–36, 309–310; farmers 309–310
Lidl 320
limited companies 319–320, 326
local: aquifers 119; communities 108–109, 292, 333; conflicts 49, 257; informants 106–107, 110; leaders 110–111; level 107–108, 135, 213, 224; management 142, 144; markets 61, 215; observations 133; peoples 108, 119; players 178; populations 108, 131, 166; production 213, 226, 229
Lokere catchments 104–105, 113
London University Joint Research Project 4, 34
Longoromit dam 109–110
long-term trends *338*
Lowi, M. 151
Lukes, S. 149

main channel 131, 140–141, 143; navigable 138–139
major channel 136, 141, 143–144
management: asymmetries 176–177; committee 109–110
managers 35, 74, 111, 176, 311
markets 4, 6, 11, 170, 186–187, 192–200, 210, 252–253, 258, 326, 338–339, 342, 347; conditions 195, 209; environmentalism 345; failures 346–347; modes 291–292; outcomes 224–225; sector 338–339; to value water 6, 192
median lines 129, 131, 133–134, 138
median upstream cropland cover 241, 247
Mediterranean *63*
Mekong Spirit 153
Mekonnen, M. M. 13
Middle East 2, 4, 37–39, 41, 49, 52–54, 119, 155, 209, 215, 217–218, 220, 253, 293
Middle East and North Africa (MENA) countries 34, 49–50, 219, 252, 258–259, 264
The Middle East Water Question 39, 122, 148
minimum storage distances 328n9
mobile water 200
Mohtar, R. H. 167
Morrisons 320

national economy 68, 74, 230
National Water Resources Plan (NWRP) 25, 59–61, **60**, 68, 71, 74
natural/naturalized: boundaries 127, 129–130, 135; capital 291, 296, 298, 331, 344, 346; capital assets 297–298; catchments 23–24; environment 182, 289; flows *62*; gas 285; land 233, 242, 247; landcover 232, 246; landscapes 19, 205; resources 2, 285, 345; vegetation 205, 230, 246–248, 340; water bodies 19–20
nature-based solutions 297–298
negative evaporative water footprints 247–248
negative water footprints 242, 244, 248
net water consumers *22*
new business model 167–168, 171
new coalitions 8, 331, 333–334
new geopolitics of food 6–7, 253, 255, 257, 261
new lands 67–69
new water structures 102, 107
New Zealand 343
nexus concepts 6, 284, 288
Nile Basin Initiative (NBI) 71, 152, 155
Nile, River 5, 53, 56–57, 59, 61, 68–74, 78, 82, 117, 148–149, 151, 154–156, 193; Basin 5, 14, 93, 102, 113, 156; flow per capita *70*; flows 59, 61, 69, 74, 91, 265; system 62–63; waters 4, 7, 90–91, 155, 268; Waters Agreement 89–92
non-agricultural land 233, 247
non-consumptive use 26–28
non-food water 81, 299
non-governmental organizations (NGOs) 315, 333
non-native irrigator 88
non-renewable groundwater 34, 117, 118, 124, 208, 209; irrigation 224
non-renewable resources 208–209
North Africa 34, 49, 215, 219, 242, 252, 264

old lands 68–69
operating policies 93
optimum cropping pattern 67
Oregon State University 42, 49
Organisation for Economic Cooperation and Development (OECD) countries 169, 277, 338, 343; economies 342, 349

organizational cultures 181–183, *184*, 189
Ouse, River 324
overhead irrigation 236, 248
Oxford handbook of food 6, 292
Oxford University Press 166

particular discourse 43–44, 46
pasture 20, 110, 197, 230, 232–233, 235–236, 239–241, 247, 309–310, 326
Pauly, D. 83
percolation water 21
performance indicators 27, 81, 83–84, 86
Permanent Court of Arbitration 136, 142–143
permitted water 197–199
Perry, C. J. 194, 325
physical: change 127, 129, 138–139, 143; water 264, 269–270; water-use efficiency 197–198, 200
pixels 15, 20, 22–24, 232–233, 235–236, 239–240, 242, 246
policy: analyses 81, 84, 86, 93; analysts 5, 81–82, 84, 86, 93; change 182, 333; intervention 81–87, 93; process 87
political: assessments 82; decisions 165, 199, 206; parties 73, 121, 123; power 193, 199; pressures 61, 103, 111; processes 206, 253; shocks 280; systems 165, 333, 347
political economy 2, 4, 111, 122, 157, 201, 266, 291, 302, 337, 339, 347–348; framework 103; of water security 4, 35
politicians 6, 8, 49, 120, 166–167, 288, 331, 333–335, 337, 343, 345
ponded irrigation water 236, 242
population, growing 165, 263, 266, 278, 313
positive footprints 243, 248; water 242–243, 247–248
poultry farms 316–317
power 3–5, 8, 37, 43–46, 52–53, 73, 101–104, 111–113, 132, 148–151, 155–156, 253, 345–347; of hegemonic imaginings 4, 45, 47; and politics 111; relations 8, 44, 148, 151, 155, 291, 339, 342; sector 170, 182; system 185; and water 182
priced food 7–8, 344, 349
primary resources 167, 169, 172
private sector 4, 6, 123, 165, 167–172, 175–177, 179, 181–182, 184–185, 192–193, 347; engagement 181–182; investments 118–119; participation 176, 333; supplies 6, 167; in sustainable development 6, 167; and water services 6, 176
productive use 195, 248
provincial boundary 136–137, 142
public policy 122, 332
Pyle, R. M. 83

qualitative patterns, virtual-water decoupling 264, 271; in Egypt 265

rainfed: agriculture 111, 220–221, 236, 244–245, 258; areas 215; cropping scenario 245–246; production 214–215, 217, 219, 221–222
rainfed crops 20; production 214, 217, 220–221
randomized controlled trial (RCT) 83
real cost 321, 337
regenerative agriculture 288, 314, 334
regions/regional 5, 49–50, 52–54, 101–102, 107–108, 111, 113, 214, 220–221, 258, 277, 324; markets 211; planning groups 186, 189
renewable: energy 54, 170, 285–286; water resources 25, 27, 117, 123, 252
resources 5, 59, **60**, 62, 74, 102–103, 110, 168, 193–195, 198, 225, 285; accounting 302; management 110; prices 193, 197, 199, 201
restricting groundwater: irrigation 218, 223–224; use 223
retailers 313–315, 318–327, 342, 345
risk: of conflict 49; management 181, 182, *184*, 189
riverain area 142–143
river basins 13, *14*, 21, 25, 74, 85, 104, 155, 195, 199, 278, 324–325; borderscapes 129, 131; scales 13, 15, 18, 22, 297–298; simulation models 86
river boundaries 127, 129, 131–134; delimitation 5, 126–127, 139, 143; protocol 139, 143
rivers 23–24, 49, 71–72, 84, 116–117, 126–127, 129–140, 142–143, 155, 205–206, 208, 301, 313, 315–317; bank 5, 126–127, 129, 131, 133, 135, 137, 139, 141, 143, 147; catchments 184–185, 316; flow 14, 23–26, 315; systems 22, 27; valleys 126, 132
Riyadh and Qassim regions 119, 123
Rosner, M. *338*
Rougé, C. 195
Russia–Ukraine war 280

sanctioned discourse 4, 43–47, 149, 151, 312
Saudi Arabia 57, 116–123, 125, 155, 207, 220, 253, 256, 258
Scalable Water Balances from Earth Observations (SWEO) 4, 13, **15**, 15–28, *23*, **25**, **27**, 33
scale effect 217
scarce: economic resources 117, 122; resource 121, 198; water resources 27, 193, 257
School of Oriental and African Studies (SOAS) 148
services: exports 270–271; sectors 74, 270
shadow water 41–42
shallow: groundwater 17, 19, 109, 208; water tables 20, 301
Shardara reservoir *25*
shared water challenges 318, 329n12
Shatt al-Arab 127, 132, 138–139, 143
Shilpakar, R. L. 14
significant harm 71, 88, 91, 93
small-scale irrigation schemes 103, 109

social power 44–45
society 5–8, 13, *14*, 166, 168, 192–194, 291–292, 318–319, 337, 339, 343, 345, 347–348
socio-economic measures 68, 74
socio-political contexts 276–277, 281
soil: factors 65; management 231, 318; moisture 15, 206, 208; moisture storage 20, 206; water 17–18, 21, 24, 154, 192, 340; water balances 24, 154; and water management 313, 322
Soil Moisture and Ocean Salinity (SMOS) 17
Sojamo, S. 154
South Africa 214, 218, 220
South Asia 220, 239, 254
South Sudan 57, 70, 102, 111
spatial distribution 215–216, 231, 236
spatial footprint of areas 242, 245
state power 44, 149
static model 61, 67
status quo conditions 82–83, 85–87, 89, 91–92
stock markets 319
streams and groundwater 340
strong economic growth 268
structural power 150, 155
study period 263, 266, 268, 270–272, 276–278, 281
subsidises farmers 342, 345
substitution effect 217
Sudan 5, 56–57, 60–62, 69–73, 76, 89–90, 92–93, 129, 136–137, 143, 155, 253, 258; and Egypt 56, 70, 72, 76, 93
Suez Canal 72–73, 78, 270
supplemental data online 16, 21, 23, 213, 217, 220–221, 223
suppliers 78, 277–278, 280–281, 314, 319–321, 323, 325–326
supply chains 6, 167–168, 170–172, 230–231, 253, 285–286, 292, 317, 322, 325–327, 337, 348
support, farmers 314, 322
surface: areas 18, 234; and groundwater 209, 215; irrigation systems 205; runoff 15, 20, 22, 26
surface water 19, 24, 49, 65, 142, 208, 211, 213–214, 216, 225; irrigation 216, 224; substitution 219, 223; use 209, 213, 217, 223; withdrawals 25, 213
sustainability/sustainable 168, 179, 205, 209–210, 287, 289, 295–298, 314–315, 318, 321, 339; agriculture 324; biophysical determinants of 210; business 319–320; development 6, 167–169, 171–172, 175, 193, 295, 346; farming 8, 310, 313, 315, 318, 322–323, 327; food systems 323; groundwater use 205–206, 215, 223–225; level of abstraction 208–209; policies 213, 215, 224–225; production 172, 323; restrictions 225; soil 315, 326; water management 323–324; water use 226
Sustainable Development Goals (SDGs) 284
system-action-transformation (SAT) **303**
system-based approach 331
system coordination 181, 183, 185, 187, 189, 191
tame problems 184, 186
targeted areas 217
Taskforce for Nature-related Financial Disclosure (TNFD) 297
Tayia, A. 264
temperature 86, 208, 236–237, 243, 246, 248
terrestrial water balances 14
total fertility rate (TFR) *76*
trade 42, 48, 165, 204, 214, 223, 253–255, 257, 263–264, 266–267, 277, 286, 292; decoupling 272, 277; patterns 265–266; policies 263
transboundary flows **150**
transboundary waters 3, 49, 148, 153, 157; analysis of 148, 157; arrangements 148–155; governance 91; interaction 153; managing resources 82; resources 53, 82, 93; resources management 5, 81, 83, 85, 87, 89, 91, 93
Transboundary Waters Interaction Nexus (TWINS) 153
transfer 37, 129, 136, 139, 183, 204, 310
transformation 169, 171, 285, 299
transition 54, 75, 170, 172, 286, 296
treatment effect 82, 84, 86
tree cover 102, 235–237, 242–243, 245, 247–248
Tripartite National Committee 89, 91
Tropical Rainfall Measurement Mission (TRMM) 16
Turkana 109–110

UK food: major retailers 315, 318, 326; retailers 315, 318; retail statistics 314; system 328
underpaid people 288–289, 292
under-priced food 288–289, 292, 310
unexamined baselines 5, 82–84
United Nations 27, 70, 136, 138–140, 142, 198
United States 57, 88, 131, 169, 209, 215, 220, 225, 252, 277–278, 285, 333–334; Department Health 82, 84; Environmental Protection Agency 82
unsustainable groundwater 215, 217, 224
upper riparians 69, 71, 73, 78
upper soil water balance 20
upstream countries 78, 155
upstream states 151–152, 155
Ureta, S. 86

value/valuing: of food imports 121, 270; soil 8, 336; of water 6, 67, 192, 193, 195, 197–198, 200, 285–286
vegetation cover 17–18, 236, 242–243
vertical soil water balance 15, 20, *21*
virtual: blue water exports 220; hydro-hegemony 149, 154; trade 220, 224
virtual-water 2, 4–7, 41–42, 48–57, 116–117, 120–121, 154–155, 165, 204–205, 213–215, 217, 219, 221, 223, 225, 229–230, 252–255, 257, 259, 263, 285–286, 289, 291–293; concept 4, 48, 229–230, 248, 300; considerations 7,

255, 257; exports 123, 155, *266*, *274*, 277, 291; flows 157, 257, 286–287, *341*; hegemony 293; paradigm 252–253, 257, 259; policies, decoupling 263, 270, 276, 278, 280–281; recoupling 264, 268–272, 276, 281; recoupling conceptual framework *265*; resources *342*; rivers 5, 154; trade 27, 49, 120, 213–214, 220, 224–225, 258–259, 265, 280; transfers 154
virtual-water decoupling 262–265, 267–281, 283; in Egypt 263–266, 270–272, 276, 280–281; role of 262–263, 265, 267, 269, 271, 273, 275, 277, 279, 283
virtual-water imports 120, 258, 262, *266*, 269, 271–272, *273*, 276, 278, 281; share of 269, 272, 276
vulnerable groundwater exporters 154, 291

Wada, Y. 291
Warner, J. 150, 151
water 2–8, 21–22, 41–42, 48–50, 52–54, 65–69, 99–105, 107–111, 115–122, 148–157, 192–201, 229–238, 262–263, 284–289, 291–293, 299–303, 309–312, 322–325, 339–343, 345–348; abstraction 107–108, 111; accounting 7–8, 13–15, 27–28, 231; accounts 26, 301, 303; allocation 14, 257; authorities *14*, 196; availability 77, 165, 225; balance 16–17, 21, 24, 27, 60–61, 65, 67, 154, 234–235; bodies 18, *19*, 24, 235; and complex problemsheds in Karamoja 99; conflicts 48–49; consumption 194, 205, 215, 232, 248, 304, 322, 325, 345, 347; costs 229, 248; deficit 68, 78; demands 68, 188, 266; development 2, 113; diplomacy 152–153, 157; distribution system *64*; efficiency 257, 322; endowments 35, 257; and energy 284, 286; environment 286; flows 101, 117, 131, 155, 300, 303, 340; framework 293; governance 6, 102, 181, 183, 185, 187, 189, 191, 305, 331; infrastructure development 5, 101; insecurity 99; intakes 233, 241–242, 247; markets 194, 196–201; negotiations 49, 152; policy 121, 186, 194, 230, 310; productivity 27, 310; quality 63, 65, 74, 181, 185, 189–190, 231, 235, 285, 288, 315–317, 325; quantity 231, 235, 325; reallocation 88, 206; requirements 67–68; risks 322, 325; sector 4, 5, 35, 68, 177, 180, 189; services 6, 169, 176–177, 179, 200; shortages 91–92; and society 6, 292; storage 18, 196; supply 20–21, 24, 56, 59, 74, 198, 285, 288, 301, 322, 345; systems 13, 185–186, 289, 302; table 118, 165, 205–206; use 13–14, 74, 87, 91, 151, 195, 217, 225, 230–231, 247, 284, 314; volume of 18, 35, 52–53, 65, 141, 155, 196, 231, 263, 299–300, 324; wars 4, 48–49, 51–53, 55, 72, 204, 257–258, 293
Water Balance Model 208, 213
water footprint/footprinting 6, 13–14, 27, 48, 229–231, 233, 235, 237, 239, 241, 243, 245–249, 251, 285–286, 305, 309, 314; analysis 14, 16, 26, 305; calculating 230, 246; concept 13, 286
Water Framework Directive (WFD) 181
water in the national economy (WINE) *74*
water management 12, 19, 111, 113, 192–194, 230, 304–305, 311, 313–315, 322–323, 326–327; processes 27, 111; zones 104
water-poor: countries 117; economies 248
water-related public policies 325, 327
water-resource decoupling 264; mechanism 264; policy analysis framework 264
water resources 5–7, 13, 34–35, 41–42, 59, 87–88, 194–196, 199, 209–210, 233, 317, 337; development 103, 113, 206; management 12, 26, 103, 107, 197–198, 286; planning 101; South East 187–189
water, sanitation and hygiene (WASH) plans 177
water-scarce/scarcity 5, 13, *14*, 49–50, 68, 70–71, 74, 101, 120, 153, 194, 199, 258, 292; countries 257, 278, 281; economies 262–263; in Egypt 71, 74; environments 291; managing 68; mitigating 6, 252, 257
water security 4, 7, 35, 69, 169, 262–265, 267, 269, 271, 273, 275, 277, 279, 283, 286–287, 347; in Egypt 264; policymaking 262, 280
water stewardship 313–315, 317–318, 322, 325–327; uptake of 318, 326
water-stressed areas 285, 322
water-use efficiency 27, 59, 309–311, 325
water-users 188, 193, 196, 200, 322, 325; associations 199–200
WaterWorld 233–235, 239; land use 236; model 231–232; scenarios 235
weaponized interdependence 255–256, 259
wheat: imports *77*; prices *338*; production 118, 121
Wichelns, D. 305
wicked problems 183–186, 193
willingness to pay (WTP) 88
world: markets 54, 258; regions 211
World Bank 12, 42, 53, 75, 109, 266, 287, 309
World Economic Forum 13, 167, 182, 193, 284
World Food Programme 254, 258
World Resources Institute 295
Wye and Usk Foundation (WUF) 316
Wye Nutrient Management Board 317
Wyre catchment project **188**

Zeitoun, M. 155